T0331812

T. WOLFRAM AND S. ELLİALTIOĞLU

Electronic and Optical Properties of *d*-band Perovskites

The perovskite family of oxides includes a vast array of insulators, metals, and semiconductors. Current intense scientific interest stems from the large number of diverse phenomena exhibited by these materials including pseudo-two-dimensional electronic energy bands, high-temperature superconductivity, metal–insulator transitions, piezoelectricity, magnetism, photochromic and catalytic activity.

This book is the first text devoted to a comprehensive theory of the solid-state properties of these fascinating materials. The text includes complete descriptions of the important energy bands, photoemission, and surface states. The chapter on high-temperature superconductors explores the electronic states in typical copper oxide materials. Theoretical results are compared with experimental results and discussed throughout the book.

With problem sets included, this is a unified, logical treatment of fundamental perovskite solid-state properties, which will appeal to graduate students and researchers alike.

THOMAS WOLFRAM is a retired professor and Chairman of the Department of Physics and Astronomy at the University of Missouri-Columbia and retired Vice President and General Manager of Amoco Laser Company, ATX.

ŞİNASİ ELLİALTIOĞLU is a professor and former Chairman of the Department of Physics at the Middle East Technical University, Ankara, Turkey.

ELECTRONIC AND OPTICAL PROPERTIES OF *D*-BAND PEROVSKITES

THOMAS WOLFRAM

Formerly of University of Missouri-Columbia

ŞİNASİ ELLİALTIOĞLU

Middle East Technical University, Ankara

CAMBRIDGE
UNIVERSITY PRESS

CAMBRIDGE
UNIVERSITY PRESS

University Printing House, Cambridge CB2 8BS, United Kingdom

One Liberty Plaza, 20th Floor, New York, NY 10006, USA

477 Williamstown Road, Port Melbourne, VIC 3207, Australia

314-321, 3rd Floor, Plot 3, Splendor Forum, Jasola District Centre, New Delhi - 110025, India

79 Anson Road, #06-04/06, Singapore 079906

Cambridge University Press is part of the University of Cambridge.

It furthers the University's mission by disseminating knowledge in the pursuit of education, learning and research at the highest international levels of excellence.

www.cambridge.org
Information on this title: www.cambridge.org/9780521850537

© T. Wolfram and Ş. Ellialtioğlu 2006

First published 2006

A catalogue record for this publication is available from the British Library

ISBN 978-0-521-85053-7 Hardback

Contents

Appendices

Preface

Metal oxides having the cubic (or nearly cubic), ABO_3 perovskite structure constitute a wide class of compounds that display an amazing variety of interesting properties. The perovskite family encompasses insulators, piezoelectrics, ferroelectrics,
metals, semiconductors, magnetic, and superconducting materials. So broad and
varied is this class of materials that a comprehensive treatise is virtually impossible
and certainly beyond the scope of this introductory text. In this book we treat
only those materials that possess electronic states described by energy band theory. However, a chapter is devoted to the quasiparticle-like excitations observed in
high-temperature superconducting metal oxides. Although principally dealing with
the cubic perovskites, tetragonal distortions and octahedral tilting are discussed in
the text. Strong electron correlation theories appropriate for the magnetic properties of the perovskites are not discussed. Discussions of the role of strong electron
correlation are frequent in the text, but the development of the many-electron theory crucial for magnetic insulators and high-temperature superconductors is not
included.

This book is primarily intended as an introductory textbook. The purpose is
to provide the reader with a qualitative understanding of the physics and chemistry that underlies the properties of "d-band" perovskites. It employs simple linear
combinations of atomic orbitals (LCAO) models to describe perovskite materials
that possess energy bands derived primarily from the d orbitals of the metal ions
and the p orbitals of the oxygen ions. The results are usually obtained analytically
with relatively simple mathematical tools and are compared with experimental data
whenever possible.

The book is considered appropriate for science and electrical engineering graduate students and advanced undergraduate seniors. It may be used as a primary
text for short courses or specialized topic seminars or it can serve as an auxiliary
text for courses in quantum mechanics, solid-state physics, solid-state chemistry,
materials science, or group theory. The reader will need a basic understanding of
quantum mechanics, and should have had an introductory course in solid-state
physics or solid-state chemistry. Knowledge of group theory is not required, but
some understanding of the role of symmetry in quantum mechanics would be helpful. The material covered is considered a prerequisite for understanding the results
of more complex models and numerical energy band calculations. Research scientists seeking a qualitative understanding of the electronic and optical properties of
the perovskites will also find this book useful.

The theoretical results are derived in sufficient detail to allow a typical reader
with a calculus background to reproduce the formulae and derive independent results. Because most of the results are presented in analytic form, the relationships

among the physical variables are transparent and can easily be understood and explored. Using these analytical results the reader can obtain numerical results for the electronic, optical, and surface properties of specific materials using nothing more sophisticated than a programmable hand calculator or a desk computer equipped with MS QuickBasic© software.

Many of the topics discussed in the book were originally published by the authors in research papers and were formulated in terms of Green's functions. In order to keep the material in this book as simple as possible the same results are obtained here by more rudimentary mathematical methods.

For the most part our understanding of the properties of metals is derived from various versions of the free-electron model (often with imposed periodic boundary conditions). The simplicity of this model does not diminish its applicability, and in many instances, particularly in the case of BCS (Bardeen–Cooper–Schrieffer) superconductors, the results obtained are quantitatively correct. Of equal importance is the pedagogical utility of the free-electron model, which permits scientists and students alike to make simple calculations and to develop scientific concepts and a useful intuition about the electronic and optical phenomena of metals.

In the case of compounds whose properties are dominated by the atomic orbitals of the constituent ions, the free-electron model is not particularly useful. For compounds such as the perovskites the physical and chemical properties are largely dependent upon the crystalline structure and the symmetry of the atomic orbitals involved in the valence bands and the bands near the Fermi level. The purpose of this book is to provide a relatively simple but complete description of the d-band perovskites based on atomic-like orbitals. Models of this type were developed many years ago by chemists and physicists alike using LCAO and other similar localized-orbital approaches. Later, such models were "put on the shelf" as theoretical solid-state physicists moved almost exclusively into the realm of momentum-space theories. Indeed, for some time it could be said with justification that solid-state physicists were the Fourier transform of solid-state chemists.

With the recent discovery of high-temperature superconductivity (HTSC) in the cuprate compounds interest in the science of the transition metal oxides has grown enormously. Interestingly, solid-state theorists have returned to real-space theories to look for an understanding of these materials. It is somewhat ironic that the original migration to $\vec{k}$-space was driven, to a large degree, by the success of the BCS theory in explaining (low-temperature) superconductivity in terms of a free-electron model. Now, HTSC is leading solid-state physicists back to real-space approaches. Not withstanding the extreme importance of strong electron correlations, renormalization effects, holons, and spinons, HTSC experimental data are most often discussed in terms of local atomic-like orbitals, the symmetry of the orbitals and the interactions between them. That is, the data are discussed in the jargon characteristic of LCAO models.

Although high-temperature superconductors are not, strictly speaking, per-ovskites, they share many structural and electronic features in common with the perovskites. For that reason we have included a chapter on the low-lying quasiparticle bands of these exciting, new materials.

1

Introductory discussion of the perovskites

1.1 Introduction

The mineral $CaTiO_3$ was discovered in the Ural Mountains by geologist Gustav Rose in 1839 and given the name perovskite in honor of the eminent Russian mineralogist, Count Lev Alexevich von Perovski. The name perovskite is now used to refer to any member of a very large family of compounds that has the formula ABC_3 and for which the B ion is surrounded by an octahedron of C ions. Perovskites ($MgSiO_3$ and $FeSiO_3$) are the most abundant compounds in the Earth's crust.

The compounds with the formula ABO_3, with $O = $ oxygen and $B = $ a transition metal ion, are a subclass of the transition metal oxides that belong to the perovskite family. Table 1.1 provides a brief list of some well-studied ABO_3 perovskites. Many of the perovskites are cubic or nearly cubic, but they often undergo one or more structural phase transitions, particularly at low temperatures.

The perovskite oxides are extremely interesting because of the enormous variety of solid-state phenomena they exhibit. These materials include insulators, semiconductors, metals, and superconductors. Some have delocalized energy-band states, some have localized electrons, and others display transitions between these

Table 1.1. *Some perovskite and related oxides.*

Insulating	Metallic	Magnetic	Superconducting	
			$SrTiO_3$(n-type)	
WO_3	ReO_3	$PbCrO_3$	Na_xWO_3	(t)
$NaTaO_3$	$NaWO_3$	$LaCrO_3$	K_xWO_3	(t)
$SrTiO_3$	$KMoO_3$	$CaMnO_3$	K_xWO_3	(h)
$BaTiO_3$	$SrNbO_3$	$LaMnO_3$	Rb_xWO_3	(h)
$KTaO_3$	$LaTiO_3$	$LaCoO_3$	Cs_xWO_3	(h)
$LiNbO_3$	$LaWO_3$	$LaFeO_3$	Li_xWO_3	(h)

$t = $ tetragonal, $h = $ hexagonal

1

two types of behavior. Many of the perovskites are magnetically ordered and a large variety of magnetic structures can be found.

The electronic properties of the perovskites can be altered in a controlled manner by substitution of ions into the A or B sites, or by departures from ideal stoichiometry.

The electronic energy bands of the perovskites are very unusual in that they exhibit two-dimensional behavior that leads to unique structure in properties such as the density of states, Fermi surface, dielectric function, phonon spectra and the photoemission spectra.

The perovskites are also important in numerous technological areas. They are employed in photochromic, electrochromic, and image storage devices. Their ferroelectric and piezoelectric properties are utilized in other device applications including switching, filtering, and surface acoustic wave signal processing.

Many of the perovskites are catalytically active. Development of perovskite catalyst systems for the oxidation of carbon monoxide and hydrocarbons, and the reduction of the oxides of nitrogen have been proposed. The perovskites are also employed in electrochemical applications including the photoelectrolysis of water to produce hydrogen.

Scientific studies of the perovskites date back many years. The physical properties of the tungsten bronzes were investigated as early as 1823 [1]. However, it is only in recent years that experimental and theoretical information on the electronic structure has begun to become available. Energy band calculations [2], neutron diffraction and inelastic scattering data [3], photoemission spectra [4], optical spectra [5], and transport data [6] are now available for materials such as ReO_3, WO_3, $NaWO_3$, $SrTiO_3$, $BaTiO_3$, $KMoO_3$, $KTaO_3$, $LaMnO_3$, $LaCoO_3$, and a variety of other perovskites.

Surface studies of single-crystal perovskites have been performed using photoelectron spectroscopies that indicate that the surface properties are extremely complex and interesting [7].

In this chapter we present brief discussions of some of the properties of the perovskite oxides. The discussions are qualitative and intended only to give the reader a general impression of the types of factors that must be considered. More quantitative discussions are given in later chapters.

In Section 1.2 we describe the structural features of the perovskites. Sections 1.3 through 1.6 give a qualitative discussion of the electronic states starting from a simple ionic model and then adding ligand field, covalency, and band effects. Section 1.7 deals briefly with localized d-electron states and why many perovskites do not have conventional energy bands. In Section 1.8 we touch upon the multiplet configurations of localized d electrons and their role in determining the magnetic properties. In Section 1.9 we discuss briefly superconductivity among the perovskites. The last section, 1.10, is a summary of some of the technological applications of the perovskites.

1.2 The perovskite structure

The formula unit for the cubic perovskite oxides is ABO_3 where A and B are metal cations and O indicates an oxygen anion. The structure, illustrated in Fig. 1.1, is simple cubic ($O_h^1, Pm3m$) with five atoms per unit cell. The lattice constant, $2a$, is close to $4\,\text{Å}$ for most of the perovskite oxides.

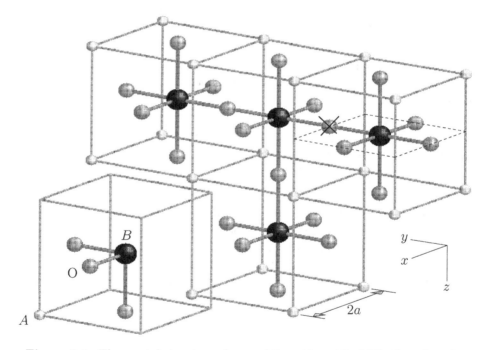

Figure 1.1. The crystal structure of perovskite oxides with ABO_3 formula unit.

The B cation is a transition metal ion such as Ti, Ni, Fe, Co, or Mn. It is located at the center of an octahedron of oxygen anions. The B site has the full cubic (O_h) point group symmetry. The A cation may be a monovalent, divalent, or trivalent metal ion such as K, Na, Li; Sr, Ba, Ca; or La, Pr, Nd. The A ion is surrounded by 12 equidistant oxygen ions. The A site also has the point group O_h.

The oxygen ions are not at sites of cubic point group symmetry. Focusing attention on the oxygen ion marked with an "$\times$" in Fig. 1.1 it may be seen that the site symmetry is D_{4h}. The B–O axis is a fourfold axis of symmetry and there are several reflection planes; the yz-plane and planes passing through the edges containing A sites. The transition metal ion (B site) will experience a cubic ligand field that lifts the fivefold degeneracy of the d-orbital energies. The oxygen ions experience an axial ligand field that splits the $2p$-orbital energies into two groups. These splittings are described in the next section.

Well-known examples of cubic perovskites are $SrTiO_3$, $KTaO_3$, and $BaTiO_3$ (above the ferroelectric transition temperature). Many of the perovskites that we shall want to include in our discussions are slightly distorted from the ideal cubic structure. If the distortions are moderate the general features are not significantly different from those of the cubic materials. $BaTiO_3$ and $SrTiO_3$ both have structural transitions to a tetragonal symmetry at certain critical temperatures. Tetragonal and orthorhombic distortions are very common among the perovskites.

Another class of compounds that we include in our discussions are the pseudo-perovskites with the formula unit BO_3. Such compounds have the perovskite structure except that the A sites are empty. Examples of pseudo-perovskites are ReO_3 and WO_3.

It is possible to form an intermediate class of perovskites from WO_3 by adding alkali ions to the empty A sites. These compounds, known as the tungsten bronzes, have the formula unit $A_x WO_3$ where x varies from 0 to 1 and A is H, Li, Na, K, Rb, or Cs. The structure is often dependent upon the value of x. WO_3 is tetragonally distorted but becomes cubic for $x > 0.5$. $NaWO_3$ is cubic.

In our discussions we shall also include substituted or mixed compounds of the form $(A_x^1 A_{1-x}^2)(B_y^1 B_{1-y}^2)O_3$ and oxygen-deficient perovskites, ABO_{3-x}. Including distorted, substituted, and non-stoichiometric compounds, the class of materials under consideration is very large. Within this broad class, examples may be found that display almost any solid-state phenomena known.

1.3 Ionic model

The perovskite oxides are highly ionic, but they also possess a significant covalent character. The ionic model is an oversimplified picture but it serves well as a starting point for thinking about the electronic properties. The ionic model assumes that the A and B cations lose electrons to the oxygen anions in sufficient numbers to produce O^{2-} ions. The usual chemical valence is assumed for the A cations; K^+, Ca^{2+}, and La^{3+}, for example. The ionic state of the transition metal ion is determined by charge neutrality. If the charge of the B ion is denoted by q_B and that of the A ion

Table 1.2. *Cations commonly found in perovskite-type oxides. In parentheses is the coordination number, Z, if the radii given are not for 12 coordination; HS and SL refer to high spin and low spin, respectively. The effective ionic radii (in Å) are from Shannon [8].*

Dodecahedral A site ($Z = 12$)			Octahedral B site ($Z = 6$)		
Ion	Electrons	Radius	Ion	Electrons	Radius
Na^+	$2p^6$	1.39	Li^+	$1s^2$	0.76
K^+	$3p^6$	1.64	Cu^{2+}	$3d^9$	0.73
Rb^+	$4p^6$	1.72	Mg^{2+}	$2p^6$	0.72
Ag^+	$2d^{10}$	1.28 (8)	Zn^{2+}	$3d^{10}$	0.74
Ca^{2+}	$3p^6$	1.34	Ti^{3+}	$3d^1$	0.67
Sr^{2+}	$4p^6$	1.44	V^{3+}	$3d^2$	0.64
Ba^{2+}	$5p^6$	1.61	Cr^{3+}	$3d^3$	0.615
Pb^{2+}	$6s^2$	1.49	Mn^{3+}(LS)	$3d^4$	0.58
La^{3+}	$4d^{10}$	1.36	Mn^{3+}(HS)	$3d^4$	0.645
Pr^{3+}	$4f^2$	1.18 (8)	Fe^{3+}(LS)	$3d^5$	0.55
Nd^{3+}	$4f^3$	1.27	Fe^{3+}(HS)	$3d^5$	0.645
Bi^{3+}	$6s^2$	1.17 (8)	Co^{3+}(LS)	$3d^6$	0.5456
Ce^{4+}	$5p^6$	1.14	Co^{3+}(HS)	$3d^6$	0.61
Th^{4+}	$6p^6$	1.21	Ni^{3+}(LS)	$3d^7$	0.56
			Ni^{3+}(HS)	$3d^7$	0.60
			Rh^{3+}	$4d^6$	0.665
			Ti^{4+}	$3p^6$	0.605
			Mn^{4+}	$3d^3$	0.53
			Ru^{4+}	$4d^4$	0.62
			Pt^{4+}	$5d^6$	0.625
			Nb^{5+}	$4p^6$	0.64
			Ta^{5+}	$5p^6$	0.64
			Mo^{6+}	$4p^6$	0.59
			W^{6+}	$5p^6$	0.60

by q_A then $q_B = 6 - q_A$ where the three oxygen ions contribute the factor of 6. A list of the common A ions and their valence states is given in Table 1.2.

Once the charge state of the B ion is determined the number of d electrons remaining is determined from the atomic electronic configuration (Table 1.2). For example, for $SrTiO_3$ we have Sr^{2+} and O^{2-} so that the titanium ion is Ti^{4+}. The electronic configuration of neutral titanium atom is [Ar] $3d^2 4s^2$. To form Ti^{4+} the outer four electrons are removed leaving the closed-shell Ar core [Ar]. Since O^{2-}

has the [Ne] configuration, all of the ions of $SrTiO_3$ have closed-shell configurations. The electronic configuration of W is [Xe] $5d^46s^2$. Thus in WO_3 the W^{6+} ion has a closed-shell [Xe] core; however, for $NaWO_3$ the W^{5+} ion has a d^1 configuration. The electronic configurations of relevant transition metal ions are given in Table 1.2.

According to the ionic model when all of the ions have closed-shell configurations the material is an insulator. If the B ion retains d electrons then the perovskite may be a metallic conductor depending on other factors to be discussed. $NaWO_3$ or ReO_3 each have d^1 configurations and are good metals. For compounds such as Na_xWO_3 it is assumed that there will be x d electrons per unit cell. That is, the Na donates its electron and the W ions donate the remaining electrons needed to form O^{2-} ions. One may imagine that there are $(1-x)$ W^{6+} and x W^{5+} ions distributed at random or on an ordered array or that each tungsten ion has an average valence of $W^{(6-x)+}$. The proper picture can not be decided from the ionic model but depends on other considerations. For Na_xWO_3 experiments show that metallic d bands are formed so that we may picture an average valency of $(6-x)+$. However, among the perovskites examples of ordered and random arrays of mixed valence B ions can also be found.

1.4 Madelung and electrostatic potentials

Starting from the ionic model, other important effects that determine the electronic properties can be added. The ionic model described above would apply to isolated or free ions. The ions are, of course, not isolated but interact in several different ways. One such interaction is through the electrostatic fields due to the charges on the ions. The most important electrostatic effect is the Madelung potential. The A and B ions are surrounded by negatively charged oxygen ions. The electrons orbiting these ions therefore experience repulsive electrostatic (Madelung) potentials. Conversely, the electrons orbiting the oxygen ions are surrounded by positively charged cations and they experience an attractive Madelung potential. The "site Madelung potentials" are defined as the electrostatic potentials at the different lattice sites due to all of the other ions. For example, the Madelung potential at a B site located at $\vec{R}_B^0$ is

$$V_M(\vec{R}_B^0) = \sum_{\vec{R}_O} \frac{e^2|q_O|}{|\vec{R}_B^0 - \vec{R}_O|} - \sum_{\vec{R}_A} \frac{e^2|q_A|}{|\vec{R}_B^0 - \vec{R}_A|} - \sum_{\vec{R}_B \neq \vec{R}_B^0} \frac{e^2|q_B|}{|\vec{R}_B^0 - \vec{R}_B|}. \qquad (1.1)$$

In (1.1), eq_O, eq_A, and eq_B are the charges on the oxygen, A, and B ions, respectively, and $\vec{R}_O, \vec{R}_A$, and $\vec{R}_B$ are the vectors for the corresponding lattice sites. The site Madelung potentials are very large for the perovskites because of the large ionic charges. Typical Madelung potentials are 30–50 eV for the B site. For

$A^{2+}B^{4+}O_3^{2-}$ perovskites the (full ionic) site potentials [9] are: $V_M(B) = +45.6\,\text{eV}$, $V_M(A) = +19.9\,\text{eV}$, and $V_M(O) = -23.8\,\text{eV}$. A table of Madelung potentials can be found in Appendix D.

The stability of the perovskite structure is largely due to the energies associated with the Madelung potentials. The attractive potential at the oxygen sites allows the oxygen ions to bind a pair of electrons. In effect the site potential adds to the electron affinity of the oxygen ion. The affinity of O^- for the second electron is actually *positive*. This means that the second electron would not be bound on a free oxygen ion. O^{2-} is stable in the lattice because of the attractive site Madelung potential. Conversely, a d electron is bound to a Ti^{4+} ion with an (ionization) energy of $-43\,\text{eV}$. In the absence of the repulsive site Madelung potential, donation of an electron from the Ti^{3+} to an O^- ion in $SrTiO_3$ would be energetically very unfavorable. The site Madelung potential adds to the ionization energy so that the d electron would have an effective binding energy of $-43 + 45.6 = +2.6\,\text{eV}$ (unbound) for $SrTiO_3$ with the full ionic charges.

Thus, it is seen that the Madelung potentials are responsible for the ionic configurations.

An orbital centered on an ion has a finite radial extent so that an electron in such an orbital would sample the electrostatic field over a distance comparable to the ionic radius. In order to determine the complete effect of the electrostatic field on the electron state we need to know the behavior of the field as a function of position near each ion site. If we use the point ion model then,

$$V(\vec{r}) = -\frac{e^2|q_B|}{|\vec{r} - \vec{R}_B^0|} + V_{es}(\vec{r}),$$

$$V_{es}(\vec{r}) = -\sum_{\vec{R}_B \neq \vec{R}_B^0} \frac{e^2|q_B|}{|\vec{r} - \vec{R}_B|} - \sum_{\vec{R}_A} \frac{e^2|q_A|}{|\vec{r} - \vec{R}_A|} + \sum_{\vec{R}_O} \frac{e^2|q_O|}{|\vec{r} - \vec{R}_O|}. \tag{1.2}$$

The potential near $\vec{R}_B^0$ can be found by expanding $V_{es}(\vec{r})$ in terms of spherical harmonics centered at $\vec{R}_B^0$. The potential $V_{es}(\vec{r})$ then takes the form of an electric multipole expansion. The monopole term is just the site Madelung potential. Thus, as we have described, the site Madelung potential produces a shift in the energy of an electron localized on the site.

The higher-order multipoles (dipole, quadrupole, etc.) create an electrostatic field (with the point group symmetry of the site) which leads to a lifting of the orbital degeneracies. The effect of the cubic electrostatic field at the B ion site is to split the fivefold degenerate d states into two groups as shown in Fig. 1.2(c). The e_g group is doubly degenerate corresponding to the d orbitals having wavefunctions with angular symmetry $(x^2 - y^2)/r^2$ and $(3z^2 - r^2)/r^2$. The threefold degenerate t_{2g} group corresponds to the states $(xy/r^2), (xz/r^2)$, and (yz/r^2).

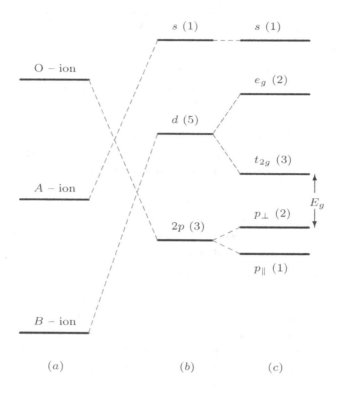

Figure 1.2. Effect of the electrostatic potentials on the ion states: (a) free ions, (b) Madelung potential, and (c) electrostatic splittings.

The oxygen $2p$ states are split by the axial electrostatic field into a doubly degenerate level denoted by $p_\perp$ and a non-degenerate $p_\parallel$ state. The notation $p_\perp$ and $p_\parallel$ refer to $2p$ orbitals oriented perpendicular and parallel to a B–O axis, respectively.

The lowest unoccupied state of the A ion is an s state. Its energy is shifted by the monopole (Madelung potential) but unaffected by the other multipole terms because it is a spatially non-degenerate function with spherical symmetry at a site of cubic symmetry.

The particular level ordering shown in Fig. 1.2 may be understood by considering the orientation of orbitals relative to the charge distributions on neighboring ions. The e_g orbitals have lobes directed along the B–O axis and directly into the negative charge clouds of oxygen ions. The t_{2g} orbitals have lobes pointed perpendicular to the B–O axis between the negative oxygen ions. As a result the e_g states experience a greater repulsion than the t_{2g} states and consequently lie at a higher energy. Similar reasoning suggests that the $p_\parallel$ states lie below the $p_\perp$ states when it is noted that B ion cores appear as positively charged centers.

In insulating perovskites such as $SrTiO_3$ the p states are completely filled while the d states are completely empty. The energy difference, E_g, between the t_{2g} and $p_\perp$ states is approximately equal to the energy gap. Metallic and semiconducting materials have the d states partially filled. $NaWO_3$ or ReO_3 have a single electron in a t_{2g} state.

In most but not all cases the energy bands involving the s state of the A ion are at energies much higher than the primary valence and conduction bands of a perovskite and therefore these bands are unoccupied. As a result the s state of the A ion usually does not play any significant role in determining the electronic properties. This is not to say that the A ion is not important. The electrostatic potentials of the A ions have a strong influence on the energy of the p–d valence and conduction bands. Furthermore, the size of the A ion is a significant factor in determining whether the crystal structure is distorted from the ideal cubic form. Nevertheless, given a particular perovskite structure and the effective electrostatic potentials acting on the B and O sites, the orbitals of the A ion may usually be omitted from electronic structure calculations. This leads to a major conceptual simplification because the electronic properties of the perovskites may be regarded as arising solely from the BO_3 part of the ABO_3 structure. This implies, for example, that the electronic *structure* of $BaTiO_3$ and $SrTiO_3$ should be essentially the same. According to the same reasoning the electronic *structure* of Na_xWO_3 should be independent of x. This does not mean that the properties are the same, but only that the available electronic states are the same. Obviously, the properties of WO_3 are completely different from those of $NaWO_3$; the former is an insulator and the latter is a metal. However, as a first approximation the only effect of the sodium is to donate electrons which occupy the t_{2g} states of the tungsten ion.

1.5 Covalent mixing

In addition to electrostatic interactions, the ions can interact because of the overlap of the electron wavefunctions. This leads to hybridization between the p and d orbitals and the formation of covalent bonds between the transition metal ions and the oxygen ions. It is frequently assumed that the covalent mixing in insulating materials such as $SrTiO_3$ is negligible. This is not correct. Nearly all of the physical and chemical properties of the perovskites are significantly affected by covalency.

To understand covalent mixing we consider a *cluster* of atoms consisting of a transition metal ion and its octahedron of oxygen ions. The wavefunctions of the cluster can be written in the form:

$$\psi^{(n)}(\vec{r}) = \sum_\alpha a_\alpha^{(n)} \varphi_{d\alpha}(\vec{r}) + \sum_{\vec{R}_i} \sum_\beta b_{i\beta}^{(n)} \varphi_{p\beta}(\vec{r} - \vec{R}_i), \tag{1.3}$$

where $\psi^{(n)}(\vec{r})$ is the cluster wavefunction for the nth eigenstate. $\varphi_{d\alpha}(\vec{r})$ is a d orbital on the B ion of α-type ($\alpha = xy, xz, \ldots$, etc.) and $\varphi_{p\beta}(\vec{r} - \vec{R}_i)$ is a p orbital centered at an oxygen ion located at $\vec{R}_i$ of the βth-type ($\beta = x$, y, or z). The coefficients $a_\alpha^{(n)}$ and $b_{i\beta}^{(n)}$ are constants which specify the amplitudes of the different orbitals which compose the nth eigenstate.

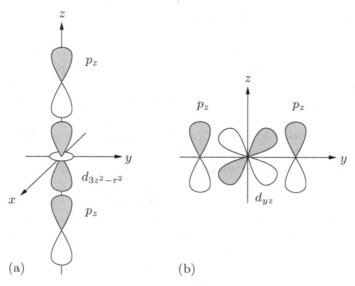

Figure 1.3. Overlap between cation d orbitals and anion p orbitals. (a) Sigma overlap and (b) pi overlap.

For the ionic model the wavefunctions are either pure d orbital ($b_{i\beta}^{(n)} = 0$) or pure p orbital ($a_\alpha^{(n)} = 0$). For the cluster the wavefunctions are still predominantly d or p orbital in character but there is a significant covalent mixing between the two (both $b_{i\beta}^{(n)}$ and $a_\alpha^{(n)} \neq 0$). The mixing comes about because of the overlap between d orbitals centered on the cation and the p orbitals on neighboring oxygen ions. There are two types of p–d overlap. The first is overlap between the d orbitals of the e_g type with p orbitals of the $p_\parallel$ type. This overlap is called "sigma" overlap. The second type, "pi" overlap occurs between t_{2g}-type d orbitals and $p_\perp$ orbitals. These two types of overlap are illustrated in Fig. 1.3. The overlap between t_{2g} and $p_\parallel$ orbitals or between e_g and $p_\perp$ orbitals vanishes by symmetry. If only the p and d orbitals are considered then there are 23 cluster states for a transition metal ion and the octahedron of oxygen ions. These 23 cluster states arise from admixtures

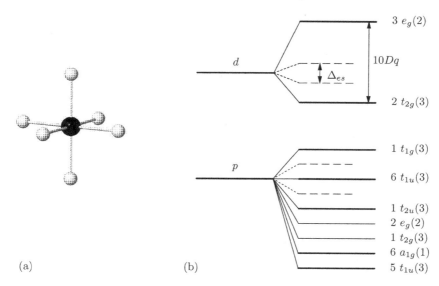

$3\,e_g(2)$

d

$10Dq$

Δ_{es}

$2\,t_{2g}(3)$

$1\,t_{1g}(3)$

p

$6\,t_{1u}(3)$

$1\,t_{2u}(3)$
$2\,e_g(2)$
$1\,t_{2g}(3)$
$6\,a_{1g}(1)$
$5\,t_{1u}(3)$

(a)   (b)

Figure 1.4. (a) BO_6 cluster and (b) the cluster levels. The dashed levels are for the electrostatic model. Δ_{es} is the electrostatic splitting.

of the 23 basis states; 5 d orbitals and 18 p orbitals, three on each of the six oxygen ions.

The cluster energy levels [10] are illustrated in Fig. 1.4. The labels given to the cluster energy levels indicate the group theoretical irreducible representations to which the wavefunctions belong. The prefix numbers are used to distinguish different levels which have the same symmetry properties. The degeneracies of the levels are indicated by the numbers in parentheses.

It is noted that the cation d orbitals are still split into the e_g and t_{2g} groups. These, so-called "ligand-field states" differ from those of the electrostatic model (Fig. 1.2) in two significant ways. First, the wavefunctions are no longer just d orbitals. They are admixtures of p and d orbitals. A second difference is that the splitting between the e_g and t_{2g} groups is much larger than for the electrostatic model. The cluster *ligand-field splitting* denoted by $10Dq$ is due to both electrostatic and covalent effects. The covalent contribution to $10Dq$ is usually much larger than the electrostatic contribution, Δ_{es}. Typically $10Dq$ is 2–3 eV in magnitude.

The ligand-field states, $3e_g$ and $2t_{2g}$, have wavefunctions in which the d orbitals combine out-of-phase with the p orbitals. The interference between the orbitals leads to a *depletion* of charge between the B and O ions. For this reason these states are called *antibonding* states. *Bonding* states are formed from in-phase combinations of the d and p orbitals. These states have wavefunctions that correspond to an accumulation of charge between the B and O ions. The bonding states are the $2e_g$

and $1t_{2g}$ levels (shown in Fig. 1.4). These states have hybridized wavefunctions, typically 70% p orbital and 30% d orbital. The percentage d-orbital admixture is a measure of the covalent bonding.

The remaining cluster levels have wavefunctions that are combinations of p orbitals located on the six oxygen ions. They do not hybridize with the d orbitals and therefore they do not contribute to the metal–oxygen bonding. Such states are called *non-bonding* states. Wavefunctions of the three types of cluster states are illustrated in Fig. 1.5.

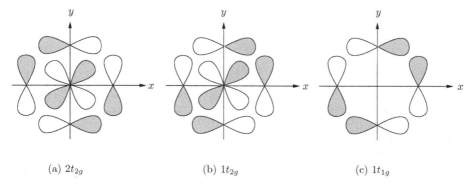

(a) $2t_{2g}$ (b) $1t_{2g}$ (c) $1t_{1g}$

Figure 1.5. Cluster states: (a) antibonding, (b) bonding, and (c) non-bonding.

It is important to note that electrons occupy d orbitals on the cation even when the $3e_g$ and $2t_{2g}$ levels are unoccupied. This is because of the covalent mixing of the d orbitals into the filled valence states below the $2t_{2g}$ level. This covalency effect is significant even for "ionic" insulators such as $SrTiO_3$. The ionic model implies that the titanium ion is Ti^{4+} with a d^0 configuration. Cluster models would give an effective valence such as $Ti^{3+}(d^1)$.

1.6 Energy bands

In the preceding section we considered a cluster model for the perovskites in which the transition metal ion interacts with the nearest-neighbor oxygen ions. The covalent mixing between the cation and anion wavefunctions leads to a partial occupation of d orbitals which, in the ionic model, were empty. A mechanistic interpretation of the covalent mixing is that the overlap between cation and anion wavefunctions provides a means of transferring electrons back and forth between the ions. Clearly, for an extended crystal structure the same mechanism will allow electrons to be shared between cations in adjacent clusters. Each oxygen of a given cluster is shared by adjacent cations. Cations can interact with each other through the intervening oxygen ion. An electron on a cation may be transferred to the oxygen ion and then from the oxygen ion to the second cation. When such processes occur the electrons become delocalized and electron energy bands are formed. It is important to note that the formation of d-electron bands requires two independent electron transfer processes. The delocalization of d electrons therefore is second order in the p–d overlap (or the probability of p to d electron transfer). This is quite different from a typical monatomic metal where delocalization is first order in the atomic overlap. For cubic perovskites the cation–cation separation is nearly 4 Å. This is too large for a significant direct overlap between cation orbitals and therefore band formation occurs by transfer of electrons between cations and anions whose separation is only about 2 Å.

In considering the energy bands of a perovskite it is appropriate to divide the crystal into unit cells each with the formula unit ABO_3. (The unit cell is shown in Fig. 1.1.) As discussed previously, the s states of the A ion can be neglected. Therefore, there will be 14 energy bands corresponding to the five d orbitals and nine p orbitals of each unit cell. The wavefunctions of the band states are characterized by a wavevector $\vec{k}$ and are of the form

$$\Psi_{\vec{k}}(\vec{r}) = \sum_{\vec{R}_d} \sum_{\alpha} a_\alpha(\vec{k}) \, e^{i\vec{k}\cdot\vec{R}_d} \, \varphi_{d\alpha}(\vec{r} - \vec{R}_d) + \sum_{\vec{R}_p} \sum_{\beta} b_\beta(\vec{k}) \, e^{i\vec{k}\cdot\vec{R}_p} \, \varphi_{p\beta}(\vec{r} - \vec{R}_p). \quad (1.4)$$

In (1.4), $a_\alpha(\vec{k}) \, e^{i\vec{k}\cdot\vec{R}_d}$ and $b_\beta(\vec{k}) \, e^{i\vec{k}\cdot\vec{R}_p}$ are respectively the amplitudes of the d and p orbitals of symmetries α and β located at the lattice sites $\vec{R}_d$ and $\vec{R}_p$.

An energy band diagram for a typical perovskite is shown in Fig. 1.6 for a model which includes only the interactions between nearest-neighbor ions [11]. For this simple model the energy bands divide into a set of sigma bands and a set of pi bands. The sigma bands involve only the e_g d orbitals and the $p_\parallel$ oxygen orbitals. The pi bands involve only the t_{2g} d orbitals and the $p_\perp$ oxygen orbitals.

The sigma bands have five branches: two distinct σ-type valence (bonding) bands, two distinct σ^*-type conduction (antibonding) bands and a single σ^0-type non-bonding band. The pi bands have nine branches: three equivalent π-type valence (bonding) bands, three equivalent π^*-type conduction (antibonding) bands, and three equivalent π^0-type non-bonding bands.

The bonding and antibonding (σ, σ^*, π, π^*) bands have wavefunctions whose p–d admixture varies as a function of the wavevector $\vec{k}$. At $\Gamma(\vec{k}=0)$ in the first Brillouin zone (see the inset in Fig. 1.6) the wavefunctions are pure p or pure d orbital in composition. The states at Γ have no covalent character and therefore correspond to the levels derived from the ionic model including the electrostatic potentials (Fig. 1.2(c)). As $\vec{k}$ varies along $\Gamma \to X \to M \to R$ the covalent mixture of the p and d orbitals increases. It is maximum at the point R, at the corner of the Brillouin zone. The states at R are very similar to the "g" states of the cluster

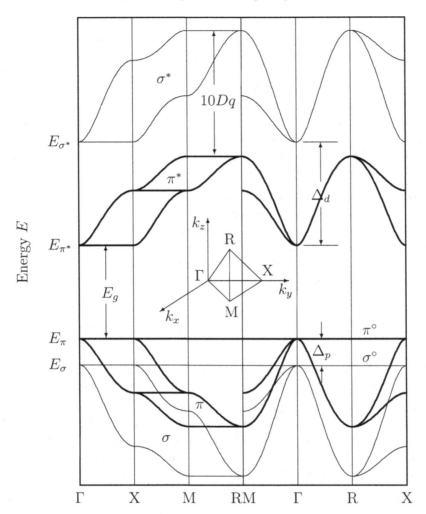

Figure 1.6. Energy bands for a typical perovskite showing the dispersion for $\vec{k}$-vectors along various lines in the Brillouin zone (inset) according to the LCAO model with nearest-neighbor interactions. The lighter curves are the pi bands and the darker curves are the sigma bands. The energies, E_g, $10Dq$, Δ_d, and Δ_p are the band gap, total (cluster) ligand-field splitting, d-orbital ligand-field splitting, and the p-orbital ligand-field splitting, respectively.

model (i.e., $2t_{2g}, 3e_g$, etc.). Thus the ionic model underestimates the covalency and the cluster model overestimates the covalency of the perovskites. The separation between the σ^* and π^* bands at Γ, $\Delta_{es}(d)$, corresponds to the electrostatic contribution to the ligand-field splitting. The separation at R is the total ligand-field band splitting and is approximately equal to $10Dq$.

The non-bonding band states for σ^0 and π^0 involve only oxygen $2p$ orbitals and therefore do not involve metal–oxygen covalent mixing. The band and cluster models produce similar non-bonding states.

The energy separation between the π^* and π^0 bands at Γ is the fundamental band gap, E_g. It varies between 1 and 4 eV and is largest for the insulating perovskites. Covalent mixing decreases with increasing band gap. The magnitude of the band gap is a measure of the ionicity of a perovskite. For example, the band gap of $SrTiO_3$ is 3.25 eV and that of ReO_3 is about 1 eV. This means that $SrTiO_3$ is much more ionic than ReO_3.

Insulating perovskites (e.g., $SrTiO_3$, $BaTiO_3$, or WO_3) have filled valence bands; that is, the σ, π, σ^0, and π^0 bands are completely occupied with electrons. The conduction bands (σ^* and π^*) are empty. Metallic perovskites such as $NaWO_3$ or ReO_3 have one electron per unit cell in the π^* conduction band. Examples of metallic compounds with two electrons in the π^* band are $CaMoO_3$, $BaMoO_3$, and $SrMoO_3$. Perovskites with more than two d electrons tend to form localized-states similar to those of the cluster model rather than delocalized band states.

Insulating perovskites can be rendered semiconducting or metallic by several means. Reduction in a hydrogen atmosphere produces oxygen vacancies. The vacancies act as donor centers; two electrons being donated by each vacancy (hydrogen itself may also remain in the lattice and act as a donor). Electron concentrations in the range of 10^{16}–10^{20} electrons/cm^3 can be produced in this way. Reduced insulating perovskites are n-type semiconductors with the Fermi level very near to the bottom of the π^* conduction band. n-type $SrTiO_3$ has been found to be a superconductor at temperatures below 0.3 K [12].

Insulating perovskites can also be doped by substituting appropriate ions into either the B or A sites. The tungsten bronzes Na_xWO_3, K_xWO_3, Li_xWO_3, and H_xWO_3 are special cases in which donor ions are substituted into the empty A sites of insulating WO_3. Electron concentrations of the order of 10^{22} electrons/cm^3 are obtained in this case. Many of the bronze compositions are superconductors.

One of the reasons perovskites are particularly valuable for research is that the electronic properties can be varied in a controlled fashion to produce almost any desired feature. The Fermi level in $SrTiO_3$ can be varied over a 3 eV range by going from cation- to anion-deficient compositions. The basic band structure does not change appreciably so the properties of such compositions are easily understood and interpreted in terms of a fixed band structure; that is the "rigid-band" approximation is valid. The rigid-band model is also applicable to the tungsten bronzes, and mixed compounds of the $A_x^{(1)}A_{1-x}^{(2)}BO_3$ type where $A^{(1)}$ and $A^{(2)}$ are different cations.

1.7 Localized d electrons

In the preceding section we indicated how the localized cluster states are delocalized because of the overlap of wavefunctions between adjacent clusters. The d-band formation is due to the transfer of electrons between cations via intervening oxygen ions. These electrons become delocalized and have an equal probability (proportional to $|e^{i\vec{k}\cdot\vec{R}}|^2=1$) of being found at any cation site. The band model neglects any possible spatial correlation between d electrons. The potential experienced by a given electron is assumed to be the same at every lattice site and equal to the average potential of the ion core and all other electrons. The usual one-electron band model explicitly ignores the fact that at any given instant of time a non-average number of electrons may be occupying the orbital of an ion. However, during the lifetime of the "non-average" ionic state the electrons on the site will experience a non-average potential. In particular, the intra-atomic Coulomb repulsion of an electron on a non-average site will be different from that at an average site.

Consider the situation in which we start with two metal ions each having n electrons. The electron–electron repulsion energy among the n electrons at each site is $\frac{1}{2}Un(n-1)$ where U is the Coulomb integral. If we transfer an electron from one site to the other there will be $n-1$ electrons on one site and $n+1$ on the other. The electron–electron repulsion energy will be $\frac{1}{2}Un(n+1)$ on the site with the extra electron and $\frac{1}{2}U(n-2)(n-1)$ on the other site. There is a change in the repulsion energy at one site of $\frac{1}{2}U[n(n+1)-n(n-1)]=nU$. At the other site the change in energy is $\frac{1}{2}U[(n-2)(n-1)-n(n-1)]=-Un+U$. Therefore, the net change is an additional repulsive energy equal to U. Thus, there is a Coulomb energy barrier to the creation of non-average ionic states.

Band formation is favorable because the delocalization of an electron reduces its kinetic energy (provided that the electron can occupy a state near the bottom of the band). For such a case the reduction in kinetic energy increases as the band width increases.

It is clear from what has been said that energy band formation will only be favorable if the reduction in kinetic energy is larger than the increase in the Coulomb energy. A variety of models which include a form of the Coulomb correlation energy have been used to find a criterion for the validity of the band model [13]. In general it is found that band theory applies when $W \gtrsim U$ where W is the band width. For W less than U, localized d-electron states are energetically favored. The precise criterion is model-dependent.

The localized electron criterion leads to interesting possibilities for the perovskites. The band width of the σ^* band is substantially larger than that of the π^* band and consequently, for a number of perovskites, the t_{2g} states are localized while the e_g states form σ and σ^* energy bands; $LaNiO_3$ with filled t_{2g} states and a single electron in the σ^* band is an example [14].

1.8 Magnetism in the perovskites

The occurrence of magnetism in the perovskites is closely connected to the existence of localized d electrons. In almost all cases where magnetism exists the d electrons are localized and possess localized spins. In such cases the local electronic configuration becomes an important consideration. One must be concerned with the multiplet structure. The tendency toward the formation of a multiplet configuration with a net spin arises from intra-atomic exchange and correlation. In atomic theory, Hund's rule states that the lowest-energy configuration corresponds to the state of maximum multiplicity or maximum spin and orbital angular momentum. Hund's rule is qualitatively applicable to the perovskites with localized d electrons. There are, however, some significant differences between atomic theory and the theory applicable to ions of the solid. The major differences between free ions and the cations in a solid perovskite are:

(1) the fivefold degenerate d states are split into the e_g and t_{2g} groups with a splitting of $10Dq$;
(2) the energy differences between different electronic configurations are not as widely separated as for the free ions;
(3) there is significant covalent mixing between the d-ion orbitals and the neighboring oxygen ion p orbitals.

As a consequence of (1) and (3) the electronic configuration of the cation should be specified in terms of the one-electron cluster states $3e_g$ and $2t_{2g}$. For simplicity the numerical descriptors of these states may be omitted. The d-electron configuration may then be specified by $(t_{2g}^n e_g^m)$, where n and m are the occupations of the $2t_{2g}$ and $3e_g$ levels, respectively.

The effect of (2) is that different valence states and different electronic configurations of the cation are closer in energy to each other than for the free ion. This is a result of polarization and electron screening of the Coulomb interactions. On applying Hund's rule to a perovskite cation the ligand-field splitting must be taken into account. When the number of d electrons, $m+n$, is between 4 and 7, Hund's rule can be violated if the ligand-field splitting is greater than the intra-atomic exchange energy. Consider, for example, $LaMnO_3$ which has Mn^{3+} ions with four d electrons. The intra-atomic exchange favors the "high-spin" configuration $^5E_g = (t_{2g} \uparrow^3 e_g \uparrow)$. However, occupying the e_g state involves a loss of binding energy equal to the ligand-field splitting. Therefore, the "low-spin" configuration $^3T_{2g} = (t_{2g} \uparrow^3 t_{2g} \downarrow)$ is competitive. Assuming a constant exchange, J, between parallel spin electrons, the intra-atomic exchange involves

$$E_{\text{ex}} = -J \sum_{i>j} \vec{s}_i \cdot \vec{s}_j,$$

where $\vec{s}_i$ and $\vec{s}_j$ are the spins of the occupied states. The ^5E_g has an exchange energy $-\frac{3}{2}J$ while for the $^3T_{2g}$, $E_{\text{ex}} = -\frac{3}{4}J$. However, the $^3T_{2g}$ has a ligand-field energy of $10Dq$. Therefore, the difference in the energies of the two configurations is

$$E(^5E_g) - E(^3T_{2g}) = -\frac{3}{4}J + 10Dq \equiv \Delta E.$$

When $\Delta E < 0$ the high spin state ^5E_g (spin $= 2$) is lower in energy than the low spin state $^3T_{2g}$ (spin $= 1$). If $\Delta E > 0$ then the low spin state is favored. Experiments on d^4 ions in perovskites show that the low spin state is usually favored. This indicates that the ligand-field splitting is larger than the intra-atomic exchange and Hund's rule does not apply.

When the cations possess localized spins, then long-range magnetic ordering can occur. The principal mechanism of spin–spin interactions is superexchange. Superexchange involves the antiferromagnetic coupling between nearest-neighbor cations by exchange of electrons with the intervening oxygen ion.

Examples of magnetically ordered perovskites are $LaCrO_3$, $PbCrO_3$, $CaMnO_3$, $LaFeO_3$, and many others. Those named above form the simple G-type magnetic cell in which the spins of nearest-neighbor cations are antiparallel. Many other types of magnetic ordering also occur among the magnetic perovskites.

As a final comment on localized d electrons we mention the importance of the Jahn–Teller effect. This effect is the spontaneous distortion of a cubic structure such as that of perovskites. When the cation electronic configuration is orbitally degenerate, the ground state will in some cases, be unstable to small distortional displacements. This Jahn–Teller distortion occurs because the electronic energy decreases linearly with displacement while the elastic energy increases as the square of the displacement. A minimum in the total energy always occurs for a small but finite distortional displacement.

1.9 Superconductivity

Superconductivity has been observed for n-type $SrTiO_3$ and for many of the compositions of the tungsten bronzes: Li_xWO_3, Na_xWO_3, K_xWO_3, Rb_xWO_3, and Cs_xWO_3. The occurrence of superconductivity in compounds whose elements are not superconducting and for which more than three-fifths of the atoms are oxygen is truly remarkable.

WO_3 is an insulator with a tetragonally distorted perovskite structure. With the addition of alkali ions to the empty A site a variety of metallic bronzes can be formed. The tungsten bronzes occur with cubic, hexagonal, and two different tetragonal perovskite-like structures [15]. For $Na_x WO_3$ the tetragonal I phase occurs in the range $0.2 < x < 0.5$. For values of $x < 0.2$ the tetragonal II phase exists. For values of $x > 0.5$ the cubic perovskite structure is stable. Tetragonal I, $Na_x WO_3$ and $K_x WO_3$ are superconducting with transition temperatures of 0.57 K [16] and 1.98 K [17], respectively. The cubic and tetragonal II phases are apparently not superconducting. Except for $Na_x WO_3$ and $K_x WO_3$, superconductivity occurs for the other bronzes when they are in the hexagonal phase [17–19]. The transition temperatures of the hexagonal bronzes are close to 2 K.

It has been found that the transition temperature of the hexagonal bronzes can be raised by a factor of 2 or 3 by etching in various acids [18]. The reasons for this enhancement are not yet clear. The transition temperature of the tetragonal I sodium tungsten bronze, $Na_x WO_3$, increases rapidly as x approaches 0.2 [19]. This enhancement occurs as the composition approaches the tetragonal II phase boundary, and is presumed to be associated with a lattice instability.

More recently various alloys of barium bismuthates have been studied extensively. The highest recorded T_c for a non-layered metal oxide is about 30 K for the alloy $Ba_{1-x} K_x BiO_3$ for $x = 0.38$. This superconducting material displays a transition to an insulating state at $x < 0.38$, but is a cubic, superconducting metal for $0.38 < x < 0.6$. The related compound $BaPb_{1-x} Bi_x O_3$ is also a superconductor with a maximum T_c of about 13 K. $BaBiO_3$ itself is an insulator even though according to conventional band theory it possesses a half-filled conduction band (antibonding Bi $6s$–O $2p$ sigma band).

In 1986 Bednorz and Müller [20] discovered a new class superconducting metal oxides ($La_2 CuO_4$ doped with Ba^{2+}, Sr^{2+}, or Ca^{2+}) one of which possessed a critical temperature, T_c, in excess of 30 K. Their discovery was followed by a worldwide research effort that turned up many other cuprate superconducting materials with even higher critical temperatures, the record high being around 166 K, a temperature that is above the boiling point of liquid nitrogen. These "high-T_c cuprate superconductors" are characterized by sets (one or more layers) of "immediately adjacent" planes of copper ions surrounded by four oxygen ions. Each set of "immediately adjacent" layers is separated from the next set by "isolation layers" (La–O planes in the case of $La_2 CuO_4$) that are poorly conducting. Despite intensive experimental and theoretical research efforts, the mechanisms underlying the high-temperature superconductivity as well as the properties of the "normal" state above T_c are not well understood. However, there seems to be agreement that the two-dimensional character of the Cu–O bonding and the resulting large density of states are important. The question of whether the mechanism of electron pairing in these materials can be explained within the framework of "conventional", BCS (phonon-mediated) superconductivity theory [21] is still open and other significantly different mechanisms have been put forward [22].

In general, these high-T_c superconductors do not have the perovskite structure. However, they share an important characteristics with the metal-oxide perovskites. Even though the perovskites are not layered in the way the high-T_c superconductors are, the electronic coupling between the transition metal d orbitals and the oxygen p orbitals is two-dimensional because of the symmetry of the interactions.

1.10 Some applications of perovskite materials

The technological uses of perovskite and perovskite-related materials are extensive and we will not attempt to review the field. In this section we shall only briefly mention some of the common applications. References are given to only a few representative papers in the vast literature.

The piezoelectric insulating perovskites such as $BaTiO_3$, $PbTiO_3$, $PbZrO_3$, $Pb(Zr_xTi_{1-x})O_3$ (PZT), and $LiNbO_3$ have been employed extensively as solid-state device materials. Some solid-state applications include switching devices, infrared detectors, and a large variety of signal processing devices [23]. These materials are employed as substrates for the generation of bulk and surface acoustic (elastic) waves. Because of their piezoelectric properties, acoustic waves are accompanied by an oscillating electric field [24, 25]. It is possible to generate acoustic waves by applying an oscillating electric field to the substrate and conversely an acoustic wave may be detected by the electric field it generates. The coupling between the elastic displacement and the electromagnetic field is nonlinear and produces second-harmonic electric fields [26]. These properties have been employed to design a number of acoustic wave signal processing devices, including time delay lines, filters, acoustic wave image devices and nonlinear convolution and correlation devices.

The nonlinear optical properties of the perovskite insulators are used for the generation of second-harmonic optical waves. The second-harmonic generation coefficient of $PbTiO_3$ is among the highest known [27]. Other applications of the insulators include photochromic, electrochromic, image storage, and display devices [27]. In the photochromic applications the transparent host materials are doped with impurity transition metal ions or rare earth ions. The impurity ions have several localized levels lying within the band gap of the substrate that correspond to different valence states. The valence state of the impurity ion can be changed by photoexcitation. Impurity ions are selected for which one valence state has an absorption band in the visible while the second does not. Colored images can be "written" by a light beam that causes photoexcitations of these metastable valence states. The images can be erased by a second light beam of a different wavelength which depopulates the metastable states. For electrochromic devices, the valence states are changed by shifting the quasi-Fermi level or by reducing the material electrochemically. A thin film of WO_3 may be changed from transparent to a deep blue color by electrochemically converting W^{6+} ions to W^{5+} ions.

There is also interest in the surface chemical properties of the perovskites. Many are excellent gas-phase catalysts and in addition several are photocatalytic and electrocatalytic.

Interest in the catalytic properties of the perovskites began with the suggestion that the rare earth cobalt oxides, $RCoO_3$ (R = rare earth ion) might prove useful as substitutes for Pt-based automotive exhaust catalysts [24, 25]. Strong catalytic activity and a high degree of selectivity have been found for a large variety of perovskite materials [28].

Many energetically favorable (exothermic) gas-phase reactions do not occur spontaneously but require a catalyst in order to occur. The detailed mechanisms of catalytic action are not known in most cases, but some general features are understood. There are two principal factors which inhibit an exothermic reaction. The first is related to the symmetry of the reactant and product states. As two molecules come together in a chemical reaction the orbitals of the complex must evolve from those of the reactants to those of the products. One can imagine a continuously changing set of hybridized molecular orbitals for the reacting molecules. It frequently happens that the occupied orbitals of the ground state of the product *can not* evolve from a hybridization of the occupied orbitals of the reactants. An example is the hydrogenation of ethylene to form ethane. The bonds associated with the hydrogen atoms of ethane evolve from a hybridization of empty antibonding states of the reacting hydrogen and ethylene molecules, and not from the occupied bonding states. Electron flow from the occupied bonding states to the empty antibonding states is forbidden by symmetry consideration. Reactions can be classified as symmetry-"allowed" or symmetry-"forbidden" in much the same way as optical translations [29]. Reactions that are "forbidden" are inhibited by a symmetry-imposed barrier.

Another barrier encountered is associated with the charge transfer involved in the reaction. As charge flows from the reactant states to the ground state of the product, the molecules must often pass through a transient polar configuration. Such polar configurations are usually energetically unfavorable because the electron affinity of a molecule is small compared to its ionization energy. The inhibition to charge transfer acts as an additional barrier to the chemical transformation.

The catalytic properties of the perovskites are directly related to the presence of coordinatively unsaturated transition metal ions on the surface. The term coordinatively unsaturated, refers to the fact that an ion on the surface will often have less than its normal complement of six oxygen ligands. Such ions provide active sites for adsorption of reactant molecules because in this way the ion can attain its normal number of ligands. The symmetry of the d orbitals is favorable for interaction with both the bonding and antibonding states of most molecules.

It is generally believed that chemisorption of one or more of the reactant molecules to form a surface complex is a precursor to a catalyzed reaction. The role of the surface complex in catalysis is twofold: d orbitals can hybridize with the reactant molecule orbitals in such a way as to provide a symmetry-allowed path for the reactions [30]. In addition, the adsorption of the reaction species greatly facilitates charge transfer processes. When molecules condense on a solid substrate the ionization energy of the molecular levels is reduced due to a process known as extra-atomic relaxation [31]. Furthermore the barrier to charge transfer is reduced by the solid-state effects of polarization and electron screening. It is also possible for charge transfer to occur via the transition metal ion. The catalyst ion acts as an intermediary to accept (donate) electrons from the reactants and to donate (accept) electrons to the product. This process involves a valence fluctuation of the cation. Such fluctuations are of low energy compared to fluctuations of charge on free molecules. The energy required for a valence fluctuation can be minimized in systems such as the mixed or non-stoichiometric perovskites since they already contain mixed valence transition metal ions.

There are several factors that make the perovskites particularly attractive as catalyst systems for research. One factor is that they form a large class of structurally similar compounds whose electronic properties can be varied in a controlled way. This permits a systematic study of the effects of variations in electronic parameters on catalytic rate, for example. (Pt is an excellent catalyst but there is little that can be done to vary its electronic state and therefore to discover why it is such a good catalyst.) A second factor making the perovskites important as catalysts is that they are highly stable at high temperatures and in hostile chemical environments.

Voorhoeve *et al.* [28] have reported extensive studies on a variety of perovskite catalysts. Co, Mn, and Ru perovskites have been investigated as catalysts for the oxidation of carbon monoxide and hydrocarbons and for the reduction of the oxides of nitrogen. (Such catalytic conversions are important in removing pollutants from auto exhaust.) Particular examples of perovskite catalysts investigated include $SrRuO_3$, $LaRuO_3$, and the substituted system $(La_xK_{1-x})(Ru_yMn_{1-y})O_3$. The catalysts are very active and highly selective in the reduction of nitrogen oxides. The use of substituted systems permits a controlled variation of valence states for the cations. The electronic properties can be tailored for a particular application.

Another important feature of perovskite catalysts is that they can be designed to simultaneously catalyze reduction and oxidation reactions. An example of such a catalyst is $(La_{0.8}Sr_{0.2})(Co_{0.9}Ru_{0.1})O_3$. Several perovskite compositions have been found to be superior or comparable to commercial catalysts.

It is also noted that surface oxygen vacancies on a perovskite can serve as catalytically active sites. NO is believed to dissociate into adsorbed N by reaction at a vacancy site on manganite catalysts. The liberated oxygen from the NO molecule can fill the empty surface oxygen site. The vacancy can be restored by reaction with an reducing agent such as CO. Such sites are useful in catalysis systems where both reduction and oxidation processes are desired.

The perovskites have not yet emerged as commercially competitive catalysts, but have proved valuable in the study of the possible mechanisms of catalysis.

As a final topic in the applications of perovskites we mention their use as electrochemical electrodes.

Materials such as $LaCoO_3$, n-type $SrTiO_3$ and the tungsten bronzes have been utilized as anode materials in electrochemical cells. They are particularly useful because of their stability in an electrolyte. Nearly ideal reversible electrode behavior has been shown for $LaCoO_3$ and related compounds such as $(La_{0.5}Sr_{0.5})CoO_3$ [32].

A particularly interesting application is the photoelectrolysis of water. The electrochemical cell consists of an n-type anode such as $SrTiO_3$, $BaTiO_3$ or some substituted perovskite and a Pt counterelectrode [33–36]. The electrolyte may be either alkaline or acid aqueous solution. The electrodes are connected through an external circuit and the electrolysis process is driven photocatalytically by photons incident on the anode surface.

In alkaline aqueous solution the anode reaction is

$$2\,p^+ + 2\,OH^- \rightarrow \frac{1}{2}\,O_2 + H_2O\,,$$

where p^+ designates a hole. Electron–hole pairs are generated in the oxide anode by absorption of incident photons with energies equal to or greater than the band gap. The electrons and holes are separated by the internal electric field of the oxide (due to band bending at the oxide–electrolyte interface). In n-type materials such as $SrTiO_3$ the band bending creates an electron depletion (hole accumulation) region at the surface. The holes combine with adsorbed hydroxyl ions to produce molecular oxygen and water as indicated by the anode reaction above. The electrons are discharged at the cathode, producing hydrogen.

The feasibility of photoelectrolysis for the production of hydrogen has been demonstrated using band-gap photons with $\hbar\omega \sim 3\,eV$ for several oxides. The first such experiments utilized n-type TiO_2 [36]. Later studies employed perovskites such as $SrTiO_3$ [33, 34]. These experiments have raised the exciting possibility of developing a solar-driven electrolysis system for the production of hydrogen fuel. The band-gap energy of $SrTiO_3$ or TiO_2 is too large for efficient solar-driven devices and therefore interest has been stimulated to search for another oxide with a smaller band gap. Methods of reducing the energy for creating electron–hole pairs in large band-gap materials are also being considered. One such method is the use of adsorbed sensitizing dye molecules. Surface states in the band-gap region offer another way for the generation of electron–hole pairs with less-than-band-gap radiation. Such surface states may also be involved in electrocatalyzing the anode reaction.

Suggested additional reading material

General survey

J. B. Goodenough, Metallic oxides. *Prog. Solid State Chem.* **5**, 145 (1971).

T. Wolfram, E. A. Kraut, and F. J. Morin, Qualitative discussion of energy bands. *Phys. Rev.* B **7**, 1677 (1973).

Reference texts

S. Sugano, Y. Tanabe, and H. Kamimura, *Multiplets of transition metal ions* (New York, Academic Press, 1970).

References

[1] F. Wohler, *Ann. Chem. Phys.* **29**, 43 (1823).
[2] L. F. Mattheiss, *Phys. Rev.* **6**, 4718 (1972).
[3] E. O. Wollan and W. C. Koehler, *Phys. Rev.* **100**, 545 (1955).
[4] F. L. Battye, H. Höchst, and A. Goldman, *Solid State Commun.* **19**, 269 (1976).
[5] M. Cardona, *Phys. Rev.* **140**, A651 (1965).
[6] P. A. Lightsey, *Phys. Rev.* B **8**, 3586 (1973).
[7] V. E. Henrich, G. Dresselhaus, and H. J. Zeiger, *Bull. Am. Phys. Soc.* **22**, 364 (1977).
[8] R. D. Shannon, *Acta Cryst.* A **32**, 751 (1976).
[9] E. A. Kraut, T. Wolfram, and W. E. Hall, *Phys. Rev.* B **6**, 1499 (1972).
[10] T. Wolfram, R. A. Hurst, and F. J. Morin, *Phys. Rev.* B **15**, 1151, (1977).
[11] T. Wolfram, E. A. Kraut, and F. J. Morin, *Phys. Rev.* B **7**, 1677 (1973).
[12] J. F. Schooley, W. R. Hosler, E. Ambler, J. H. Becker, M. L. Cohen, and O. S. Koonce, *Phys. Rev. Lett.* **14**, 305 (1965).
[13] J. Hubbard, *Proc. Roy. Soc. (London)* A **276**, 238 (1963).

[14] J. B. Goodenough and P. M. Roccah, *J. Appl. Phys.* **36**, 1031 (1965).

[15] A. D. Wadsley, *Non-stoichiometric compounds*, ed. L. Mandalcorn (New York, Academic Press, 1964).

[16] Ch. J. Raub, A. R. Sweedler, M. A. Jensen, S. Broadston, and B. T. Matthias, *Phys. Rev. Lett.* **13**, 746 (1964).

[17] A. R. Sweedler, Ch. J. Raub, and B. T. Matthias, *Phys. Lett.* **15**, 108 (1965).

[18] J. P. Remeika, T. H. Geballe, B. T. Matthias, A. B. Copper, G. W. Hull, and E. M. Kelley, *Phys. Lett.* **24A**, 565 (1967).

[19] H. R. Shanks, *Solid State Commun.* **15**, 753 (1974).

[20] J. G. Bednorz and K. A. Müller, *Z. Phys.* B **64**, 189 (1986).

[21] J. Bardeen, L. N. Cooper, and J. R. Schrieffer, *Phys. Rev.* **108**, 1175 (1957).

[22] See for example, P. W. Anderson, *The theory of superconductivity in the high-T_c cuprates* (Princeton, NJ, Princeton University Press, 1997).

[23] A collection of papers on the applications may be found in the *IEEE Trans. Sonics Ultrason.* **19** (2) (1972).

[24] L. A. Pedersen and W. E. Libby, *Science* **176**, 1355 (1972).

[25] D. B. Meadowcraft, *Nature* **226**, 847 (1970).

[26] T. C. Lim, E. A. Kraut, and R. B. Thompson, *Appl. Phys. Lett.* **20**, 127 (1972).

[27] Z. J. Kiss, Photochromics. *Phys. Today* **23**, 42 (1970).

[28] R. J. H. Voorhoeve, D. W. Johnson, Jr., J. P. Remeika, and P. K. Gallagher, *Science* **195**, 827 (1977).

[29] R. B. Woodward and R. Hoffman, *The conservation of orbital symmetry* (Weinheim/Bergstrasse, Verlag-Chemic, 1970).

[30] T. Wolfram and F. J. Morin, *Appl. Phys.* **8**, 125 (1970).

[31] D. A. Shirley, *J. Vac. Sci. Technol.* **12**, 280 (1975).

[32] A. C. C. Tseung and H. L. Beven, *J. Electroanal. Chem. Interfacial Electrochem.* **45**, 429 (1973).

[33] J. G. Mavrodies, D. I. Tchernev, J. A. Kafalas, and D. F. Kolesar, *Mater. Res. Bull.* **10**, 1023 (1975).

[34] M. S. Wrighton, A. B. Ellis, P. T. Wolczanski, D. L. Morse, H. B. Abrahamson, and D. S. Ginley, *J. Amer. Chem. Soc.* **98**, 2774 (1976).

[35] R. D. Nasby and R. K. Quinn, *Mater. Res. Bull.* **11**, 985 (1976).

[36] A. Fujishima and K. Honda, *Nature* **238**, 37 (1972).

Problems for Chapter 1

1. The perovskite structure is often depicted as an array of octahedra. Each octahedron consists of the B ion at the center and six oxygen ions, one at each corner (shared by adjacent octahedra), along the $\pm x$, $\pm y$, and $\pm z$ axes. The A ions are located in the spaces between the octahedra. Make a sketch of the perovskite structure using an

array of octahedra and indicate the positions of the different ions.

2. Using the ionic model discuss the electronic structure expected for $KTaO_3$. What are the electronic configurations of the ions? Would you expect this material to be metallic or insulating?

3. For the perovskites why are the electronic states derived from the A ion usually less important than those of the B and O ions?

4. The notation, M:ABO_3, indicates an ABO_3 perovskite doped with M ions. Classify the following materials as n-type or p-type semiconductors: Nb:$SrTiO_3$, La:$BaTiO_3$, Na:WO_3, $KTaO_{3-\delta}$ (oxygen deficient).

5. Using information contained in the chapter what would you expect for the ionic energy gap in eV between the p- and d-levels in $BaTiO_3$? Assume covalency reduces the effective charges to 80% of their full ionic charges and that the electron "affinity" for adding an electron to the O ion is a repulsive energy of 9 eV. What effect would you expect ligand-field splitting to have on the energy gap? Explain your answer.

6. The notation "g" and "u" for the levels of the BO_6 cluster comes from the German words "gerade" and "ungerade" meaning even and odd or symmetric and unsymmetric. For a cluster with cubic symmetry the states must be either "g" or "u", and "g" and "u" functions can not be combined. The "g" cluster states are symmetric with respect to inversion through the center of the B ion. The d orbitals are all symmetric under inversion. Specify the combinations of neighboring p orbitals that will covalently mix with the d orbitals to form "g" states.

7. Energy band states are electron waves that vary in phase as $e^{i\vec{k}\cdot\vec{r}}$ where $\vec{k}$ is the wavevector for the state. For $\vec{k}=0$ the phase of an orbital is the same in all unit cells. Thus the oxygen orbitals have the same phase on either side of the B ion. Explain why the BO_6 cluster can never have a wavefunction involving p and d orbitals for which the p orbitals have the same phase on either side of the B ion.

8. The energy bands for the ABO_3 structure are illustrated in Fig. 1.6. Discuss why the parameter $10Dq$ is shown as the energy difference between the π^* and σ^* bands at M or R rather than at Γ. The components of the wavevectors for Γ, M, and R are $(0,0,0)$, $(1,1,0)$, and $(1,1,1)$, respectively, in units of $\pi/2a$.

9. The density of states, $\rho(E)$, is defined to be the number of electronic states in the energy range between E and $E+dE$. In Fig. 1.6, the energy bands have flat bands along various symmetry directions. What happens to $\rho(E)$ at an energy for which one of the bands is flat?

2

Review of the quantum mechanics of N-electron systems

This chapter is intended as a brief review of the quantum theory of N-electron systems. It also serves to introduce the linear combinations of atomic orbitals (LCAO) method.

2.1 The Hamiltonian

The total N-electron Hamiltonian, H_T, for an atomic, molecular, or solid system is of the form:

$$
H_T = \sum_A \left\{ -\frac{\hbar^2}{2M_A} \nabla_A^2 + \sum_{B<A} \frac{Z_A Z_B e^2}{|\vec{R}_A - \vec{R}_B|} \right\}
$$
$$
+ \sum_i \left\{ -\frac{\hbar^2}{2m} \nabla_i^2 + \sum_{j<i} \frac{e^2}{|\vec{r}_i - \vec{r}_j|} \right\} - \sum_A \sum_i \frac{Z_A e^2}{|\vec{r}_i - \vec{R}_A|}, \tag{2.1}
$$

where $\vec{R}_A$ and $\vec{R}_B$ are the nuclear positions and $\vec{r}_i$ and $\vec{r}_j$ are the electron coordinates. The terms in the first set of brackets are the kinetic energies of the nuclei (having mass M_A) and the Coulomb repulsions among them. The terms in the second brackets are the kinetic energies of electrons and the electron–electron repulsions. The last term is the electron–nuclear attractions. eZ_A is the charge of the nucleus located at $\vec{R}_A$, $-e$ is the charge of an electron and m is the electron mass.

For our purposes we consider the nuclei fixed at their equilibrium positions in the crystal and seek the solutions of Schrödinger's equation for the electronic wavefunction (Born–Oppenheimer approximation [1]). The electronic wavefunctions satisfy the equation

$$
H\Psi(\tau) = E\Psi(\tau) \tag{2.2}
$$

$$
H = -\sum_i \frac{\hbar^2}{2m} \nabla_i^2 + \sum_i \sum_{j<i} \frac{e^2}{|\vec{r}_i - \vec{r}_j|} - \sum_A \sum_i \frac{Z_A e^2}{|\vec{r}_i - \vec{R}_A|} \tag{2.3}
$$

where $\Psi(\tau)$ is the N-electron wavefunction with τ representing the N electron spatial and spin coordinates. The electronic energy, E, is determined by

$$E = \frac{\int \Psi^*(\tau)\, H\, \Psi(\tau)\, d\tau}{\int \Psi^*(\tau)\, \Psi(\tau)\, d\tau}\,. \tag{2.4}$$

2.2 The Slater determinant state

The N-electron wavefunction $\Psi(\tau)$,

$$\Psi(\tau) \equiv \Psi(\vec{r}_1, \vec{r}_2, \vec{r}_3, \ldots, \vec{r}_N;\ s_1, s_2, s_3, \ldots, s_N), \tag{2.5}$$

is required to be antisymmetric under the set of operations which interchange the space and spin coordinates of any pair of electrons.

The Slater determinant wavefunction is an approximate N-electron wavefunction, constructed from one-electron spin orbitals, that satisfies the antisymmetric property. Let $\phi_i(\vec{r})$, $i = 1, 2, \ldots$ be a complete set of orthonormal one-electron functions and $\chi_i(s)$ the spin functions corresponding to either "spin up" or "spin down". The spin orbitals $\psi_i(\tau)$ are the product states

$$\psi_i(\tau) \equiv \phi_i(\vec{r})\, \chi_i(s) \tag{2.6}$$

where τ represents the space and spin coordinates of an electron. The orthogonality condition is

$$\int \psi_i^*(\tau)\, \psi_j(\tau)\, d\tau = \delta_{ij}\,. \tag{2.7}$$

The Slater determinant wavefunction for an N-electron system is

$$\Delta_\nu^N = \frac{1}{\sqrt{N!}} \begin{vmatrix} \psi_1(1) & \psi_2(1) & \cdots & \psi_N(1) \\ \psi_1(2) & \psi_2(2) & \cdots & \psi_N(2) \\ & \vdots & & \\ \psi_1(N) & \psi_2(N) & \cdots & \psi_N(N) \end{vmatrix} \tag{2.8}$$

where ν symbolizes the set of N spin orbitals used in constructing Δ_ν^N. The factor $1/\sqrt{N!}$ ensures that the wavefunction is normalized so that $\int (\Delta_\nu^N)^*\, \Delta_\nu^N\, d\tau_1\, d\tau_2 \cdots d\tau_N = 1$. A different Slater determinant can be constructed from each different set of N spin orbitals. The wavefunction of an N-electron system can be approximated by a linear combination of such Slater determinants

$$\Psi = \sum_\nu a_\nu \Delta_\nu^N \tag{2.9}$$

where a_ν are the constant coefficients specifying the amplitude of different "configurations" comprising the total wavefunction, Ψ. Use of more than one slater determinant is referred to as configuration interaction. In many applications to solids and molecules, configuration mixing is omitted and Ψ is approximated by a single Slater determinant. We shall limit this discussion to a single Slater determinant wavefunction.

The electronic Hamiltonian, H, includes one-electron operators $f(i)$,

$$f(i) = -\frac{\hbar^2}{2m}\,\nabla_i^2 - \sum_A \frac{Z_A\,e^2}{|\vec{r}_i - \vec{R}_A|}, \tag{2.10}$$

and two-electron operators g_{ij},

$$g_{ij} = \frac{e^2}{|\vec{r}_i - \vec{r}_j|}. \tag{2.11}$$

The expectation or average value of the energy of a particular N-electron Slater determinant wavefunction is

$$\langle E_\nu^N \rangle = \int (\Delta_\nu^N)^* H \Delta_\nu^N \, d\tau$$

$$= \sum_i \int \phi_i(1)^* f(1)\,\phi_i(1)\,d\vec{r}_1$$

$$+ \frac{1}{2}\sum_{ij} \iint \psi_i(1)^* \psi_j(2)^* g_{12}\,(1 - P_{12})\,\psi_j(2)\,\psi_i(1)\,d\tau_1\,d\tau_2 \tag{2.12}$$

where the sums are over all N states appearing in the set ν. The operator, P_{12} is the exchange operator defined by

$$P_{12}\,\psi_j(2)\,\psi_i(1) = \psi_j(1)\,\psi_i(2). \tag{2.13}$$

2.3 Koopman's theorem

Consider two Slater determinant wavefunctions Δ_ν^N and $\Delta_{\nu(k)}^{N-1}$ where the set of $(N-1)$ spin orbitals in $\nu(k)$ is the same as those of ν except that the state ψ_k does not appear.

We want to compare the average energy of these two approximate wavefunctions. The energy difference, according to (2.12), is

$$\varepsilon_k^N \equiv \langle E_\nu^N \rangle - \langle E_{\nu(k)}^{N-1} \rangle$$

$$= \int \phi_k(1)^* f(1)\,\phi_k(1)\,d\vec{r}_1$$

$$+ \sum_j \iint \psi_k(1)^* \psi_j(2)^* g_{12}\,(1 - P_{12})\,\psi_j(2)\,\psi_k(1)\,d\tau_1\,d\tau_2. \tag{2.14}$$

ε_k^N is the energy difference between the N and $(N-1)$-electron states having $(N-1)$ common spin orbitals. As we shall see ε_k^N is closely related to, but not equal to, the binding energy of an electron of the N-electron system.

2.4 Hartree–Fock equations

The Hartree–Fock equations are obtained by application of the variational principle to a Slater N-electron wavefunction. Given an approximate wavefunction constructed from one-electron spin orbitals, the best choices of basis orbitals are those that minimize the average energy. We therefore seek the minimum value of $\langle E_\nu^N \rangle$ with variations in the spin-orbital functions. The variations are constrained by the requirement that the states remain orthonormal:

$$\int d\tau \, \psi_i(\tau)^* \, \psi_j(\tau) = \delta_{ij} \, .$$

This constraint may be imposed by the method of Lagrange multipliers. Thus we minimize the function F, where

$$F \equiv \langle E_\nu^N \rangle - \sum_i \sum_{j<i} \lambda_{ij}^N \left\{ \int d\tau \, \psi_i(\tau)^* \, \psi_j(\tau) - \delta_{ij} \right\} . \qquad (2.15)$$

The constants, λ_{ij}^N, are the Lagrange multipliers. We require

$$\frac{\delta F}{\delta \phi_k^*} = 0$$

and find

$$\left\{ f(1) + \left[\sum_j \int d\tau_2 \, \psi_j(2)^* g_{12} \, (1 - P_{12}) \, \psi_j(2) \right] \right\} \psi_i(1) = \sum_j \lambda_{ij}^N \, \psi_j(1) \, . \qquad (2.16)$$

Equations (2.16) are known as the Hartree–Fock equations and it is convenient to transform them to a representation in which λ_{ij}^N is diagonal. The set of numbers λ_{ij}^N form a matrix that can be diagonalized by a unitary transformation:

$$\varepsilon_i^N \delta_{ij} = \sum_{k\ell} C_{ki}^* \, \lambda_{k\ell}^N \, C_{\ell j} . \qquad (2.17)$$

The unitary properties of the transformation are expressed by the relations:

$$C_{ik}^+ = C_{ki}^* \, , \qquad (2.18)$$

$$\sum_k C_{ki}^* \, C_{kj} = \delta_{ij} \, . \qquad (2.19)$$

We define the transformation of the spatial part of the spin orbitals by

$$\phi_i(1) = \sum_k C_{ik}\, \phi'_k(1), \tag{2.20}$$

$$\phi_i(1)^* = \sum_k C_{ik}^+\, \phi'_k(1)^* = \sum_k C_{ki}^*\, \phi'_k(1)^*. \tag{2.21}$$

We note that the term

$$\sum_j \int d\tau_2\, \psi_j(2)^*\, g_{12}\,(1-P_{12})\,\psi_j(2) = \sum_j \int d\tau_2\, \psi'_j(2)\, g_{12}\,(1-P_{12})\,\psi'_j(2) \tag{2.22}$$

where

$$\psi'_j(2) = \phi'_j(r_2)\,\chi_j(s_2)\,.$$

Substituting (2.20) and (2.21) into (2.16) gives

$$\sum_k C_{ik}\left\{ f(1) + \left[\sum_j \int d\tau_2\, \psi'_j(2)^*\, g_{12}(1-P_{12})\,\psi'_j(2)\right]\right\}\psi'_k(1)$$

$$= \sum_{kj} C_{jk}\, \lambda^N_{ij}\, \psi'_k(1). \tag{2.23}$$

Multiplying (2.23) by $C^*_{\ell i}$ and summing over all i gives

$$\left\{ f(1) + \left[\sum_j \int d\tau_2\, \psi'_j(2)^*\, g_{12}\,(1-P_{12})\,\psi'_j(2)\right]\right\}\psi'_\ell(1) = \varepsilon^N_\ell\, \psi'_\ell(1). \tag{2.24}$$

Equation (2.24) is the diagonal form of the Hartree–Fock equations. If we multiply (2.24) by $\psi'_\ell(1)^*$ and integrate over τ_1 then we obtain

$$\int d\tau_1\, \psi'_\ell(1)^*\, f(1)\, \psi'_\ell(1)$$

$$+ \sum_j \int d\tau_1 \int d\tau_2\, \psi'_\ell(1)^*\, \psi'_j(2)^*\, g_{12}\,(1-P_{12})\,\psi'_j(2)\, \psi'_\ell(1) = \varepsilon^N_\ell. \tag{2.25}$$

Comparison of (2.25) with (2.14) shows that

$$\varepsilon^N_\ell = \langle E^N_\nu \rangle - \langle E^{(N-1)}_{\nu(\ell)} \rangle. \tag{2.26}$$

Thus the eigenvalues of the Hartree–Fock equations are, by Koopman's theorem, equal to the difference between the energies of the N and $N{-}1$ electron systems having $(N{-}1)$ common spin orbitals. There are several reasons why ε^N_ℓ is *not* the true binding energy of an electron in the state ψ'_ℓ. First, we note that the Hartree–Fock equations are approximate. However, even if the Hartree–Fock equations were exact, ε^N_ℓ would still not be the correct binding energy because the set of spin orbitals $\{\nu(k)\}$ are not those which minimize the energy of the $(N-1)$-electron system. If we solved the Hartree–Fock equations for the N and $(N-1)$-electron systems separately we would find that the spin orbitals which result are different. For small systems such as atoms and molecules this difference can be large. For larger systems such as solids, ε_ℓ^N is a better approximation to the binding energy. We shall return to the question of the relation of ε_ℓ^N to the binding energy in a later chapter where we discuss the interpretation of photoemission experiments.

2.5 Hartree–Fock potential

The Hartree–Fock equations, (2.24), may be written in the form

$$\left[\frac{-\hbar^2}{2m}\nabla_i^2 + V_N(\vec{r}) + V_C(\vec{r})\right]\phi_i(\vec{r}) - \int V_{\text{ex}}^{(i)}(\vec{r},\vec{r}')\phi_i(\vec{r}')\, d\vec{r}' = \varepsilon_i^N\,\phi_i(\vec{r})$$

(2.27)

where $V_N(\vec{r})$ is the nuclear attraction potential, $V_C(\vec{r})$ is the direct Coulomb potential and $V_{\text{ex}}^{(i)}(\vec{r},\vec{r}')$ is the non-local exchange potential. These are defined by

$$V_N(\vec{r}) = -\sum_{\vec{R}_A}\frac{Z_A\,e^2}{|\vec{r}-\vec{R}_A|}\,,$$

(2.28)

$$V_C(\vec{r}) = \sum_j e^2\int d\vec{r}'\frac{\phi_j(\vec{r}')^*\phi_j(\vec{r}')}{|\vec{r}-\vec{r}'|}\,,$$

(2.29)

and

$$V_{\text{ex}}^{(i)}(\vec{r},\vec{r}') = \sum_j e^2\frac{\phi_j(\vec{r}')^*\phi_j(\vec{r})}{|\vec{r}-\vec{r}'|}\chi(s_j)\,\chi(s_i)\,.$$

(2.30)

The charge density at $\vec{r}'$ due to an electron in the orbital ϕ_j is given by $e\,\phi_j(\vec{r}')^*\phi_j(\vec{r}')$. The total charge density, $\rho(\vec{r}')$, due to all occupied states is the sum of the individual charge densities. The Coulomb potential may therefore be written as

$$V_C(\vec{r}) = e\int d\vec{r}'\frac{\rho(\vec{r}')}{|\vec{r}-\vec{r}'|}$$

(2.31)

which is recognized as the potential arising from a continuous charge distribution. The Coulomb potential entering (2.27) (defined in (2.29)) includes the interaction of an electron with itself. This self-interaction, however, is exactly canceled by one of the terms in the exchange potential.

The exchange potential, $V_{\text{ex}}^{(i)}(\vec{r},\vec{r}')$, is seen to be a non-local potential acting between electrons with parallel spins. The dependence on the superscript "i" is only that the states contributing to the exchange must have the same spin state as χ_i. The potential is therefore the same for all states having the same spin and is not dependent on the spatial orbital. The exchange potential may be written in the form of a local potential,

$$\int V_{\text{ex}}^{(i)}(\vec{r}, \vec{r}')\, \phi_i(\vec{r}')\, d\vec{r}' = v_{\text{ex}}^{(i)}(\vec{r})\, \phi_i(\vec{r}) \tag{2.32}$$

where

$$v_{\text{ex}}^{(i)}(\vec{r}) \equiv \sum_j \int d\vec{r}' \left\{ \frac{\phi_j^*(\vec{r}')\, \phi_j(\vec{r})\, \phi_i(\vec{r}')}{|\vec{r} - \vec{r}'|\, \phi_i(\vec{r})} \right\} (\chi(s_j),\, \chi(s_i)). \tag{2.33}$$

The local potential, $v_{ex}^{(i)}(\vec{r})$, depends explicitly on the spatial orbital $\phi_i(\vec{r})$.

For parallel spins the exchange reduces the Coulomb repulsion between two electrons. This comes about because the antisymmetric wavefunction must vanish whenever two electrons with parallel spins are at the same point. This may be seen by noting that when the two coordinates of two electrons $\vec{r}_n$ and $\vec{r}_m$ are equal then two of the rows of the Slater determinant are equal and the determinant vanishes. This effect is expressed by the Pauli exclusion principle which prevents parallel spin electrons from occupying the same point in space. The exchange potential is a form of electron–electron correlation. The probability that one electron is at $\vec{r}_n$ and a second electron is at $\vec{r}_m$ is

$$\Gamma(\tau_n, \tau_m) = \int d\tau' \Delta_\nu^{N*} \Delta_\nu^N \tag{2.34}$$

where the integration over $d\tau'$ is over all τ except τ_n and τ_m. For *parallel* spin electrons the probability is found to be

$$\Gamma_{\text{p}}(\vec{r}_n, \vec{r}_m) = \frac{1}{N!} \sum_i \sum_{j \neq i} \left\{ |\phi_i(\vec{r}_n)|^2 |\phi_j(\vec{r}_m)|^2 - \phi_i^*(\vec{r}_n)\, \phi_j^*(\vec{r}_m)\, \phi_i(\vec{r}_m)\, \phi_j(\vec{r}_n) \right\}. \tag{2.35}$$

For *antiparallel* spin electrons

$$\Gamma_{\text{a}}(\vec{r}_n, \vec{r}_m) = \frac{1}{N!} \sum_i \sum_{j \neq i} |\phi_i(\vec{r}_n)|^2 |\phi_j(\vec{r}_m)|^2. \tag{2.36}$$

It is seen that the probability of antiparallel spin electrons at $\vec{r}_n$ and $\vec{r}_m$ is equal to the product of the individual probabilities. Thus there is no correlation between antiparallel electrons. However, the parallel spin electron probability has an interference due to exchange-correlation. When $\vec{r}_m = \vec{r}_n$, the probability of parallel spin electrons vanishes. If we fix one electron at $\vec{r}_n$ then the probability of finding another (parallel spin) electron near $\vec{r}_n$ is small. The depletion of the probability due to the second term in (2.35) is called the "exchange hole". As an electron moves through space it is always surrounded by the exchange hole which is the result of the correlated motion of same spin electrons as they avoid occupying the same point in space.

Returning to the Hartree–Fock equations, (2.27), it is clear that the potential is not known *a priori*. To construct the potential one must have the orbitals, but to obtain the orbitals one must have the potential. Some type of self-consistent procedure is required in order to obtain the solutions of the Hartree–Fock equations. In practice the equations are solved iteratively. A starting potential V^0 is assumed and the orbitals are determined by solution of the eigenvalue equation. These orbitals are then used to construct a new potential, V^1. The process is iterated until V^n is (sufficiently close) equal to V^{n+1}. It is assumed that such self-consistent solutions are unique.

2.6 Approximate exchange potential

The Hartree–Fock equations are difficult to solve self-consistently because of the complex, non-local exchange potential. As a result an approximate form of the exchange potential is desirable. One approximation, known as the "Xα" approximation, has proved to be particularly convenient and it has been employed extensively in solid-state calculations. The essential idea comes from the Fermi–Thomas [1–3] model where the exchange energy is found to be proportional to the cube root of the charge density. Slater [4] suggested that a semiempirical potential, $V_{X\alpha}$, be used in place of the Hartree–Fock exchange potential, where

$$V_{X\alpha}(\vec{r}) = \frac{9}{2}\,\alpha\left(\frac{3\rho(\vec{r})}{8\pi}\right)^{1/3}. \tag{2.37}$$

Here the quantity α is a scaling parameter which is generally in the range $\frac{2}{3} \leq \alpha \leq 1$. The X in X$\alpha$ is shorthand for exchange. The Xα approximation replaces the non-local Hartree–Fock exchange potential by a local potential that is proportional to the cube root of the charge density. The form in (2.37) is used for non-magnetic systems. For magnetic systems spin dependent forms are used which are defined so that their sum is equal to (2.37) when the number of spin-up and spin-down electrons is equal. The form used is

$$V_{X\alpha}^{(s)}(\vec{r}) = \frac{9}{2}\,\alpha\left(\frac{3\rho_s(\vec{r})}{4\pi}\right)^{1/3} \tag{2.38}$$

where $\rho_s(\vec{r})$ is the charge density for $s = \uparrow$ or $s = \downarrow$. When self-consistent solutions are determined, the degree of spin polarization must also be determined self-consistently. The optimum scaling parameter, α, appearing in (2.38) is usually not the same as that in (2.37).

There are two points worth noting with regard to the use of the Xα approximation. The first is that the exact cancelation mentioned in the preceding section between the Coulomb and exchange self-interaction terms is lost. The second point is that Koopman's theorem no longer applies to the eigenvalues. That is, the eigenvalues ε_ℓ^N of the Hartree–Fock equations with V_{ex} replaced by $V_{X\alpha}$ do not correspond to the energy difference between the Slater determinant states Δ_ν^N and $\Delta_{\nu(\ell)}^{(N-1)}$. On the other hand, the Hartree–Fock eigenvalues do not correspond to the true binding energy of an electron because of the neglect of the relaxation of the orbitals of the $(N-1)$ electron system. In many cases the eigenvalues found for the $X\alpha$ approximation compare reasonably well with experimental ionization energies.

Orbital relaxation effects are sometimes included by means of what is called the "transition state" approximation. With this method, $X\alpha$ solutions are obtained with one half of an electron assigned to the orbital whose ionization energy is sought. The ionization energy is therefore approximated by the eigenvalues of a system with $(N-\frac{1}{2})$ electrons.

The difference between an N-electron (ground state) eigenvalue ε_ℓ^N and the ionization energy I_ℓ is often found to be nearly independent of ℓ. In such a case, ground state energy differences, $\varepsilon_\ell^N - \varepsilon_k^N$, compare well with ionization energy differences, $I_\ell - I_k$.

2.7 The LCAO method

The Hartree–Fock equations with or without the $X\alpha$ approximation are the basis for many current electronic structure calculations. Many different methods are employed in the solution of Hartree–Fock equations. Each method has its advantages and disadvantages. The LCAO method for finding the solutions is particularly valuable because it provides a very simple and intuitive interpretation of the electronic structure. With the LCAO method the spatial parts of the spin orbitals comprising the Slater determinant are expressed as linear combinations of atomic orbitals. The Hartree–Fock equations are then transformed to matrix equations which determine the amplitudes of the atomic orbitals that make up an eigenstate.

The orbitals, $\phi_i(\vec{r})$, which form the basis of the Slater determinant, Δ_ν, are written in the form

$$\phi_i(\vec{r}) = \sum_{\vec{R}_n} \sum_\alpha C_{n\alpha}^{(i)} \, \varphi_\alpha(\vec{r} - \vec{R}_n) \tag{2.39}$$

where $\varphi_\alpha(\vec{r} - \vec{R}_n)$ is an atomic orbital for an atom located at $\vec{R}_n$. The subscript α labels the different atomic states; $\alpha = 1s, 2s, 2p_x, 2p_y, 2p_z, \ldots$. The coefficients, $C_{n\alpha}^{(i)}$, specify the amplitudes of the atomic orbitals for the state whose eigenvalue is $\varepsilon_i^N \equiv \varepsilon_i$. The Hartree–Fock equations with the local exchange potential $V_{X\alpha}$ are of the form

$$H \phi_i(\vec{r}) = \varepsilon_i \phi_i(\vec{r}). \qquad (2.40)$$

Using (2.39) we have

$$\sum_{n\alpha} C_{n\alpha}^{(i)} \left\{ \frac{-\hbar^2}{2m} \nabla^2 + V_N(\vec{r}) + V_C(\vec{r}) - V_{X\alpha}(\vec{r}) \right\} \varphi_\alpha(\vec{r} - \vec{R}_n)$$

$$= \varepsilon_i \sum_{n\alpha} C_{n\alpha}^{(i)} \varphi_\alpha(\vec{r} - \vec{R}_n).$$

Multiplying (2.40) by $\varphi_\beta^*(\vec{r} - \vec{R}_m)$ and integrating over $\vec{r}$ leads to the result

$$\sum_{n\alpha} \left\{ H_{m\beta,n\alpha} - \varepsilon_i S_{m\beta,n\alpha} \right\} C_{n\alpha}^{(i)} = 0, \qquad (2.41)$$

where the matrix elements are

$$H_{m\beta,n\alpha} = \int d\vec{r}\, \varphi_\beta^*(\vec{r} - \vec{R}_m) \left[\frac{-\hbar^2}{2m} \nabla^2 + V_N(\vec{r}) + V_C(\vec{r}) - V_{X\alpha}(\vec{r}) \right] \varphi_\alpha(\vec{r} - \vec{R}_n)$$

$$(2.42)$$

and

$$S_{m\beta,n\alpha} = \int d\vec{r}\, \varphi_\beta(\vec{r} - \vec{R}_m)^* \varphi_\alpha(\vec{r} - \vec{R}_n). \qquad (2.43)$$

The atomic orbitals belonging to a given atom are orthonormal, but orbitals on different atoms have overlap that is specified by $S_{m\beta,n\alpha}$. If we denote the matrix of the Hamiltonian whose elements are $H_{m\beta,n\alpha}$ by $\mathbb{H}$ and the overlap matrix by $\mathbb{S}$ then (2.41) is

$$(\mathbb{H} - \varepsilon_i \mathbb{S})\, \vec{C}^{(i)} = 0 \qquad (2.44)$$

where the components of the vector $\vec{C}^{(i)}$ are the $C_{n\alpha}^{(i)}$. Because of the overlap matrix, $\mathbb{S}$, the eigenvectors of (2.44) are *not* orthogonal. Instead they satisfy the condition

$$(\vec{C}^{(i)}, \mathbb{S}\,\vec{C}^{(j)}) = \sum_{n\alpha} \sum_{m\beta} C_{n\alpha}^{(i)} S_{n\alpha,m\beta} C_{m\beta}^{(j)} = \delta_{ij}. \qquad (2.45)$$

The diagonal elements of $\mathbb{H}$ are of the order of the ionization energy of the corresponding atomic state. The off-diagonal matrix elements are called transfer or resonance integrals or simply LCAO integrals.

In the use of the LCAO-Xα approach, the atomic orbitals are usually taken from prior calculations which are available for most atoms. The "Herman–Skillman" orbitals [5] are frequently used. To find self-consistent solutions some type of iterative procedure has to be employed. Very often, the charge density of the non-interacting atoms is used to generate initial potentials for V_C and $V_{X\alpha}$. Each iteration produces a set of $C_{n\alpha}^{(i)}$'s which may be used to calculate a new charge density for the interacting atoms.

The charge density is given by

$$\rho(\vec{r}) = \sum_{\vec{k}} \Psi_{\vec{k}}^* \Psi_{\vec{k}} = \sum_{\vec{k}} \sum_{i\alpha} \sum_{j\beta} C_{i\alpha}^* \, C_{j\beta} \, \varphi_\alpha^*(\vec{r} - \vec{R}_i) \, \varphi_\beta(\vec{r} - \vec{R}_j)$$

$$\int d\vec{r}\, \rho(\vec{r}) = \sum_{\vec{k}(\text{occ})} (\vec{C}^{(k)}, \mathbb{S}\,\vec{C}^{(k)}) = N \tag{2.46}$$

where (occ) means that the sum is over the occupied eigenstates. For an N-electron system the N lowest-energy eigenstates are the occupied states.

The LCAO-Xα self-consistent method is as rigorous as any other method employing the Xα approximation. The advantages of the method are twofold. First, the wavefunctions are very easy to conceptualize and the interpretation of the results in terms of elementary chemical concepts is immediate. A second advantage is the absence of the necessity to employ artificial boundary conditions such as those employed in most other methods. For example, the "multiple scattering" [6] or linear combinations of "muffin-tin" orbitals (LCMTO) [7] methods currently employed use artificial spherical boundaries about the atoms and surrounding the molecule itself. These methods are not well suited for systems such as planar molecules.

The disadvantage of the LCAO-Xα method seems to be principally the amount of computational time required to obtain accurate numerical solutions. On the other hand, and of primary importance in our discussions, is the fact that the LCAO-Xα method is ideally suited as a basis for the development of simpler, empirical models.

The LCAO method provides a rigorous solution to the self-consistent problem only if the atomic orbital basis set includes *all* of the atomic states. In practice, the set of atomic states employed is finite and restricted to only a few atomic states beyond those occupied for a free atom. This introduces an error which is difficult to assess. All methods which express the orbitals $\phi_i(\vec{r})$ in terms of a finite set of basis states suffer from this type of "truncation error".

2.8 Orthogonalized atomic orbitals

It is often convenient to work with localized orbitals that are orthogonal in order to eliminate the overlap between orbitals localized on different atomic sites. This is accomplished by the transformation:

$$\vec{D}^{(k)} = \mathbb{S}^{1/2} \, \vec{C}^{(k)} \,, \tag{2.47}$$

$$\mathbb{H}' = \mathbb{S}^{-1/2} \mathbb{H} \, \mathbb{S}^{-1/2} \,, \tag{2.48}$$

which leads to the new matrix equation

$$[\mathbb{H}' - \varepsilon_k \mathbb{I}]\, \vec{D}^{(k)} = 0\,. \tag{2.49}$$

It is always possible to make this transformation because the eigenvalues of $\mathbb{S}$ are real, positive numbers and therefore $\mathbb{S}^{1/2}$ and $\mathbb{S}^{-1/2}$ can always be constructed.

Equation (2.49) is in the standard eigenvalue form and the eigenvectors, $\vec{D}^{(k)}$, form an orthonormal set

$$(\vec{D}^{(k)}, \vec{D}^{(\ell)}) = \delta_{k\ell}\,. \tag{2.50}$$

The new localized orbitals corresponding to the transformation (2.47) are called Löwdin orbitals [8]. They are related to the atomic orbitals by the relation

$$\xi_\alpha(\vec{r} - \vec{R}_k) = \sum_{j\beta}(\mathbb{S}^{-1/2})_{j\beta,k\alpha}\, \varphi_\beta(\vec{r} - \vec{R}_j)\,. \tag{2.51}$$

The Löwdin orbital, $\xi_\alpha(\vec{r} - \vec{R}_k)$, is localized near $\vec{R}_k$ but it is somewhat more extended than the atomic orbital $\varphi_\alpha(\vec{r} - \vec{R}_k)$.

An important property of the transformation (2.51) is that it preserves the symmetry properties. That is, $\xi_\alpha(\vec{r} - \vec{R}_k)$ has symmetry transformation properties that are *identical* to those of $\varphi_\alpha(\vec{r} - \vec{R}_k)$. Proof of this property will be given in Chapter 3.

Suggested reference texts

C. J. Ballhausen and H. B. Gray, *Molecular orbital theory* (New York, W. A. Benjamin Inc., 1965).

J. C. Slater, *The self-consistent field for molecules and solids*, Vol. **4**, Quantum theory of molecules and solids (New York, McGraw-Hill, 1974).

G. A. Segal, ed., *Semiempirical methods of electronic structure, Part A: Methods* (New York, Plenum Press, 1977).

References

[1] L. H. Thomas, *Proc. Camb. Phil. Soc.* **23**, 542 (1927).

[2] E. Fermi, *Z. Physik* **48**, 73 (1928).

[3] P. A. M. Dirac, *Proc. Camb. Phil. Soc.* **26**, 376 (1930).

[4] J. C. Slater, *Phys. Rev.* **81**, 385 (1951).

[5] F. Herman and S. Skillman, *Atomic structure calculations* (Englewood Cliffs, NJ, Prentice-Hall, 1963).

[6] K. H. Johnson, *J. Chem. Phys.* **45**, 3085 (1966), *Int. J. Quantum Chem.* **51**, 361 (1967); *ibid* **52**, 223 (1968).

[7] O. K. Andersen, *Phys. Rev.* B **12**, 3060 (1975).

[8] P.-O. Löwdin, *J. Chem. Phys.* **18**, 365 (1950).

Problems for Chapter 2

1. Show that the Slater determinant for an N-electron system vanishes if $\psi_i = \psi_j$ for any $i \neq j$. Explain how this result is related to the Pauli exclusion principle.

2. In x-ray photoelectron spectroscopy a photon impinges on a solid surface. An electron in an initial state of energy E_i absorbs the photon and is ejected from the solid into a high-energy electronic state E_f. Discuss the connection between the Hartree–Fock energies and Koopman's theorem in this photoexcitation process? Under what conditions is the Born–Oppenheimer approximation valid?

3. Explain the following terms:
   (a) overlap integral
   (b) transfer or resonance integral
   (c) Xα approximation
   (d) exchange hole.

4. Consider two Slater determinant wavefunctions, Δ_a^N and Δ_b^N, with different configurations. Assume the orthogonal basis orbitals for Δ_a^N are $(\Psi_a, \Psi_2, \cdots, \Psi_N)$ and those for Δ_b^N are $(\Psi_b, \Psi_2, \cdots, \Psi_N)$, with $\Psi_b \neq \Psi_a$. Show that Δ_a^N is orthogonal Δ_b^N.

5. Empirical LCAO models use adjustable parameters for the diagonal and two-center interaction integrals between neighboring atoms lying within some cutoff radius R_0. Interactions beyond R_0 are assumed to be negligible. In addition, overlap integrals are often ignored. Give reasons why both of these assumptions may be approximately valid.

6. Consider two empirical models, model I and model II. Assume I and II use orbitals with the same symmetry properties. Model I assumes the overlap integrals between

orbitals on different sites vanish. Model II uses the overlap integrals between different sites as adjustable parameters. As a result, model II has many more adjustable parameters than model I. Which model is capable of the most accurate representation of the electronic states?

3

Empirical LCAO model

The LCAO method described in the previous chapter forms the basis for a number of empirical or qualitative models. In such models the LCAO matrix elements are treated as "fitting" parameters to be determined from experiment or in some empirical way. Such models have provided a great deal of physical insight into the electronic properties of molecules and solids.

One of the first and simplest LCAO models was used by Hückel [1] to discuss the general qualitative features of conjugated molecules. Later, Slater and Koster [2] introduced an LCAO method for the analysis of the energy bands of solids. The Slater–Koster LCAO model has been used extensively as an interpolation scheme.

The LCAO parameters are determined by choosing the model parameters to give results that approximate those of more accurate numerical energy band calculations at a few points in the Brillouin zone. Once the parameters are determined the LCAO model gives approximate energies at any point in the Brillouin zone.

LCAO models have been remarkably useful for ordered solids and molecules having a high degree of symmetry. The reason for this is that in many cases the electronic structure is qualitatively determined by symmetry or group theoretical considerations. The group theoretical properties of a system are preserved in LCAO models and therefore they are able to correctly represent the general features of the electronic states.

3.1 LCAO matrix elements

The LCAO matrix elements (see (2.42)) were derived in Chapter 2. They are of the form:

$$H_{k\alpha,j\beta} \equiv \int \varphi_\alpha(\vec{r} - \vec{R}_k)^* H(\vec{r}) \, \varphi_\beta(\vec{r} - \vec{R}_j) \, d\vec{r} \,, \tag{3.1}$$

$$H(\vec{r}) = -\frac{\hbar^2}{2m} \nabla^2 + V^{\mathrm{T}}(\vec{r}) \,, \tag{3.2}$$

where φ_α may be taken as an atomic orbital or a Löwdin orbital and where $V^T(\vec{r})$ consists of the nuclear attraction, Coulomb, and exchange potentials. V^T may be expressed in terms of a sum of potentials localized at each atomic site,

$$V^T(\vec{r}) = \sum_{\vec{R}} v(\vec{r} - \vec{R}). \qquad (3.3)$$

Using (3.2), the LCAO matrix element may be decomposed into a kinetic energy matrix element and a potential energy matrix element:

$$H_{k\alpha,j\beta} = T_{k\alpha,j\beta} + V^T_{k\alpha,j\beta}, \qquad (3.4)$$

$$T_{k\alpha,j\beta} = \int \varphi_\alpha(\vec{r} - \vec{R}_k)^* \left(-\frac{\hbar^2}{2m}\nabla^2 \right) \varphi_\beta(\vec{r} - \vec{R}_j)\, d\vec{r}, \qquad (3.5)$$

$$V^T_{k\alpha,j\beta} = \int \varphi_\alpha(\vec{r} - \vec{R}_k)^* V^T(r)\, \varphi_\beta(\vec{r} - \vec{R}_j)\, d\vec{r}. \qquad (3.6)$$

Using (3.3) $V^T_{k\alpha,j\beta}$ takes the form

$$V^T_{k\alpha,j\beta} = \sum_{R} \int \varphi_\alpha(\vec{r} - \vec{R}_k)^* v(\vec{r} - \vec{R})\, \varphi_\beta(\vec{r} - \vec{R}_j)\, d\vec{r}. \qquad (3.7)$$

The kinetic energy matrix elements in (3.5) have two types of integrals. The matrix elements for $\vec{R}_k = \vec{R}_j$ are "one-center" integrals, while for $\vec{R}_k \neq \vec{R}_j$ they are "two-center" integrals. The potential energy matrix elements have three possible types of integrals; one-center integrals for $\vec{R}_k = \vec{R} = \vec{R}_j$, two-center integrals when two of the position vectors are the same, and three-center integrals for $\vec{R}_k \neq \vec{R} \neq \vec{R}_j$. Since the amplitudes of atomic orbitals decrease exponentially with distance from the nucleus, it is clear that the integrals involved in the matrix elements will decrease rapidly with increasing $|\vec{R}_k - \vec{R}_j|$ and may therefore be neglected beyond some cutoff distance. In a similar fashion, the localized potential $v(\vec{r} - \vec{R})$ will decrease with distance away from $\vec{R}$. Therefore the integrals also decrease rapidly as either $|\vec{R} - \vec{R}_k|$ or $|\vec{R} - \vec{R}_j|$ increases. In general then, one expects the one-center integrals to be the largest, followed by the two-center integrals and with the three-center integrals being the smallest. However, there are many three-center contributions and for accurate calculations they must be retained.

3.2 Slater–Koster model

In order to reduce the number of integrals involved in calculating the LCAO matrix elements certain approximations must be employed. One approximation already mentioned is to neglect matrix elements for which $|\vec{R}_k - \vec{R}|, |\vec{R}_j - \vec{R}|$, or $|\vec{R}_k - \vec{R}_j|$ exceed some chosen distance.

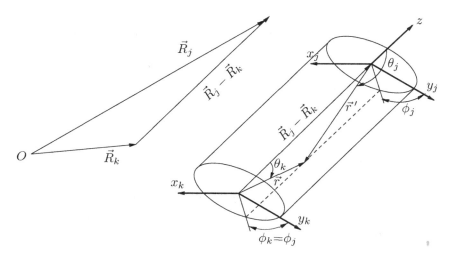

Figure 3.1. Coordinate system with the z-axis along the internuclear axis of atoms located at $\vec{R}_j$ and $\vec{R}_k$ showing that $\phi_k = \phi_j$.

Slater and Koster suggested that the three-center contributions be neglected altogether for the purpose of establishing a workable empirical model. As a further simplification they assumed the local potentials to be spherically symmetric about the atomic center so that

$$v(\vec{r} - \vec{R}) = v(|\vec{r} - \vec{R}|) . \tag{3.8}$$

This approximation is not accurate because the charge density around a given ion in a molecule or solid is seldom spherical. Nevertheless it is a reasonable starting point. Use of this approximation does not mean that the final, resulting charge density will be spherically symmetric about the atomic sites. Instead, the charge density calculated from the resulting wavefunctions will reflect the bonding symmetry between the neighboring atomic orbitals. Therefore, this approximation can be considered as the first step of a self-consistent procedure. With these assumptions we may express the matrix elements of H with $\vec{R}_k \neq \vec{R}_j$ as

$$H_{k\alpha,j\beta} = T_{k\alpha,j\beta} + \sum_R \int \varphi_\alpha(\vec{r} - \vec{R}_k)^* v(\vec{r} - \vec{R}) \varphi_\beta(\vec{r} - \vec{R}_j) \, d\vec{r}$$

$$\simeq \int \varphi_\alpha(\vec{r} - \vec{R}_k)^* \left[\frac{-\hbar^2}{2m} \nabla^2 + v(|\vec{r} - \vec{R}_k|) + v(|\vec{r} - \vec{R}_j|) \right] \varphi_\beta(\vec{r} - \vec{R}_j) \, d\vec{r}. \tag{3.9}$$

The atomic orbitals will be of the form

$$\varphi_\alpha(\vec{r} - \vec{R}_k) = R_{n_\alpha}(r) \, P_{\ell_\alpha}^{m_\alpha}(\cos \theta_k) \, e^{im_\alpha \phi_k} \tag{3.10}$$

where n_α, ℓ_α, and m_α are the principal, orbital, and magnetic quantum numbers, respectively, $R_{n_\alpha}(r)$ is the radial part of the wavefunction, and the angular part, $P_{\ell_\alpha}^{m_\alpha}(\cos\theta_k)$, is an associated Legendre polynomial. If the z-axis is chosen along the line joining two atoms, $\vec{R}_j - \vec{R}_k$, as shown in Fig. 3.1, then it is clear that the ϕ_k is equal to ϕ_j. Therefore, the integral of (3.9) contains a part,

$$\int_0^{2\pi} e^{-i(m_\alpha - m_\beta)\phi}\, d\phi = \delta_{m_\alpha m_\beta}.$$

This means that the only non-vanishing two-center integrals are those for which the orbitals have the same symmetry about the internuclear axis. That is, the matrix elements are non-zero only if there is a non-vanishing overlap between the two orbitals involved. Some pictorial examples are shown in Fig. 3.2 to illustrate this principle. In practice it is a trivial matter to sketch the angular parts of the orbitals and to determine whether the matrix element vanishes by symmetry.

A useful nomenclature for describing the nature of the overlap of two atomic orbitals has been developed. Non-zero overlap exists when $m_\alpha = m_\beta = m$. For these cases the overlap is called σ, π, and δ for $m=0$, 1, or 2, respectively. Examples of these are illustrated in Fig. 3.3. The atomic orbitals generally employed are chosen to be real by taking linear combinations of atomic states. Thus $p_x \propto \cos\phi$ and $p_y \propto \sin\phi$ are employed instead of functions involving $e^{im\phi}$. In all cases, the functions combined have the same $|m|$. Thus, when we use these real atomic orbitals, the rule for deciding whether the overlap vanishes becomes $|m_\alpha| = |m_\beta|$, where $|m_\alpha|$ is the magnitude of the magnetic quantum number associated with the functions that make up the real atomic orbital φ_α.

The overlap integrals are denoted by $S(\beta\alpha t)$. For example, $S(pd\pi)$ represents the overlap between a p orbital with a d orbital where each has $|m| = 1$. The interaction matrix elements themselves are represented by $(\beta\alpha t)$ alone. Both overlap and interaction matrix elements are defined by convention for the specific configurations

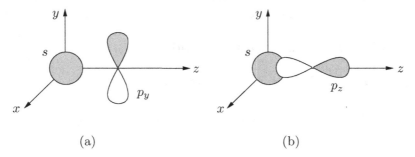

(a)   (b)

Figure 3.2. (a) Overlap between an s orbital and a p orbital. The overlap vanishes by symmetry $m_s = 0$, $m_{p_y} = 1$. (b) Non-zero σ overlap between an s orbital and a p_z orbital: $m_s = m_{p_z} = 0$.

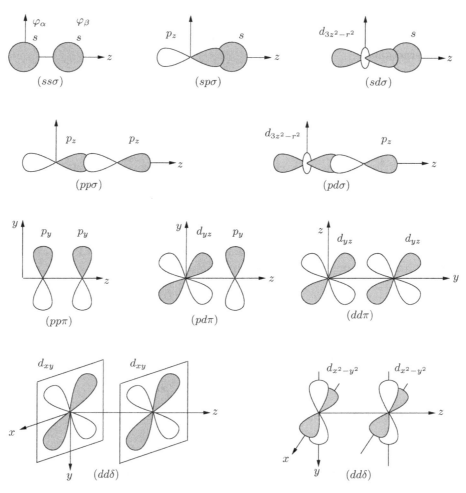

and derived from the above definitions by symmetry:

Figure 3.3. Fundamental types of overlaps (or matrix elements).

as shown in Fig. 3.3. Other configurations are related to these basic interactions. For example, a $(pd\pi)$ overlap changes sign if the positions of the orbitals are interchanged as shown in Fig. 3.4. The relative signs of particular interactions are usually obvious. The relation, between the definitions in Fig. 3.3 and the various basic interactions is

$$(\beta\alpha t) \equiv \int \varphi_\alpha(\vec{r} - \vec{R}_i)^* \, H \, \varphi_\beta(\vec{r} - \vec{R}_j) \, d\vec{r} \tag{3.11}$$

$$\mathbb{S}(\beta\alpha t) \equiv \int \varphi_\alpha(\vec{r} - \vec{R}_i)^* \varphi_\beta(\vec{r} - \vec{R}_j) \, d\vec{r} \tag{3.12}$$

$$(t = \sigma, \pi, \text{ or } \delta)$$

where $\vec{R}_j - \vec{R}_i$ is a vector with components along the *positive* z-axis. The matrix element for $d_{x^2-y^2}$ with s or p, shown in Fig. 3.3, have been deduced from the basic matrix elements by rotating the coordinate system and then re-expressing the transformed orbitals in terms of those which have basic definitions. To do this one must pay attention to the fact that the normalization of the angular part of the wavefunctions for different orbitals is sometimes different. Table 3.1 lists the forms of the orbitals. For compactness we shall often use the notation d_{x^2} for $d_{x^2-y^2}$ and d_{z^2} for $d_{3z^2-r^2}$.

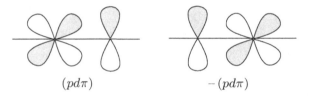

$$(pd\pi) \qquad\qquad\qquad -(pd\pi)$$

Figure 3.4. Definition of $(pd\pi)$ overlap and interaction integral.

Consider the overlap of the $d_{x^2-y^2}$ with the s orbital as shown in Fig. 3.3. We can relabel the x-axis as z' and $y = y'$. Then the relevant angular part becomes

$$d_{x^2} = \sqrt{\frac{15}{16\pi}} \left(\frac{x^2 - y^2}{r^2} \right)$$

$$\rightarrow \sqrt{\frac{15}{16\pi}} \left(\frac{z'^2 - y'^2}{r'^2} \right)$$

$$= \sqrt{\frac{15}{16\pi}} \left[\frac{1}{2} \left(\frac{3z'^2 - r'^2}{r'^2} \right) + \frac{1}{2} \left(\frac{x'^2 - y'^2}{r'^2} \right) \right]$$

$$= \frac{\sqrt{3}}{2} d_{z'^2} + \frac{1}{2} d_{x'^2} .$$

The s orbital has no angular variation and so it remains unchanged by the coordinate transformation. Then

$$\int d^*_{x^2} \; s \; d\vec{r} = \frac{\sqrt{3}}{2} \int d^*_{z'^2} \; s \; d\vec{r} + \frac{1}{2} \int d^*_{x'^2} \; s \; d\vec{r} = \frac{\sqrt{3}}{2}(sd\sigma). \tag{3.13}$$

The second integral on the right-hand side of (3.13) vanishes by symmetry.

Table 3.1. *Forms of atomic orbitals.*

Orbital $q\#$	Angular function	Radial function	
ns	$\sqrt{\frac{1}{4\pi}}$	$R_s^{(n)}(r)$	
np_x	$\sqrt{\frac{3}{4\pi}}\left(\frac{x}{r}\right)$	$rR_p^{(n)}(r)$	
np_y	$\sqrt{\frac{3}{4\pi}}\left(\frac{y}{r}\right)$	$rR_p^{(n)}(r)$	
np_z	$\sqrt{\frac{3}{4\pi}}\left(\frac{z}{r}\right)$	$rR_p^{(n)}(r)$	
nd_{z^2}	$\sqrt{\frac{5}{16\pi}}\left(\frac{3z^2-r^2}{r^2}\right)$	$r^2R_d^{(n)}(r)$	$\left.\rule{0pt}{30pt}\right\}\; e_g$
nd_{x^2}	$\sqrt{\frac{15}{16\pi}}\left(\frac{x^2-y^2}{r^2}\right)$	$r^2R_d^{(n)}(r)$	
nd_{xy}	$\sqrt{\frac{15}{4\pi}}\left(\frac{xy}{r^2}\right)$	$r^2R_d^{(n)}(r)$	$\left.\rule{0pt}{44pt}\right\}\; t_{2g}$
nd_{yz}	$\sqrt{\frac{15}{4\pi}}\left(\frac{yz}{r^2}\right)$	$r^2R_d^{(n)}(r)$	
nd_{zx}	$\sqrt{\frac{15}{4\pi}}\left(\frac{zx}{r^2}\right)$	$r^2R_d^{(n)}(r)$	

In more general cases we shall have LCAO integrals between orbitals displaced from one another in an arbitrary direction. It becomes tedious to perform the transformations of the orbitals. To ease this discomfort, Slater and Koster worked out a table which gives matrix elements for displaced orbitals in terms of the fundamental integrals. Their results are given in Table 3.2. To use the table to evaluate the interaction integral of the form:

$$E_{\beta\alpha} = \int \varphi_\alpha(\vec{r} - \vec{R}_i)^* H \; \varphi_\beta(\vec{r} - \vec{R}_j) \, d\vec{r}, \tag{3.14}$$

one need only calculate the direction cosines $\ell, m,$ and n of the vector $\vec{R}_{ji} = \vec{R}_j - \vec{R}_i$;

$$\ell = \frac{x_j - x_i}{|\vec{R}_{ji}|}, \qquad m = \frac{y_j - y_i}{|\vec{R}_{ji}|}, \qquad n = \frac{z_j - z_i}{|\vec{R}_{ji}|}. \tag{3.15}$$

For example, if $\varphi_\beta(\vec{r} - \vec{R}_j)$ is a p_x orbital with $\vec{R}_j = a(1, -1, 1)$ and $\varphi_\alpha(\vec{r} - \vec{R}_i)$ is a d_{z^2} orbital at $\vec{R}_i = a(-1, 1, 2)$ then $|\vec{R}_{ji}| = 3a$, $\ell = +\frac{2}{3}$, $m = -\frac{2}{3}$, and $n = -\frac{1}{3}$.

In Table 3.2 we find the line labeled $E_{x,3z^2-r^2}$ and the integral is

$$\int d_{z^2}\, H\, p_x\, d\vec{r} = \ell\left[n^2 - \frac{1}{2}(\ell^2 + m^2)\right](pd\sigma) - \sqrt{3}\ell n^2 (pd\pi)$$

$$= \frac{2}{3}\left[\frac{1}{9} - \frac{1}{2}\left(\frac{4}{9} + \frac{4}{9}\right)\right](pd\sigma) - \sqrt{3}\left(\frac{2}{3}\cdot\frac{1}{9}\right)(pd\pi)$$

$$= -\frac{2}{9}\left[(pd\sigma) + \frac{(pd\pi)}{\sqrt{3}}\right]. \tag{3.16}$$

3.3 Symmetry properties of the Löwdin orbitals

When using the LCAO method as an empirical model it is often convenient to use the orthogonalized Löwdin orbitals [3] discussed in Section 2.8. With the Löwdin orbital basis the overlap matrix elements between orbitals on different sites vanish and the interaction matrix elements are between Löwdin orbitals rather than atomic orbitals. The formalism described in the preceding section can then be used without change. This is allowed only because the symmetry properties of the Löwdin orbitals are *identical* to those of the corresponding atomic orbitals. In the remainder of this section we give a proof of this.

We want to show that the Löwdin orbital $\xi_\alpha(\vec{r} - \vec{R}_\ell)$ has precisely the same symmetry properties as the atomic orbital, $\varphi_\alpha(\vec{r} - \vec{R}_\ell)$. They are related by

$$\xi_\alpha(\vec{r} - \vec{R}_\ell) = \sum_{m\nu} (\mathbb{S}^{-1/2})_{m\nu,\ell\alpha}\, \varphi_\nu(\vec{r} - \vec{R}_m) \tag{3.17}$$

where $\mathbb{S}$ is the overlap matrix

$$\mathbb{S}_{m\nu,\ell\alpha} = \int \varphi_\nu(\vec{r} - \vec{R}_m)^* \varphi_\alpha(\vec{r} - \vec{R}_\ell)\, d\vec{r}. \tag{3.18}$$

The symmetry properties of φ are completely determined by their transformation properties under a symmetry operation of the group corresponding to the system in question. If $\mathbb{O}$ is a symmetry operator of the group then from group theory we know that

$$\mathbb{O}\,\varphi_\alpha(\vec{r} - \vec{R}_\ell) = \sum_{k\beta} \Gamma(\mathbb{O})_{k\beta,\ell\alpha}\, \varphi_\beta(\vec{r} - \vec{R}_k) \tag{3.19}$$

where $\Gamma(\mathbb{O})$ is a unitary representation matrix for $\mathbb{O}$; $\Gamma(\mathbb{O})^{-1}_{\ell\alpha,k\beta} = \Gamma(\mathbb{O})^*_{k\beta,\ell\alpha}$. For the Löwdin orbitals to have precisely the same symmetry properties as the atomic

Table 3.2. *LCAO two-center integrals [2] for orbitals centered at $\vec{R}_i$ and $\vec{R}_j$. The variables ℓ, m, and n are the direction cosines of $(\vec{R}_j - \vec{R}_i)$.*

$E_{s,s}$	$(ss\sigma)$
$E_{s,x}$	$\ell(sp\sigma)$
$E_{x,x}$	$\ell^2(pp\sigma) + (1-\ell^2)(pp\pi)$
$E_{x,y}$	$\ell m(pp\sigma) - \ell m(pp\pi)$
$E_{x,z}$	$\ell n(pp\sigma) - \ell n(pp\pi)$
$E_{s,xy}$	$\sqrt{3}\ell m(sd\sigma)$
E_{s,x^2-y^2}	$\frac{1}{2}\sqrt{3}(\ell^2 - m^2)(sd\sigma)$
$E_{s,3z^2-r^2}$	$[n^2 - \frac{1}{2}(\ell^2 + m^2)](sd\sigma)$
$E_{x,xy}$	$\sqrt{3}\ell^2 m(pd\sigma) + m(1-2\ell^2)(pd\pi)$
$E_{x,yz}$	$\sqrt{3}\ell mn(pd\sigma) - 2\ell mn(pd\pi)$
$E_{x,zx}$	$\sqrt{3}\ell^2 n(pd\sigma) + n(1-2\ell^2)(pd\pi)$
E_{x,x^2-y^2}	$\frac{1}{2}\sqrt{3}\ell(\ell^2 - m^2)(pd\sigma) + \ell(1-\ell^2+m^2)(pd\pi)$
E_{y,x^2-y^2}	$\frac{1}{2}\sqrt{3}m(\ell^2 - m^2)(pd\sigma) - m(1+\ell^2-m^2)(pd\pi)$
E_{z,x^2-y^2}	$\frac{1}{2}\sqrt{3}n(\ell^2 - m^2)(pd\sigma) - n(\ell^2 - m^2)(pd\pi)$
$E_{x,3z^2-r^2}$	$\ell[n^2 - \frac{1}{2}(\ell^2 + m^2)](pd\sigma) - \sqrt{3}\ell n^2(pd\pi)$
$E_{y,3z^2-r^2}$	$m[n^2 - \frac{1}{2}(\ell^2 + m^2)](pd\sigma) - \sqrt{3}mn^2(pd\pi)$
$E_{z,3z^2-r^2}$	$n[n^2 - \frac{1}{2}(\ell^2 + m^2)](pd\sigma) + \sqrt{3}n(\ell^2 + m^2)(pd\pi)$
$E_{xy,xy}$	$3\ell^2 m^2(dd\sigma) + (\ell^2 + m^2 - 4\ell^2 m^2)(dd\pi) + (n^2 + \ell^2 m^2)(dd\delta)$
$E_{xy,yz}$	$3\ell m^2 n(dd\sigma) + \ell n(1-4m^2)(dd\pi) + \ell n(m^2 - 1)(dd\delta)$
$E_{xy,zx}$	$3\ell^2 mn(dd\sigma) + mn(1-4\ell^2)(dd\pi) + mn(\ell^2 - 1)(dd\delta)$
E_{xy,x^2-y^2}	$\frac{3}{2}\ell m(\ell^2 - m^2)(dd\sigma) + 2\ell m(m^2 - \ell^2)(dd\pi) + \frac{1}{2}\ell m(\ell^2 - m^2)(dd\delta)$
E_{yz,x^2-y^2}	$\frac{3}{2}mn(\ell^2 - m^2)(dd\sigma) - mn[1 + 2(\ell^2 - m^2)](dd\pi) + mn[1 + \frac{1}{2}(\ell^2 - m^2)](dd\delta)$
E_{zx,x^2-y^2}	$\frac{3}{2}n\ell(\ell^2 - m^2)(dd\sigma) + n\ell[1 - 2(\ell^2 - m^2)](dd\pi) - n\ell[1 - \frac{1}{2}(\ell^2 - m^2)](dd\delta)$
$E_{xy,3z^2-r^2}$	$\sqrt{3}\ell m[n^2 - \frac{1}{2}(\ell^2 + m^2)](dd\sigma) - 2\sqrt{3}\ell mn^2(dd\pi) + \frac{1}{2}\sqrt{3}\ell m(1+n^2)(dd\delta)$
$E_{yz,3z^2-r^2}$	$\sqrt{3}mn[n^2 - \frac{1}{2}(\ell^2 + m^2)](dd\sigma) + \sqrt{3}mn(\ell^2 + m^2 - n^2)(dd\pi)$
	$\quad - \frac{1}{2}\sqrt{3}mn(\ell^2 + m^2)(dd\delta)$
$E_{zx,3z^2-r^2}$	$\sqrt{3}\ell n[n^2 - \frac{1}{2}(\ell^2 + m^2)](dd\sigma) - 2\sqrt{3}\ell n(\ell^2 + m^2 - n^2)(dd\pi)$
	$\quad + \frac{1}{2}\sqrt{3}\ell n(\ell^2 + m^2)(dd\delta)$
$E_{x^2-y^2,x^2-y^2}$	$\frac{3}{4}(\ell^2 - m^2)^2(dd\sigma) + [\ell^2 + m^2 - (\ell^2 - m^2)^2](dd\pi) + [n^2 + \frac{1}{4}(\ell^2 - m^2)^2](dd\delta)$
$E_{x^2-y^2,3z^2-r^2}$	$\frac{1}{2}\sqrt{3}(\ell^2 - m^2)[n^2 - \frac{1}{2}(\ell^2 + m^2)](dd\sigma) + \sqrt{3}n^2(m^2 - \ell^2)(dd\pi)$
	$\quad + \frac{1}{4}\sqrt{3}(1+n^2)(\ell^2 - m^2)(dd\delta)$
$E_{3z^2-r^2,3z^2-r^2}$	$[n^2 - \frac{1}{2}(\ell^2 + m^2)](dd\sigma) + 3n^2(\ell^2 + m^2)(dd\pi) + \frac{3}{4}(\ell^2 + m^2)^2(dd\delta)$

orbitals, we require that

$$\mathbb{O}\,\xi_\alpha(\vec{r} - \vec{R}_\ell) = \sum_{k\beta} \Gamma(\mathbb{O})_{k\beta,\ell\alpha}\,\xi_\beta(\vec{r} - \vec{R}_k). \tag{3.20}$$

However, from (3.17) we know

$$\mathbb{O}\,\xi_\alpha(\vec{r} - \vec{R}_\ell) = \sum_{m\nu} (\mathbb{S}^{-1/2})_{m\nu,\ell\alpha}\,\mathbb{O}\,\varphi_\nu(\vec{r} - \vec{R}_m)$$

$$= \sum_{m\nu} (\mathbb{S}^{-1/2})_{m\nu,\ell\alpha} \sum_{k\beta} \Gamma(\mathbb{O})_{k\beta,m\nu}\,\varphi_\beta(\vec{r} - \vec{R}_k)$$

$$= \sum_{k\beta} \left[\Gamma(\mathbb{O})\mathbb{S}^{-1/2} \right]_{k\beta,\ell\alpha}\,\varphi_\beta(\vec{r} - \vec{R}_k)\,. \tag{3.21}$$

On the other hand, substitution of (3.17) into (3.20) gives

$$\mathbb{O}\,\xi_\alpha(\vec{r} - \vec{R}_\ell) = \sum_{k\beta} \Gamma(\mathbb{O})_{k\beta,\ell\alpha} \left(\sum_{m\nu} (\mathbb{S}^{-1/2})_{m\nu,k\beta}\,\varphi_\nu(\vec{r} - \vec{R}_m) \right)$$

$$= \sum_{m\nu} \left[\mathbb{S}^{-1/2}\Gamma(\mathbb{O}) \right]_{m\nu,\ell\alpha}\,\varphi_\nu(\vec{r} - \vec{R}_m)$$

$$= \sum_{k\beta} \left[\mathbb{S}^{-1/2}\Gamma(\mathbb{O}) \right]_{k\beta,\ell\alpha}\,\varphi_\beta(\vec{r} - \vec{R}_k)\,. \tag{3.22}$$

Now comparison of (3.21) with (3.22) shows that the equations are compatible only if

$$\mathbb{S}^{-1/2}\Gamma(\mathbb{O}) = \Gamma(\mathbb{O})\mathbb{S}^{-1/2}$$

or

$$\Gamma(\mathbb{O})^{-1}\mathbb{S}^{-1/2}\Gamma(\mathbb{O}) = \mathbb{S}^{-1/2}. \tag{3.23}$$

According to (3.23) the Löwdin orbitals can possess the same symmetry as the atomic orbitals only if $\mathbb{S}^{-1/2}$ is invariant under the unitary transformation, Γ. We shall prove that this is true for physical systems.

The overlap matrix $\mathbb{S}$ has the form,

$$\mathbb{S} = \mathbb{I} + \Delta \tag{3.24}$$

where $\mathbb{I}$ is the unit matrix and Δ has zero diagonal elements. This form results because the atomic orbitals are normalized and therefore the overlap of the orbital with itself is unity. The non-vanishing off-diagonal elements of Δ correspond to the overlap between atomic orbitals centered on different atoms. These overlap integrals are not necessarily small but they are necessarily less than the diagonal overlap for any atomic orbitals. That is,

$$|\Delta_{k\beta,\ell\alpha}| < 1\,. \tag{3.25}$$

Therefore, it is possible to expand $\mathbb{S}^{-1/2}$ in a convergent power-series as follows:

$$\mathbb{S}^{-1/2} = (\mathbb{I} + \Delta)^{-1/2} = \mathbb{I} - \frac{1}{2}\Delta + \frac{3}{8}\Delta^2 - \frac{5}{16}\Delta^3 + \cdots .$$

If we can prove that every power of Δ is invariant under transformation by $\Gamma(\mathbb{O})$ then it will follow that $\mathbb{S}^{-1/2}$ is also invariant. The proof is easy:

$$\left[\Gamma^{-1}(\mathbb{O})\Delta\Gamma(\mathbb{O})\right]_{k\alpha,j\beta} = \sum_{m\nu}\sum_{n\gamma}\Gamma^{-1}(\mathbb{O})_{k\alpha,m\nu}\Delta_{m\nu,n\gamma}\Gamma(\mathbb{O})_{n\gamma,j\beta}$$

$$= \int d\vec{r}\left(\sum_{m\nu}\Gamma^{-1}(\mathbb{O})_{k\alpha,m\nu}\ \varphi_\nu^*(\vec{r}-\vec{R}_m)\right)\left(\sum_{n\gamma}\Gamma(\mathbb{O})_{n\gamma,j\beta}\ \varphi_\gamma(\vec{r}-\vec{R}_n)\right)$$

$$= \int \left[\mathbb{O}\varphi_\alpha(\vec{r}-\vec{R}_k)\right]^*\left[\mathbb{O}\varphi_\beta(\vec{r}-\vec{R}_j)\right]\ d\vec{r}$$

$$= \int \varphi_\alpha(\vec{r}-\vec{R}_k)^*\varphi_\beta(\vec{r}-\vec{R}_j)\ d\vec{r} = \Delta_{k\alpha,j\beta}\ . \tag{3.26}$$

Equation (3.26) shows that Δ is invariant, and therefore so is every power of Δ, since

$$\Gamma^{-1}(\mathbb{O})\Delta^N\Gamma(\mathbb{O}) = \left[\Gamma^{-1}(\mathbb{O})\Delta\Gamma(\mathbb{O})\right]\left[\Gamma^{-1}(\mathbb{O})\Delta\Gamma(\mathbb{O})\right]\cdots\left[\Gamma^{-1}(\mathbb{O})\Delta\Gamma(\mathbb{O})\right]$$

$$= \Delta^N.$$

This completes the proof.

A few comments on the use of the LCAO method for building an *empirical* model are now in order. In particular, we note that it is not necessary to employ atomic orbitals as the basis functions. Any set of orbitals that possess the symmetry of the atomic orbitals can be used. If the interactions are to be treated as empirical parameters, the results are independent of the actual basis orbitals (assuming the same number of basis orbitals are employed). The symmetry types and degeneracies of the electronic states do not depend on the actual basis orbitals, provided they possess the same transformation properties as the atomic orbitals. For example, s–p hybrids are often used when deriving molecular electronic states. The symmetry and degeneracies of the resulting electronic states are no different from those obtained when one employs atomic p orbitals or Löwdin orbitals. Including overlap integrals in an empirical model does not improve the results even though it appears there are more empirical parameters (the overlap integrals). The two models (one without overlap integrals and one with overlap integrals) are equivalent because they are related by a unitary transformation.

Suggested reference texts

C. J. Ballhausen and H. B. Gray, *Molecular Orbital Theory* (New York, W. A. Benjamin, 1965).

W. G. Richards and J. A. Horsley, *Ab initio Molecular Orbital Calculations for Chemists* (Oxford, Clarendon Press, 1970).

References

[1] E. Hückel, *Z. Physik* **70**, 204 (1931); *ibid* **72**, 310 (1932); *ibid* **76**, 628 (1932).
[2] J. C. Slater and G. F. Koster, *Phys. Rev.* **94**, 1498 (1954).
[3] P.-O. Löwdin, *J. Chem. Phys.* **18**, 365 (1950).

Problems for Chapter 3

1. Construct the matrix eigenvalue equation for a $d_{x^2-y^2}$ orbital at $(0,0,0)$ interacting with p orbitals at $(a,0,0)$ and $(-a,0,0)$.
   (a) Find the eigenvalues and eigenvectors. Classify the resulting states as bonding, non-bonding, or antibonding. Use E_d and E_p for the diagonal energies.
   (b) Determine the amount of covalent mixing, that is, the ratio of the d to p amplitudes squared for the various states.
   (c) Using $E_p = -9\,\text{eV}$, $E_d = -5\,\text{eV}$ and $(pd\sigma) = 1\,\text{eV}$, calculate the eigenvalues and d to p ratios of the amplitudes squared.

2. Show that the angular function for nd_{xy} in Table 3.1 is properly normalized.

3. Using the definitions of the angular functions in Table 3.1 express the orbital function $3y^2 - r^2$ as a linear combination of the two orbitals d_{z^2} and d_{x^2}.

4. Using Table 3.2 express the interactions of the following orbitals in terms of the Slater–Koster parameters:
   (a) p_x orbital located at $(0,a,0)$ with a p_y orbital located at $(a,0,0)$,
   (b) p_x orbital at $(0,a,0)$ with a $d_{x^2-y^2}$ orbital at $(0,0,0)$,
   (c) d_{xz} at $(0,0,0)$ with a d_{xz} at $(0,a,0)$.

5. Derive the Slater–Koster formula for E_{x,x^2-y^2} shown in Table 3.2.

4

LCAO energy band model for cubic perovskites

In Chapter 1 a qualitative description of the energy bands of perovskites was developed starting from a simple ionic model. It was argued that the essential electronic structure of the perovskites is derived from the BO_3 ions of the ABO_3 compound. The A ion was shown to be important in determining the ionic state of the B ion. Also, the A ion contributes to the electrostatic potentials. However, the energy bands associated with the outer s orbitals of the A ion were found to be far removed in energy from the lowest empty d bands and were unoccupied. As a result of these considerations the electronic structure of the A ion can be neglected in discussing the principal features of perovskite energy bands.

The energy bands that are electronically and chemically active are derived from the d orbitals of the transition metal (B) cation and the $2p$ orbitals of the oxygen anions. The deep core states of the ions produce atomic-like levels at energies far below the valence bands and may also be omitted in our discussion.

One of the first energy band calculations for perovskites was carried out by Kahn and Leyendecker [1] for SrTiO$_3$. They employed a semiempirical approach based on the method of Slater and Koster [2] (described in Section 3.2). The model presented in this chapter follows closely the work of Kahn and Leyendecker.

4.1 The unit cell and Brillouin zone

The BO_3 unit cell is shown in Fig. 4.1. The B ion is located at the origin and the three oxygen ions are located at a distance, a, along the three coordinate axes. The A ion (not shown in Fig. 4.1) is located at (a, a, a).

Let $\vec{e}_x$, $\vec{e}_y$, and $\vec{e}_z$ represent unit vectors along the x, y, and z axes, respectively, then the perovskite lattice can be described as a single cubic lattice of unit cells with lattice vectors

$$\vec{R}_B(\hat{n}) = 2a(n_x\vec{e}_x + n_y\vec{e}_y + n_z\vec{e}_z).\tag{4.1}$$

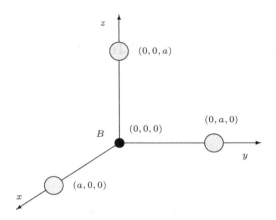

Figure 4.1. BO_3 unit cell for a perovskite.

The symbol $\hat{n}$ represents the three integers n_x, n_y, and n_z, which may be positive, negative, or zero. The oxygen ions are located at

$$\vec{R}_O^j(\hat{n}) = \vec{R}_B(\hat{n}) + a\vec{e}_j \qquad (j = x, \, y, \, \text{or} \, z) \qquad (4.2)$$

where $\vec{e}_j$ is one of the unit vectors, $\vec{e}_x$, $\vec{e}_y$, or $\vec{e}_z$. The A ions are located at the positions

$$\vec{R}_A(\hat{n}) = \vec{R}_B(\hat{n}) + (a\vec{e}_x + a\vec{e}_y + a\vec{e}_z). \qquad (4.3)$$

For each unit cell we shall consider 14 basis states; five d orbitals centered at $\vec{R}_B(\hat{n})$ and three $2p$ orbitals centered on each of the three oxygen ions. These 14 basis states produce 14 energy bands. Each energy band has N states, where N is the number of unit cells in the solid. Each of the $14N$ energy band states is specified by a band index, ν, and a wavevector, $\vec{k}$. The wavevector is chosen to lie in the first Brillouin zone of the perovskite structure. This Brillouin zone is a cube in $\vec{k}$-space as shown in Fig. 4.2.

We can assume that the $\vec{k}$-vectors corresponding to the N states of each band lie on a cubic lattice obtained by dividing the first Brillouin zone into N equal sized cubes.

In the limit as $N \to \infty$, the spacing between these $\vec{k}$ vectors becomes arbitrarily small and $\vec{k}$ may be treated as a continuous variable.

The points of high symmetry in the zone are shown in Fig. 4.2(a). These points

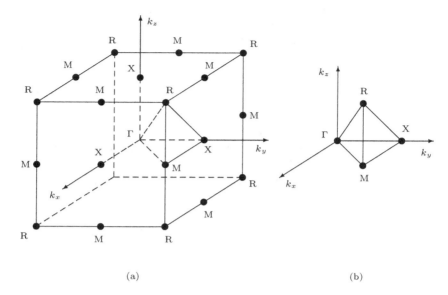

(a)   (b)

Figure 4.2. (a) Brillouin zone for a simple cubic perovskite showing the points of high symmetry, (b) 1/48 segment of the Brillouin zone.

have the coordinates in $\vec{k}$-space as follows:

$$\Gamma = \left(\frac{\pi}{2a}\right)(0,0,0)$$

$$X = \left(\frac{\pi}{2a}\right)(\pm 1, 0, 0); \qquad \left(\frac{\pi}{2a}\right)(0, \pm 1, 0); \qquad \left(\frac{\pi}{2a}\right)(0, 0, \pm 1)$$

$$M = \left(\frac{\pi}{2a}\right)(\pm 1, \pm 1, 0); \qquad \left(\frac{\pi}{2a}\right)(\pm 1, 0, \pm 1); \qquad \left(\frac{\pi}{2a}\right)(0, \pm 1, \pm 1)$$

$$R = \left(\frac{\pi}{2a}\right)(\pm 1, \pm 1, \pm 1). \tag{4.4}$$

Figure 4.2(b) shows a representative segment of the first Brillouin zone having 1/48 of the volume of the full zone. The entire zone can be constructed from 48 of these segments joined in an appropriate manner. If the 48 operations of the cubic point group (O_h) are performed on a segment with Γ as the origin, then the segment will be rotated, reflected, or inverted in such a way as to map out the entire first Brillouin zone. This means that every $\vec{k}$-vector point *inside* a segment is related to 47 equivalent points in the 47 other segments by a (cubic) symmetry operation. Consequently, it is necessary to find the energy band solutions only for the $\vec{k}$-vectors which lie inside and on the surface of one of the segments of the Brillouin zone. Solutions for symmetry-equivalent $\vec{k}$-vectors can then be obtained by performing the symmetry operation of O_h on the wavefunctions from the segment. The energies of symmetry-equivalent $\vec{k}$-vector states are the same.

4.2 LCAO matrix equation for an infinite lattice

In Section 2.8 we obtained the LCAO matrix eigenvalue equation:

$$(\mathbb{H}' - E_i\,\mathbb{I})\vec{D}^{(i)} = 0 \tag{4.5}$$

where the matrix elements of $\mathbb{H}'$ are integrals of a one-electron Hamiltonian between Löwdin (orthogonalized) orbital basis states.

The Löwdin orbitals are specified by a symmetry index, α, and a position vector. For compounds such as the perovskites it is convenient to specify the atomic position by a lattice vector $\vec{R}_m$ which locates the unit cell and a vector $\vec{\tau}_j$ which locates the position of the jth atom within the unit cell. Thus, we write the Löwdin orbitals as

$$\xi_\alpha(\vec{r} - \vec{R}_m - \vec{\tau}_j) = \xi_\alpha(\vec{r} - \vec{R}_{mj})\,, \tag{4.6}$$

where

$$\vec{R}_{mj} \equiv \vec{R}_m + \vec{\tau}_j\,.$$

For a solid having N unit cells, each unit cell having n_s basis orbitals, (4.5) is an $n_sN \times n_sN$ matrix equation. The eigenvector $\vec{D}^{(i)}$ has n_sN components, $d^{(i)}_{mj\alpha}$, which specify the amplitudes of the Löwdin orbitals comprising the ith eigenstate. In an infinite, periodic solid the modulus $|d^{(i)}_{mj\alpha}|^2$ must be the same for every equivalent atomic position. Therefore the amplitudes for a particular symmetry type orbital of equivalent atoms can differ at most by a phase factor. According to Bloch's theorem for periodic systems the amplitudes can be taken in the form

$$d^{(i)}_{mj\alpha} = \frac{1}{\sqrt{N}}e^{i\vec{k}\cdot\vec{R}_m}d_{j\alpha}(\vec{k}, \nu) \qquad (i = \vec{k}, \nu)\,. \tag{4.7}$$

For convenience we also introduce a phase factor within the unit cell and write $d_{j\alpha}(\vec{k}, \nu) = e^{i\vec{k}\cdot\vec{\tau}_j}a_{j\alpha}$. The eigenstates are characterized by the wavevector, $\vec{k}$, and a band index ν and therefore on the right-hand side of (4.7) we have replaced the eigenstate index i by $(\vec{k}, \nu)$.

The LCAO Bloch wavefunction then assumes the form

$$\psi_{\vec{k}\nu}(\vec{r}) = \frac{1}{\sqrt{N}}\sum_m\sum_{j\alpha}e^{i\vec{k}\cdot\vec{R}_{mj}}a_{j\alpha}(\vec{k}, \nu)\xi_\alpha(\vec{r} - \vec{R}_{mj})\,. \tag{4.8}$$

For a periodic solid the matrix elements of the Hamiltonian between a given pair of orbital types depend only upon the difference in the position vectors locating the orbitals so that

$$[\mathbb{H}]_{mj\beta,ni\alpha} = H_{j\beta,i\alpha}(\vec{R}_n - \vec{R}_m)\,. \tag{4.9}$$

Equation (4.5) may be reduced to a matrix of $n_s \times n_s$ dimensions by using

(4.7) and (4.9), which reflect the translational invariance of the solid. One has for the matrix eigenvalue equation:

$$\frac{1}{\sqrt{N}} \sum_n \sum_{i\alpha} \left\{ H_{j\beta,i\alpha}(\vec{R}_n - \vec{R}_m) - E_{\vec{k}\nu} \delta_{\alpha\beta} \delta_{nm} \delta_{ij} \right\} e^{i\vec{k}\cdot\vec{R}_{ni}} a_{i\alpha}(\vec{k},\nu) = 0.$$

(4.10)

We multiply (4.10) by $\frac{1}{\sqrt{N}} e^{-i\vec{k}\cdot\vec{R}_{mj}}$ and sum over all $\vec{R}_m$ to obtain

$$\frac{1}{\sqrt{N}} \sum_{i\alpha} \left\{ \sum_m \sum_n H_{j\beta,i\alpha}(\vec{R}_n - \vec{R}_m) - E_{\vec{k}\nu} \delta_{\alpha\beta} \delta_{nm} \delta_{ij} \right\} e^{i\vec{k}\cdot(\vec{R}_{ni} - \vec{R}_{mj})} a_{i\alpha}(\vec{k},\nu)$$

$$= \sum_{i\alpha} \left\{ h_{j\beta,i\alpha}(\vec{k}) - E_{\vec{k}\nu} \delta_{\alpha\beta} \delta_{ij} \right\} a_{i\alpha}(\vec{k},\nu) = 0.$$

(4.11)

In obtaining (4.11) we have defined

$$h_{j\beta,i\alpha}(\vec{k}) \equiv \sum_p e^{i\vec{k}\cdot\vec{R}_p} e^{i\vec{k}\cdot(\vec{\tau}_i - \vec{\tau}_j)} H_{j\beta,i\alpha}(\vec{R}_p)$$

(4.12)

and used the relations

$$\sum_n \sum_m e^{i\vec{k}\cdot(\vec{R}_n - \vec{R}_m)} H_{j\beta,i\alpha}(\vec{R}_n - \vec{R}_m) = N \sum_p e^{i\vec{k}\cdot(\vec{R}_p)} H_{j\beta,i\alpha}(\vec{R}_p),$$

$$\sum_n \sum_m e^{i\vec{k}\cdot(\vec{R}_n - \vec{R}_m)} \delta_{nm} = N,$$

$$\delta_{ij} e^{i\vec{k}\cdot(\vec{\tau}_i - \vec{\tau}_j)} = \delta_{ij}.$$

Equation (4.11) is the desired result. It shows that the energies and wavefunctions are determined by an $n_s \times n_s$ secular equation. In this form the matrix elements, $h_{j\beta,i\alpha}(\vec{k})$ are the lattice Fourier transforms of the LCAO integrals.

4.3 LCAO matrix elements for the perovskite

In order to solve (4.11), we must specify the elements $h_{j\beta,i\alpha}(\vec{k})$ and hence $H_{j\beta,i\alpha}(\vec{R}_p)$. The lattice-space matrix elements can be parameterized by using the Slater–Koster method described in Section 3.2. We must consider the matrix elements between the 14 basis states within a unit cell and in neighboring unit cells.

For the perovskites an excellent model is obtained if only matrix elements between first and second nearest neighbors are retained. With this approximation, cation–anion (nearest-neighbor) interactions and anion–anion (second-nearest-neighbor) interactions between adjacent oxygen ions are retained and all other interactions are neglected.

The Hamiltonian is of the form

$$H(\vec{r}) = -\frac{\hbar^2}{2m}\nabla^2 + v_{\text{eff}}(\vec{r}),\tag{4.13}$$

where v_{eff} includes the nuclear, Coulomb, and exchange potentials. For our purpose here it is not necessary to know the explicit form of v_{eff}. We need only know that it possesses certain symmetry properties. In particular, $v_{\text{eff}}(\vec{r})$ must be invariant under the operations of the cubic point group O_h when the origin for $\vec{r}$ is at a B-ion site. Similarly, when the origin is at an oxygen site then $v_{\text{eff}}(\vec{r})$ must be invariant to the operations of the point group C_{4v}.

(a) Diagonal LCAO matrix elements

The symmetry properties of the Löwdin orbitals are identical to those of the corresponding atomic orbitals as was proved in Section 3.3. As a consequence, the forms of the atomic wavefunctions listed in Table 3.1 will also be the forms of the Löwdin orbitals. Table 3.2, for the LCAO two-center integrals, may therefore be used without change for either atomic or Löwdin basis orbitals.

The functions listed in Table 3.1 are the linear combinations of the spherical harmonics appropriate for cubic symmetry. According to group theory, the e_g and t_{2g} type d orbitals belong to different irreducible representations of the O_h point group. This means that the matrix elements of a Hamiltonian (which is invariant under O_h) between e_g and t_{2g} orbitals *centered on the same B ion site* must vanish. Furthermore, the two e_g orbitals (three t_{2g} orbitals) belong to *different rows* of the same irreducible representation. This means that the matrix elements of the Hamiltonian between different e_g (t_{2g}) orbitals centered on the same site must also vanish. No such symmetry restrictions apply to matrix elements between orbitals centered on different sites.

Similar symmetry considerations show that the matrix elements of the Hamiltonian between the different p orbitals centered on the same oxygen ion must also vanish.

The above discussion shows that the only non-vanishing LCAO integrals between orbitals centered on the same atomic site are the diagonal matrix elements.

The diagonal matrix elements are of the form

$$H_{\alpha\alpha}(0) = \int \varphi_\alpha(\vec{r})^* H(\vec{r})\varphi_\alpha(\vec{r})\,d\vec{r}.\tag{4.14}$$

From the discussion given in Section 1.3 we know that these integrals will be approximately the sum of an ionization energy plus a Madelung potential plus an electrostatic splitting resulting from the non-spherical part of $v_{\text{eff}}(\vec{r})$. For the d orbitals we define these elements as

$$E_d + V_{\text{M}}(B) + \eta(j)\Delta(d) \qquad (j = e_g \text{ or } t_{2g}). \qquad (4.15)$$

The energy E_d is approximately equal to the ionization energy of a d electron for the free transition metal ion having the appropriate charge state (e.g., $\text{Ti}^{3+} \rightarrow \text{Ti}^{4+}$ for SrTiO_3). The term $V_{\text{M}}(B)$ is the Madelung potential at the B-ion site and $\eta(j)$ is a numerical parameter which determines how the electrostatic splitting, $\Delta(d)$, is shared between the e_g and t_{2g} levels. According to group theory the total energy of a manifold of degenerate levels is unchanged by the non-spherical part of $v_{\text{eff}}(\vec{r})$, that is the "center of gravity" of the levels is unchanged by the electrostatic splitting of the d states. This requires that

$$2\,\eta(e_g) + 3\,\eta(t_{2g}) = 0\,,$$
$$\Delta(d)\,\eta(e_g) - \Delta(d)\,\eta(t_{2g}) = \Delta(d)\,. \qquad (4.16)$$

Thus we find that $\eta(e_g) = \frac{3}{5}$ and $\eta(t_{2g}) = -\frac{2}{5}$.

In a similar manner, we define the diagonal matrix elements for the $2p$ oxygen orbitals by

$$E_p + V_{\text{M}}(O) + \eta(j)\,\Delta(p) \qquad (j = p_{\perp},\ p_{\|})\,, \qquad (4.17)$$

where E_p is the (fictitious) electron affinity of the O^- ion to form O^{2-}, $V_{\text{M}}(O)$ is the oxygen-site Madelung potential and $\Delta(p)$ is the electrostatic splitting due to the non-spherical part of $v_{\text{eff}}(\vec{r})$. Invariance of the center of gravity of the $2p$ manifold requires that $\eta(p_{\perp}) = \frac{1}{3}$, while $\eta(p_{\|}) = -\frac{2}{3}$ where $p_{\perp}$ and $p_{\|}$ refer to $2p$ orbitals oriented perpendicular and parallel, respectively, to the B–O internuclear axis.

The consequence of (4.15) and (4.17) is that we can characterize the LCAO diagonal matrix elements in terms of four parameters E_e, E_t, $E_{\perp}$, and $E_{\|}$ which are defined as follows:

$$E_e = E_d + V_{\text{M}}(B) + \eta(e_g)\,\Delta(d)\,,$$
$$E_t = E_d + V_{\text{M}}(B) + \eta(t_{2g})\,\Delta(d)\,,$$
$$E_{\perp} = E_p + V_{\text{M}}(O) + \eta(p_{\perp})\,\Delta(p)\,,$$
$$E_{\|} = E_p + V_{\text{M}}(O) + \eta(p_{\|})\,\Delta(p)\,. \qquad (4.18)$$

(b) Off-diagonal matrix elements

For the model we are considering, which neglects the third nearest-neighbor interactions, there are two types of off-diagonal matrix elements; cation–anion interactions and anion–anion interactions. The cation–anion matrix elements are of the type,

$$H_{\beta\alpha}(\pm a\vec{e}_j) = \int \varphi_\alpha(\vec{r})^* H \, \varphi_\beta(\vec{r} \pm a\vec{e}_j) \, d\vec{r} \quad (j = x, y, \text{ or } z).$$

$$(4.19)$$

The oxygen–oxygen off-diagonal matrix elements are of the type,

$$H_{\beta\alpha}(\pm a\vec{e}_i \pm a\vec{e}_j) = \int \varphi_\alpha(\vec{r} \pm a\vec{e}_i)^* H \, \varphi_\beta(\vec{r} \pm a\vec{e}_j) \, d\vec{r} \quad (4.20)$$

$$(i \neq j; \, i, j = x, y, \text{ or } z).$$

In (4.19), for the p–d matrix elements, φ_α is a $d(p)$ orbital and φ_β is a $p(d)$ orbital. The p–d matrix elements can be calculated in terms of the $(pd\sigma)$ and $(pd\pi)$ integrals (shown in Section 3.2) with the help of Table 3.2. The t_{2g} d orbitals interact only with the $p_\perp$-type orbitals; the t_{2g}-$p_\parallel$ LCAO integrals vanish by symmetry. Similarly, the e_g-type d orbitals have non-vanishing interactions only with the $p_\parallel$-type orbitals. For the t_{2g}-$p_\perp$ type matrix elements one finds

$$\int d_{\alpha\beta}(\vec{r})^* H \, p_\alpha(\vec{r} \mp a\vec{e}_\beta) \, d\vec{r} = \pm(pd\pi),$$

$$(4.21)$$

$$\int p_\alpha(\vec{r})^* H \, d_{\alpha\beta}(\vec{r} \pm a\vec{e}_\beta) \, d\vec{r} = \mp(pd\pi) \quad (\alpha\beta = xy, xz, \text{ or } yz).$$

$$(4.22)$$

For the e_g-$p_\parallel$ type LCAO matrix elements we have

$$\int d_{3z^2 - r^2}(\vec{r})^* H \, p_\alpha(\vec{r} \mp a\vec{e}_\alpha) \, d\vec{r} = \mp\frac{1}{2}(pd\sigma) \quad (\alpha = x \text{ or } y),$$

$$(4.23)$$

$$\int d_{3z^2 - r^2}(\vec{r})^* H \, p_z(\vec{r} \mp a\vec{e}_z) \, d\vec{r} = \pm(pd\sigma),$$

$$(4.24)$$

$$\int d_{x^2 - y^2}(\vec{r})^* H \, p_x(\vec{r} \mp a\vec{e}_x) \, d\vec{r} = \pm\frac{\sqrt{3}}{2}(pd\sigma),$$

$$(4.25)$$

$$\int d_{x^2 - y^2}(\vec{r})^* H \, p_y(\vec{r} \mp a\vec{e}_y) \, d\vec{r} = \mp\frac{\sqrt{3}}{2}(pd\sigma).$$

$$(4.26)$$

The remaining possibilities are determined by noting that

$$\int p(\vec{r})^* H \, d(\vec{r} \pm a\vec{e}_j) \, d\vec{r} = - \int d(\vec{r})^* H \, p(\vec{r} \pm a\vec{e}_j) \, d\vec{r}$$

for any p and d orbitals. It is noted that all of the possible p–d interactions are described in terms of only two LCAO integrals: $(pd\pi)$ and $(pd\sigma)$.

Next, we consider the interactions between p orbitals on adjacent oxygen ions. These interactions can be expressed in terms of the two LCAO integrals $(pp\pi)$ and $(pp\sigma)$. There are three types of integrals that give non-vanishing matrix elements. They are

$$\int p_\alpha(\vec{r} \mp a\vec{e}_\alpha)^* H \, p_\beta(\vec{r} \mp a\vec{e}_\beta) \, d\vec{r} = -(\mp)(\mp)\frac{1}{2}[(pp\pi) - (pp\sigma)] \quad (4.27)$$

$$(\alpha \neq \beta; \ \alpha, \beta = x, y, \text{ or } z),$$

where $(\mp)(\mp)$ means the product of the signs occurring in the arguments of the orbitals,

$$\int p_\alpha(\vec{r} \mp a\vec{e}_\alpha)^* H \ p_\alpha(\vec{r} \mp a\vec{e}_\beta) \ d\vec{r} = \frac{1}{2}[(pp\pi) + (pp\sigma)], \qquad (4.28)$$

$$(\alpha \neq \beta; \ \alpha, \beta = x, y, \text{ or } z),$$

$$\int p_\gamma(\vec{r} \pm a\vec{e}_\alpha)^* H \ p_\gamma(\vec{r} \pm a\vec{e}_\beta) \ d\vec{r} = (pp\pi), \qquad (4.29)$$

$$(\alpha \neq \beta \neq \gamma; \ \alpha, \beta, \gamma = x, y, \text{ or } z).$$

It is helpful to visualize the various types of interaction described by (4.21)–(4.29). Figure 4.3 shows schematic representations of some of these interactions. Equations (4.21)–(4.29) give all of the possible matrix elements for interactions between first and second nearest-neighbor ions. They are characterized entirely in terms of four basic LCAO integrals: $(pd\sigma)$, $(pd\pi)$, $(pp\sigma)$, and $(pp\pi)$. There are also four diagonal integrals: $E_e, E_t, E_\parallel$, and $E_\perp$. For this approximate model the electronic structure of the cubic perovskites is determined entirely by these eight parameters.

4.4 LCAO eigenvalue equation for the cubic perovskites

In the preceding section we determined the forms of all of the matrix elements which enter the model. To find the energy bands and wavefunctions we must solve the 14×14 matrix eigenvalue equation corresponding to (4.11).

At this point, it is convenient to make a choice of the labels for the rows and columns of the 14×14 matrix (see Table 4.1). We make the following correspondence:

$$
\begin{array}{llll}
d_{z^2}(\vec{r}) \Rightarrow 1; & p_z(\vec{r} - a\vec{e}_z) \Rightarrow 2; & d_{x^2-y^2}(\vec{r}) \Rightarrow 3; & \\
& p_x(\vec{r} - a\vec{e}_x) \Rightarrow 4; & p_y(\vec{r} - a\vec{e}_y) \Rightarrow 5 & \\
d_{xy}(\vec{r}) \Rightarrow 6; & p_x(\vec{r} - a\vec{e}_y) \Rightarrow 7; & p_y(\vec{r} - a\vec{e}_x) \Rightarrow 8 & (4.30) \\
d_{xz}(\vec{r}) \Rightarrow 9; & p_x(\vec{r} - a\vec{e}_z) \Rightarrow 10; & p_z(\vec{r} - a\vec{e}_x) \Rightarrow 11 & \\
d_{yz}(\vec{r}) \Rightarrow 12; & p_y(\vec{r} - a\vec{e}_z) \Rightarrow 13; & p_z(\vec{r} - a\vec{e}_y) \Rightarrow 14. &
\end{array}
$$

This choice is suggested by the fact that in the absence of oxygen–oxygen interactions, the 14×14 matrix block-diagonalizes into a 5×5 and three 3×3 matrices where rows and columns 1–5 form the 5×5, rows and columns 6–8, 9–11 and 12–14 form the three 3×3 matrices. Since the oxygen–oxygen LCAO integrals are small this choice of labels places the largest matrix elements in the diagonal blocks.

Next, we determine the matrix elements, $h_{j\beta,i\alpha}(\vec{k})$, which enter (4.11). For the

(a)

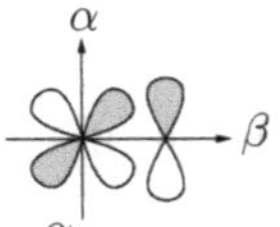

$$\int d_{\alpha\beta}(\vec{r})^* H \, p_\alpha(\vec{r} - a\vec{e}_\beta) \, d\vec{r} = (pd\pi) \qquad (4.21)$$

(b)

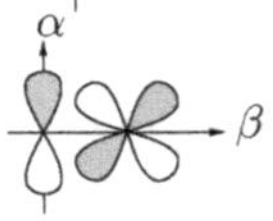

$$\int p_\alpha(\vec{r})^* H \, d_{\alpha\beta}(\vec{r} - a\vec{e}_\beta) \, d\vec{r} = -(pd\pi) \qquad (4.22)$$

(c)

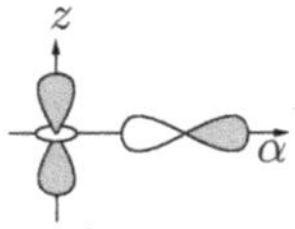

$$\int d_{3z^2-r^2}(\vec{r})^* H \, p_\alpha(\vec{r} - a\vec{e}_\alpha) \, d\vec{r} = -\tfrac{1}{2}(pd\sigma) \qquad (4.23)$$

(d)

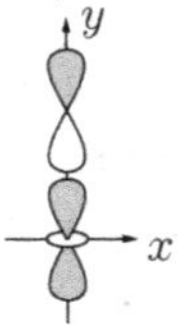

$$\int d_{3z^2-r^2}(\vec{r})^* H \, p_z(\vec{r} - a\vec{e}_z) \, d\vec{r} = (pd\sigma) \qquad (4.24)$$

(e)

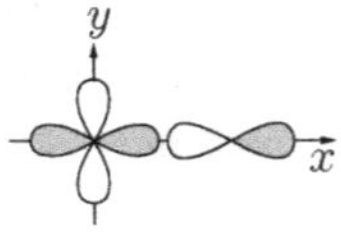

$$\int d_{x^2-y^2}(\vec{r})^* H \, p_x(\vec{r} - a\vec{e}_x) \, d\vec{r} = \tfrac{\sqrt{3}}{2}(pd\sigma) \qquad (4.25)$$

(f)

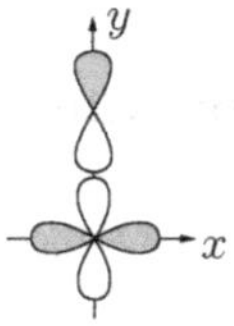

$$\int d_{x^2-y^2}(\vec{r})^* H \, p_y(\vec{r} - a\vec{e}_y) \, d\vec{r} = -\tfrac{\sqrt{3}}{2}(pd\sigma) \qquad (4.26)$$

(g)

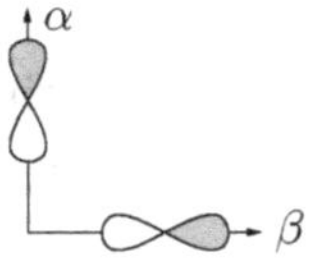

$$\int p_\alpha(\vec{r} - a\vec{e}_\alpha)^* H \, p_\beta(\vec{r} - a\vec{e}_\beta) \, d\vec{r}$$
$$= -\tfrac{1}{2}[(pp\sigma) - (pp\pi)] \qquad (4.27)$$

(h)

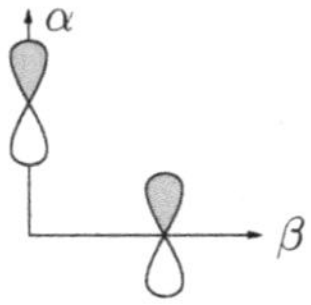

$$\int p_\alpha(\vec{r} - a\vec{e}_\alpha)^* H \, p_\alpha(\vec{r} - a\vec{e}_\beta) \, d\vec{r}$$
$$= \tfrac{1}{2}[(pp\sigma) + (pp\pi)] \qquad (4.28)$$

(i)

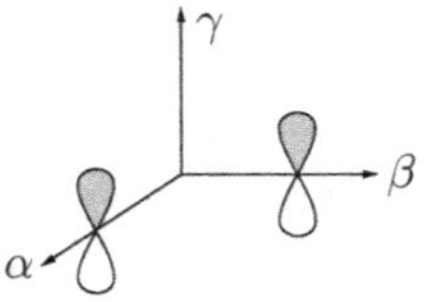

$$\int p_\gamma(\vec{r} - a\vec{e}_\alpha)^* H \, p_\gamma(\vec{r} - a\vec{e}_\beta) \, d\vec{r} = (pp\pi) \qquad (4.29)$$

Figure 4.3. Schematics of the two-center integrals.

Table 4.1. The 14×14 matrix $(\mathbb{H}-E_{\vec{k}\nu}\mathbb{I})$ with vanishing determinant.

	1	2	3	4	5	6	7	8	9	10	11	12	13	14
1	$E_c-E_{\vec{k}\nu}$	$2i(pd\sigma)S_z$	0	$-i(pd\sigma)S_x$	$-i(pd\sigma)S_y$	0	0	0	0	0	0	0	0	0
2		$E_\parallel-E_{\vec{k}\nu}$	0	$-2bS_xS_z$	$-2bS_yS_z$	0	0	0	0	0	$2cC_xC_z$	0	0	$2cC_yC_z$
3			$E_c-E_{\vec{k}\nu}$	$\sqrt{3}i(pd\sigma)S_x$	$-\sqrt{3}i(pd\sigma)S_y$	0	0	0	0	$2cC_zC_x$	0	0	$2cC_yC_z$	0
4				$E_\parallel-E_{\vec{k}\nu}$	$-2bS_xS_y$	0	0	0	0	0	0	0	0	0
5					$E_\parallel-E_{\vec{k}\nu}$	0	0	0	0	0	0	0	0	0
6						$E_t-E_{\vec{k}\nu}$	$2i(pd\pi)S_y$	$2i(pd\pi)S_x$	0	0	0	0	0	0
7							$E_\perp-E_{\vec{k}\nu}$	$-2bS_xS_y$	0	$4(pp\pi)C_yC_z$	0	0	0	0
8								$E_\perp-E_{\vec{k}\nu}$	0	0	$4(pp\pi)C_xC_z$	0	0	0
9									$E_t-E_{\vec{k}\nu}$	$2i(pd\pi)S_z$	$2i(pd\pi)S_z$	0	0	0
10										$E_\perp-E_{\vec{k}\nu}$	$-2bS_xS_z$	0	$4(pp\pi)C_zC_x$	0
11											$E_\perp-E_{\vec{k}\nu}$	0	0	$4(pp\pi)C_xC_y$
12												$E_t-E_{\vec{k}\nu}$	$2i(pd\pi)S_z$	$2i(pd\pi)S_y$
13													$E_\perp-E_{\vec{k}\nu}$	$-2bS_yS_z$
14														$E_\perp-E_{\vec{k}\nu}$

$H_{ji}=H_{ij}^*$

Parameters used are $b\equiv(pp\sigma)-(pp\pi)$, $c\equiv(pp\sigma)+(pp\pi)$, $S_\alpha\equiv\sin k_\alpha a$ and $C_\alpha\equiv\cos k_\alpha a$. $i\equiv\sqrt{-1}$.

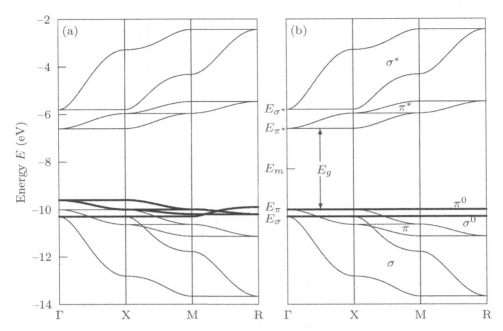

Figure 4.4. (a) Energy bands for the parameters (all in eV): $E_e = -5.8$, $E_t = -6.4$, $E_\perp = -10.0$, $E_\parallel = -10.5$, $(pd\sigma) = 2.1$, $(pd\pi) = 0.8$, $(pp\sigma) = -0.2$, $(pp\pi) = -0.1$. (b) Energy bands for the same parameters as in (a) except $(pp\sigma) = (pp\pi) = 0$.

model considered here we need only retain the terms for which

$$|\vec{R}_p + \vec{\tau}_i - \vec{\tau}_j| \leq \sqrt{2}a \simeq 2.76\,\text{Å}. \tag{4.31}$$

Using the results of (4.18) and (4.21)–(4.29) we obtain the matrix shown in Table 4.1. The eigenvalues, $E_{\vec{k}}$, are determined by the matrix eigenvalue equation

$$\det(\mathbb{H} - E_{\vec{k}}\mathbb{I}) = 0, \tag{4.32}$$

where $\mathbb{I}$ is a $n_s \times n_s$ unit matrix.

Before discussing the details of how solutions for the energy bands may be obtained, it is helpful to have a picture of the general structure of the energy bands [3]. This is best accomplished by displaying a typical set of bands obtained from (4.32).

Figure 4.4(a) shows the energy bands using the eight LCAO parameters for SrTiO$_3$ [4, 5]. Figure 4.4(b) shows a similar set of bands for the same parameters as used in calculating the bands in Fig. 4.4(a) except that the oxygen–oxygen interaction parameters, $(pp\pi)$ and $(pp\sigma)$, are taken to be zero.

The basic structure of the bands is most easily explained by beginning with

Fig. 4.4(b). There are basically three groups of bands. The first group consists of the bands labeled σ^* and π^* which lie between -7 and $-2\,$eV. These are the d-electron conduction bands. A second group, σ and π, are the mirror images of the σ^* and π^* bands, respectively. They are oxygen valence bands. The last group consists of flat bands labeled σ^0 and π^0. These are non-bonding oxygen valence bands. The wavefunctions for the σ and σ^* bands involve only the LCAO parameter $(pd\sigma)$ and those of the π and π^* involve only $(pd\pi)$. The widths of the σ and σ^* bands are determined by $(pd\sigma)$. For the π and π^* bands the widths are determined by $(pd\pi)$. The wavefunctions of the σ, σ^*, π, and π^* bands are admixtures of p and d orbitals. The non-bonding σ^0 and π^0 bands have wavefunctions that are entirely composed of oxygen $2p$ orbitals.

When the oxygen–oxygen interactions, $(pp\pi)$ and $(pp\sigma)$, are non-zero (as for the bands in Fig. 4.4(a)) the non-bonding bands are no longer flat. They are broadened into bands whose widths are controlled by $(pp\pi)$ and $(pp\sigma)$. There is also some minor changes in the σ and π valence bands, because of interaction with the non-bonding bands near crossing points. However, the general structure of Fig. 4.4(a) is not qualitatively different from that of Fig. 4.4(b) and the σ^* and π^* bands are essentially the same in both figures.

The above analysis suggests that the qualitative features of the energy bands of the perovskites are determined by nearest-neighbor cation–anion interactions and that the effects of the anion–anion interactions are small. In the remainder of this chapter we investigate the analytic solution of (4.32) when $(pp\pi)$ and $(pp\sigma)$ vanish. Discussions and solutions of the secular matrix equation including the effects of the oxygen–oxygen interactions are given in Chapter 5.

4.5 Qualitative features of the energy bands

In this section we obtain and discuss the solutions of (4.32) in the absence of oxygen–oxygen interactions.

Inspection of the matrix in Table 4.1 shows that the matrix equation block-diagonalizes into a 5×5 and three equivalent 3×3 blocks when $(pp\pi)$ and $(pp\sigma)$ are set to zero.

(a) Pi bands

The 3×3 blocks involve only E_t and the $(pd\pi)$ two-center integral and therefore we refer to these bonds as the "pi bands". Consider the 3×3 block obtained from the rows and columns 6–8. The other two 3×3 blocks are equivalent since they may be obtained from the first by permutation of the coordinate axis labels; substitution of z for y in the first 3×3 block gives the second and substitution of z for x gives the last 3×3 block.

The secular equation for the 3×3 blocks is of the form

$$\begin{pmatrix} (E_t - E_{\vec{k}\nu}) & 2i(pd\pi)S_\beta & 2i(pd\pi)S_\alpha \\ -2i(pd\pi)S_\beta & (E_\perp - E_{\vec{k}\nu}) & 0 \\ -2i(pd\pi)S_\alpha & 0 & (E_\perp - E_{\vec{k}\nu}) \end{pmatrix} \begin{pmatrix} a_{\alpha\beta} \\ a_\alpha \\ a_\beta \end{pmatrix} = 0 \qquad (4.33)$$

where $\alpha\beta = xy, xz$, or yz. The coefficients $a_{\alpha\beta}$, a_α, and a_β specify the amplitudes of the orbitals $d_{\alpha\beta}(\vec{r})$, $p_\alpha(\vec{r} - a\vec{e}_\beta)$, and $p_\beta(\vec{r} - a\vec{e}_\alpha)$ making up the eigenstates. Requiring the determinant of the coefficients to vanish gives the eigenvalue condition (secular equation),

$$(E_t - E_{\vec{k}\nu})(E_\perp - E_{\vec{k}\nu})^2 - 4(pd\pi)^2(S_\alpha^2 + S_\beta^2)(E_\perp - E_{\vec{k}\nu}) = 0. \qquad (4.34)$$

Since the term $(E_\perp - E_{\vec{k}\nu})$ can be factored out from (4.34), one eigenvalue is

$$E_{\vec{k}\nu} = E_\perp = E_{\vec{k}\pi^0}. \qquad (4.35)$$

The eigenvector for this energy is easily seen to be

$$\frac{1}{\sqrt{S_\alpha^2 + S_\beta^2}} \begin{pmatrix} 0 \\ S_\alpha \\ -S_\beta \end{pmatrix}. \qquad (4.36)$$

The real-space wavefunction corresponding to (4.36) is

$$\Psi_{\vec{k}\pi^0}(\vec{r}) = \frac{1}{\sqrt{N(S_\alpha^2 + S_\beta^2)}} \sum_m \left\{ S_\alpha \, p_\alpha(\vec{r} - \vec{R}_m - a\vec{e}_\beta) \, e^{i\vec{k}\cdot(\vec{R}_m + a\vec{e}_\beta)} \right.$$

$$\left. - S_\beta \, p_\beta(\vec{r} - \vec{R}_m - a\vec{e}_\alpha) \, e^{i\vec{k}\cdot(\vec{R}_m + a\vec{e}_\alpha)} \right\} \qquad (4.37)$$

where the sum is over the lattice vectors, $\vec{R}_m$, for the locations of the unit cells only. It is clear from (4.37) that the wavefunction involves only the oxygen orbitals and contains no d-orbital mixture. This solution correspond to the three symmetry equivalent π^0 non-bonding bands. Their energy, $E_{\vec{k}\nu} = E_\perp$ is independent of the $\vec{k}$-vector so the bands are "flat", that is without dispersion.

The remaining two eigenvalues of (4.34) are

$$E_{\vec{k}\binom{\pi^*}{\pi}}^{\alpha\beta} = \frac{1}{2}(E_\perp + E_t) \pm \sqrt{\left[\frac{1}{2}(E_t - E_\perp)\right]^2 + 4(pd\pi)^2(S_\alpha^2 + S_\beta^2)}. \qquad (4.38)$$

For $\vec{k} = 0$ (at Γ), $E_{\vec{k}\pi^*} = E_t$ and $E_{\vec{k}\pi} = E_\perp$. Therefore, we see that the energy gap between the π (or π^0) valence band and the π^* conduction band is

$$E_g^0(\Gamma) = E_t - E_\perp, \qquad (4.39)$$

and the "mid-gap" energy, $E_m^0(\Gamma)$, is

$$E_m^0(\Gamma) = \frac{1}{2}(E_t + E_\perp). \tag{4.40}$$

The energies, $E_g^0(\Gamma)$ and $E_m^0(\Gamma)$, shown in Fig. 4.4(b), allow the energies of the π and π^* bands to be expressed in a more physical form; namely,

$$E_{\vec{k}(\pi^*)}^{\alpha\beta} = E_m^0(\Gamma) \pm \sqrt{\left[\frac{1}{2}E_g^0(\Gamma)\right]^2 + 4(pd\pi)^2(S_\alpha^2 + S_\beta^2)}. \tag{4.41}$$

For simplicity, in the remainder of this book, we shall use the notation, E_g and E_m for $E_g^0(\Gamma)$ and $E_m^0(\Gamma)$, respectively. The center of gravity of the two bands is at the mid-gap and the bands are mirror reflections of one another with the "mirror" located at mid-gap. It is also to be noted that these bands have a two-dimensional character in that each band depends only on two components of $\vec{k}$. Thus $E_{\vec{k}\pi^*}^{\alpha\beta}$ and $E_{\vec{k}\pi}^{\alpha\beta}$ are flat along the γ direction in the Brillouin zone, where $\gamma \neq \alpha \neq \beta$.

The eigenvectors for these bands are

$$C_{\vec{k}\nu}\begin{pmatrix} (E_\perp - E_{\vec{k}\nu}^{\alpha\beta}) \\ 2i(pd\pi)S_\beta \\ 2i(pd\pi)S_\alpha \end{pmatrix} ; \qquad (\nu = \pi^* \text{ or } \pi) \tag{4.42}$$

where

$$C_{\vec{k}\nu}^{-1} = \sqrt{(E_\perp - E_{\vec{k}\nu}^{\alpha\beta})^2 + 4(pd\pi)^2(S_\alpha^2 + S_\beta^2)},$$

$$= \sqrt{2(E - E_m)(E - E_\perp)}. \tag{4.43}$$

The validity of (4.42) is easily verified by substitution into (4.33) and use of (4.34). The real-space wavefunctions are given by

$$\Psi_{\vec{k}\nu}^{\alpha\beta}(\vec{r}) = \frac{C_{\vec{k}\nu}}{\sqrt{N}} \sum_m e^{i\vec{k}\cdot\vec{R}_m}\Big\{(E_\perp - E_{\vec{k}\nu}^{\alpha\beta}) \, d_{\alpha\beta}(\vec{r}-\vec{R}_m)$$

$$+2i(pd\pi)S_\beta \, p_\alpha(\vec{r}-\vec{R}_m-a\vec{e}_\beta) \, e^{ik_\beta a}$$

$$+2i(pd\pi)S_\alpha \, p_\beta(\vec{r}-\vec{R}_m-a\vec{e}_\alpha) \, e^{ik_\alpha a}\Big\}. \tag{4.44}$$

Except at Γ the π^* and π band wavefunctions are admixtures of d and p orbitals. The admixture varies with $\vec{k}$. At Γ, $E_{\vec{k}\pi^*} = E_t$ and $E_{\vec{k}\pi} = E_\perp$. Equation (4.42) shows that the eigenvector for the π^* band is pure d orbital since $S_\alpha = S_\beta = 0$. For the π bands we must determine eigenvectors components by a limiting process because $(E_\perp - E_{\vec{k}\pi}) \to 0$ as $\vec{k} \to 0$, so all the components of (4.42) are tending to zero. It is easily shown that $(E_\perp - E_{\vec{k}\pi})$ tends to zero quadratically for small $|\vec{k}|$

and thus the eigenvector tends to the form

$$\frac{1}{\sqrt{4(S_\alpha^2 + S_\beta^2)}} \begin{pmatrix} 0 \\ 2iS_\alpha \\ 2iS_\beta \end{pmatrix} \xrightarrow{\vec{k} \to 0} \begin{pmatrix} 0 \\ \cos\phi \\ \sin\phi \end{pmatrix}, \tag{4.45}$$

where $\cos\phi = k_\alpha/\sqrt{k_\alpha^2 + k_\beta^2}$. Therefore, the π bands become pure p orbital at Γ. We conclude that there is no mixing between the p and d orbitals at Γ. This turns out to be a general result (as we shall show later) and is not associated with the approximations made in the secular equation. The eigenvalues of the three symmetry-equivalent blocks are the same so each eigenvalue is triply degenerate. As a result, arbitrary linear combinations of symmetry-equivalent eigenvectors are also eigenvectors.

At other points in the Brillouin zone the wavefunctions are admixtures of p and d orbitals. The mixing increases as $|\vec{k}|$ increases and is maximum at the point R in the Brillouin zone.

The admixture, r_d, is easily calculated. For the π valence band the d-orbital probability $r_d(E)$, at $E_{\vec{k}\nu}^{\alpha\beta} = E$ is according to (4.42) given by

$$r_d(E) = \frac{|a_{\alpha\beta}|^2}{|a_{\alpha\beta}|^2 + |a_\alpha|^2 + |a_\beta|^2} = \frac{(E_\perp - E)^2}{(E_\perp - E)^2 + 4(pd\pi)^2(S_\alpha^2 + S_\beta^2)}$$

$$= \frac{(E - E_\perp)}{2(E - E_m)} \tag{4.46}$$

where $E_m = \frac{1}{2}(E_t + E_\perp)$. If we define $E' \equiv E - E_m$ the result is

$$r_d(E') = \frac{1}{2} + \frac{E_g}{4E'}. \tag{4.47}$$

Equation (4.47) is valid for energies within the range of the valence band, that is for $-E_B \le E' \le -E_g/2$ with $E_B = \sqrt{(E_g/2)^2 + 8(pd\pi)^2}$. The result shows that for a given E' the amount of mixing of the d orbitals into the valence band is dependent only on the band gap, E_g, and the p–d interaction, $(pd\pi)$.

A rough measure of the d-orbital probability averaged over the valence band, $\langle r_d \rangle$, can be obtained if the density of states is taken to be constant (see Chapter 6 for detailed calculations of the density of states). For this approximation,

$$\langle r_d \rangle \equiv \frac{1}{(E_\perp - E_m + E_B)} \int_{E_m - E_B}^{E_\perp} dE \frac{(E - E_\perp)}{2(E - E_m)} \tag{4.48}$$

$$= \frac{1}{(-E_g/2 + E_B)} \int_{-E_B}^{-E_g/2} dE' \left(\frac{1}{2} + \frac{E_g}{4E'}\right) \tag{4.49}$$

$$= \frac{1}{2} + \frac{1}{2} \frac{\eta}{1 - \eta} \ln \eta, \qquad \text{with} \qquad \eta = \frac{E_g}{2E_B}. \tag{4.50}$$

$\langle r_d \rangle$ depends only on the ratio of the band gap, E_g, to E_B, a measure of band width. The ratio, η, lies between 0 and 1. Small values of η correspond to strong covalent mixing. The behavior of $\langle r_d \rangle$ versus E_g for several values of $(pd\pi)$ is shown in Fig. 4.5(a). Figure 4.5(b) shows $\langle r_d \rangle$ as a function of $(pd\pi)$ for several values of E_g.

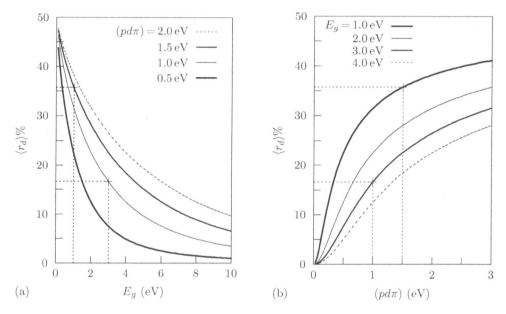

Figure 4.5. The average d-orbital mixing in the π valence band as functions of E_g and $(pd\pi)$. The two examples solved in the text are marked on the graphs.

For perovskites such as $SrTiO_3$ or $BaTiO_3$, E_g is approximately $3\,eV$ and $(pd\pi)$ about $1\,eV$ so that

$$E_B = \sqrt{(E_g/2)^2 + 8(pd\pi)^2} = 3.2016 \text{ eV} \tag{4.51}$$

$$\eta = \frac{E_g}{2E_B} = 0.4685 \tag{4.52}$$

$$\langle r_d \rangle = \frac{1}{2} + \frac{1}{2} \frac{\eta}{1 - \eta} \ln \eta = 0.1658. \tag{4.53}$$

This result means that the average valence-band wavefunction (for $SrTiO_3$ or $BaTiO_3$) is about 17% d orbital and 83% p orbital in content. For the average π^* conduction-band wavefunction it is just the reverse: 17% p orbital and 83% d orbital in content.

For the metallic perovskite ReO_3, the band gap is only about $1\,eV$ while $(pd\pi) \simeq 1.5\,eV$. In this case, $E_B = 4.2720\,eV$, $\eta = 0.1170$, and $\langle r_d \rangle = 0.3578$. Thus the average valence-band wavefunction has about 36% d orbital and 64% p orbital. Clearly, ReO_3 is much more covalent than $SrTiO_3$ or $BaTiO_3$.

(b) Sigma bands

The upper 5×5 block in (4.32) determines the sigma bands. With $(pp\sigma) = (pp\pi) = 0$ the remaining three parameters involved are E_e, $E_\parallel$, and $(pd\sigma)$. The matrix equation is

$$
\begin{pmatrix}
(E_e - E_{\vec{k}\nu}) & 2i(pd\sigma)S_z & 0 & -i(pd\sigma)S_x & -i(pd\sigma)S_y \\
-2i(pd\sigma)S_z & (E_\parallel - E_{\vec{k}\nu}) & 0 & 0 & 0 \\
0 & 0 & (E_e - E_{\vec{k}\nu}) & \sqrt{3}i(pd\sigma)S_x & -\sqrt{3}i(pd\sigma)S_y \\
i(pd\sigma)S_x & 0 & -\sqrt{3}i(pd\sigma)S_x & (E_\parallel - E_{\vec{k}\nu}) & 0 \\
i(pd\sigma)S_y & 0 & \sqrt{3}i(pd\sigma)S_y & 0 & (E_\parallel - E_{\vec{k}\nu})
\end{pmatrix}
\begin{pmatrix}
a_{z^2} \\
a_z \\
a_{x^2} \\
a_x \\
a_y
\end{pmatrix}
= 0.
$$

$$(4.54)$$

The coefficients specify the amplitudes of the orbitals as follows:

$$
\begin{aligned}
a_{z^2} &\Rightarrow d_{z^2}(\vec{r}) \\
a_z &\Rightarrow p_z(\vec{r} - a\vec{e}_z) \\
a_{x^2} &\Rightarrow d_{x^2-y^2}(\vec{r}) \\
a_x &\Rightarrow p_x(\vec{r} - a\vec{e}_x) \\
a_y &\Rightarrow p_y(\vec{r} - a\vec{e}_y).
\end{aligned}
$$

$$(4.55)$$

This 5×5 matrix equation can be solved exactly. The eigenvalue equation obtained from the determinant of the matrix is

$$
\{(E_e - E_{\vec{k}\nu})^2(E_\parallel - E_{\vec{k}\nu})^2 - 4(pd\sigma)^2(E_e - E_{\vec{k}\nu})(E_\parallel - E_{\vec{k}\nu})(S_x^2 + S_y^2 + S_z^2)
$$
$$
+12(pd\sigma)^4(S_x^2 S_y^2 + S_y^2 S_z^2 + S_z^2 S_x^2)\}(E_\parallel - E_{\vec{k}\nu}) = 0.
$$

$$(4.56)$$

Since the term $(E_\parallel - E_{\vec{k}\nu})$ can be factored from the result one obvious eigenvalue is

$$
E_{\vec{k}\nu} = E_\parallel.
$$

$$(4.57)$$

This flat, non-bonding sigma band will be denoted by $E_{\vec{k}\sigma^0}$. The eigenvector is easily found to be

$$\frac{1}{\sqrt{S_x^2 S_y^2 + S_y^2 S_z^2 + S_z^2 S_x^2}} \begin{pmatrix} 0 \\ S_x S_y \\ 0 \\ S_y S_z \\ S_z S_x \end{pmatrix}. \tag{4.58}$$

The corresponding real-space wavefunction is

$$\Psi_{\vec{k}\sigma^0}(\vec{r}) = \frac{1}{\sqrt{N(S_x^2 S_y^2 + S_y^2 S_z^2 + S_z^2 S_x^2)}} \sum_m e^{i\vec{k}\cdot\vec{R}_m} \left\{ S_x S_y \, p_z(\vec{r} - \vec{R}_m - a\vec{e}_z) \, e^{ik_z a} \right.$$

$$\left. + S_y S_z \, p_x(\vec{r} - \vec{R}_m - a\vec{e}_x) \, e^{ik_x a} + S_z S_x \, p_y(\vec{r} - \vec{R}_m - a\vec{e}_y) \, e^{ik_y a} \right\}. \tag{4.59}$$

Equation (4.59) shows that the wavefunction for the σ^0 band is composed entirely of p orbitals.

Returning to the eigenvalue equation (4.56), we see that after factoring out the term $(E_\| - E_{\vec{k}\nu})$ the remaining expression is quadratic in the variable $(E_e - E_{\vec{k}\nu})(E_\| - E_{\vec{k}\nu})$. This allows an immediate solution with the result that

$$(E_e - E_{\vec{k}\nu})(E_\| - E_{\vec{k}\nu}) = 2(pd\sigma)^2 \Big\{ (S_x^2 + S_y^2 + S_z^2)$$

$$\pm \sqrt{(S_x^4 + S_y^4 + S_z^4) - (S_x^2 S_y^2 + S_y^2 S_z^2 + S_z^2 S_x^2)} \Big\}. \tag{4.60}$$

Equation (4.60) is quadratic in the variable $E_{\vec{k}\nu}$ so it can be solved to give the eigenvalues. They are:

$$E_{\vec{k}\sigma^*}^{(\pm)} = \frac{1}{2}[E_e + E_\|] + \sqrt{\left[\frac{1}{2}(E_e - E_\|)\right]^2 + 2(pd\sigma)^2 \left[(S_x^2 + S_y^2 + S_z^2) \pm S^2\right]} \tag{4.61}$$

$$E_{\vec{k}\sigma}^{(\pm)} = \frac{1}{2}(E_e + E_\|) - \sqrt{\left[\frac{1}{2}(E_e - E_\|)\right]^2 + 2(pd\sigma)^2 \left[(S_x^2 + S_y^2 + S_z^2) \pm S^2\right]} \tag{4.62}$$

$$S^2 \equiv \sqrt{(S_x^4 + S_y^4 + S_z^4) - (S_x^2 S_y^2 + S_y^2 S_z^2 + S_z^2 S_x^2)}. \tag{4.63}$$

In general $E_{\vec{k}\sigma^*}^{(\pm)}$ and $E_{\vec{k}\sigma}^{(\pm)}$ depend on all three components of $\vec{k}$. However, $E_{\vec{k}\sigma^*}^{(-)}$ and $E_{\vec{k}\sigma}^{(-)}$ are flat along any Γ to X direction. To see this let $\vec{k} = k_\alpha \vec{e}_\alpha$, where $\alpha = x, y$, or z, then $S^2 = S_\alpha^2$ and $E_{\vec{k}\sigma}^{(-)} = E_\|$; $E_{\vec{k}\sigma^*}^{(-)} = E_e$ independent of the magnitude of $\vec{k}$.

The eigenvectors are:

$$C_{\vec{k}\nu}\begin{pmatrix} (E_{\parallel} - E_{\vec{k}\nu})X_\nu/(pd\sigma) \\ 2iS_zX_\nu \\ -\sqrt{3}(E_{\parallel} - E_{\vec{k}\nu})(S_x^2 - S_y^2)/(pd\sigma) \\ -iS_x[X_\nu + 3(S_x^2 - S_y^2)] \\ -iS_y[X_\nu - 3(S_x^2 - S_y^2)] \end{pmatrix}, \tag{4.64}$$

with $\nu = \sigma^*(\pm)$ or $\sigma(\pm)$ and

$$X_\nu = (E_{\parallel} - E_{\vec{k}\nu})(E_e - E_{\vec{k}\nu})/(pd\sigma)^2 - 3(S_x^2 + S_y^2),$$
$$= 2S_z^2 - S_x^2 - S_y^2 \pm 2S^2,$$

where the factor, $C_{\vec{k}\nu}$, the normalization coefficient is equal to the inverse square root of the sum of the squares of the components in (4.64), namely

$$C_{\vec{k}\nu} = \left\{ X_\nu^2[\eta_{\parallel}^2 + S_x^2 + S_y^2 + 4S_z^2] + (S_x^2 - S_y^2)^2[6X_\nu + 3\eta_{\parallel}^2 + 9(S_x^2 + S_y^2)] \right\}^{-1/2} \tag{4.65}$$

where $\eta_{\parallel} = (E_{\parallel} - E_{\vec{k}\nu})/(pd\sigma)$. The real-space wavefunction is

$$\Psi_{\vec{k}\nu}(\vec{r}) = \frac{C_{\vec{k}\nu}}{\sqrt{N}} \sum_m e^{i\vec{k}\cdot\vec{R}} \left\{ a_{z^2}d_{z^2}(\vec{r} - \vec{R}_m) + a_z e^{ik_z a}p_z(\vec{r} - \vec{R}_m - a\vec{e}_z) \right.$$

$$\left. + a_x e^{ik_z a}p_x(\vec{r} - \vec{R}_m - a\vec{e}_x) + a_y e^{ik_y a}p_y(\vec{r} - \vec{R}_m - a\vec{e}_y) \right\}. \tag{4.66}$$

The form of the eigenvector given by (4.64) is not convenient to use when only one of the components of $\vec{k}$ is non-zero (that is along Γ–X) because all of the components vanish and a limiting process must be used. It is more convenient to return to the original matrix equation, (4.54) and recalculate the eigenvectors.

If $k_x = k_y = 0$ and $k_z \neq 0$, then (4.54) becomes

$$\begin{pmatrix} (E_e - E_{k_z\nu}) & 2i(pd\sigma)S_z & 0 & 0 & 0 \\ -2i(pd\sigma)S_z & (E_{\parallel} - E_{k_z\nu}) & 0 & 0 & 0 \\ 0 & 0 & (E_e - E_{k_z\nu}) & 0 & 0 \\ 0 & 0 & 0 & (E_{\parallel} - E_{k_z\nu}) & 0 \\ 0 & 0 & 0 & 0 & (E_{\parallel} - E_{k_z\nu}) \end{pmatrix} \begin{pmatrix} a_{z^2} \\ a_z \\ a_{x^2} \\ a_x \\ a_y \end{pmatrix} = 0. \tag{4.67}$$

It is immediately seen that two of the valence-band states have eigenvalues $E_{\parallel}$ and

that $\begin{pmatrix} 0 \\ 0 \\ 0 \\ 1 \\ 0 \end{pmatrix}$ and $\begin{pmatrix} 0 \\ 0 \\ 0 \\ 0 \\ 1 \end{pmatrix}$ are the eigenvectors. Similarly, one of the conduction bands is flat with energy $E^{(-)}_{k_z\sigma*} = E_e$ and eigenvector $\begin{pmatrix} 0 \\ 0 \\ 1 \\ 0 \\ 0 \end{pmatrix}$. The two orbitals $d_{z^2}(\vec{r})$ and

$p_z(\vec{r} - a\vec{e}_z)$ are mixed. The 2×2 matrix may be diagonalized to obtain:

$$E^{(+)}_{k_z\sigma*} = \frac{1}{2}(E_e + E_{\parallel}) + \sqrt{\left[\frac{1}{2}(E_e - E_{\parallel})\right]^2 + 4(pd\sigma)^2 S_z^2} \xrightarrow{k_z \to 0} E_e + \frac{4(pd\pi)^2(k_za)^2}{E_g},$$
(4.68a)

$$E^{(+)}_{k_z\sigma} = \frac{1}{2}(E_e + E_{\parallel}) - \sqrt{\left[\frac{1}{2}(E_e - E_{\parallel})\right]^2 + 4(pd\sigma)^2 S_z^2} \xrightarrow{k_z \to 0} E_{\parallel} - \frac{4(pd\pi)^2(k_za)^2}{E_g}.$$
(4.68b)

The amplitudes a_{z^2} and a_z are the only non-zero components of the eigenvectors. For the conduction band and the valence band the eigenvectors, respectively, are

$$\frac{1}{\sqrt{(E_{\parallel} - E^{(+)}_{k_z\sigma*})^2 + 4(pd\sigma)^2 S_z^2}} \begin{pmatrix} \dfrac{(E_{\parallel} - E^{(+)}_{k_z\sigma*})}{2i(pd\sigma)S_z} \\ 0 \\ 0 \\ 0 \\ 0 \end{pmatrix} \xrightarrow{k_z \to 0} \begin{pmatrix} 1 \\ 0 \\ 0 \\ 0 \\ 0 \end{pmatrix}, \quad (4.69a)$$

$$\frac{1}{\sqrt{(E_e - E^{(+)}_{k_z\sigma})^2 + 4(pd\sigma)^2 S_z^2}} \begin{pmatrix} 2i(pd\sigma)S_z \\ -(E_e - E^{(+)}_{k_z\sigma}) \\ 0 \\ 0 \\ 0 \end{pmatrix} \xrightarrow{k_z \to 0} \begin{pmatrix} 0 \\ 1 \\ 0 \\ 0 \\ 0 \end{pmatrix}. \quad (4.69b)$$

It follows from (4.68) and (4.69) that the mixing between the p and d orbitals vanishes at Γ and increases linearly with $|\vec{k}|$ away from Γ along the vector $(0, 0, k_z)$. Similar results can easily be obtained for other symmetry directions.

4.6 Summary of the chapter results

In the preceding sections of this chapter we formulated a 14×14 matrix energy band problem for a model which includes cation–anion interactions (nearest-neighbor) and anion–anion (second-nearest-neighbor) interactions. The cation–anion interactions involved the LCAO two-center integrals $(pd\sigma)$ and $(pd\pi)$. For the perovskites, $(pd\pi)$ is usually in the range of 1.0–1.5 eV and $(pd\sigma)$ is usually two to three times larger than $(pd\pi)$. The anion–anion interactions involve the LCAO parameters $(pp\sigma)$ and $(pp\pi)$. These parameters are usually 5–10 times smaller than the p–d two-center integrals. Because of this, the qualitative features of the energy bands can still be obtained even when $(pp\sigma)$ and $(pp\pi)$ are neglected.[1] This point is demonstrated in Fig. 4.4 which shows that the major effect of $(pp\sigma)$ and $(pp\pi)$ is to broaden the flat non-bonding oxygen bands into bands of narrow widths.

The LCAO energy bands model was solved analytically for the approximation $(pp\sigma) = (pp\pi) = 0$. For this approximation the matrix equation is block-diagonal. A 5×5 block involves only the e_g-type d orbitals and the $2p$ orbitals oriented parallel to the B–O axis. These orbitals interact only through LCAO integral $(pd\sigma)$. The five bands resulting from the 5×5 block were designated as the "sigma bands". One band, a flat non-bonding band called σ^0, involves only the $2p$ orbitals. Two conduction bands, termed σ^* bands, are formed whose width depends on $(pd\sigma)$. Two valence bands, called σ bands, are also formed. The σ bands are the mirror reflection of the σ^* bands with the mirror located at the mean energy $(E_e + E_{\parallel})/2$.

The mixing between the p and d orbitals was shown to vanish at Γ, and to increase with $|\vec{k}|$ near Γ. The maximum covalent mixing occurs at R in the Brillouin zone. The remaining 9×9 of the secular matrix is block-diagonal and consists of three equivalent 3×3 blocks. Each block involves one of the three t_{2g}-type d orbitals, $d_{\alpha\beta}$, and the p orbitals which interact through the LCAO $(pd\pi)$ parameter. These bands were named the "pi bands". Each 3×3 yields one (π^*) conduction band, one (π) valence band and one flat non-bonding (π^0) valence band. The π and π^* bands are mirror reflections of one another with the mirror located at $(E_t + E_{\perp})/2$. The π and π^* bands were found to possess two-dimensional character. The dispersion $(E_{\vec{k}\nu}$ versus $\vec{k})$ depends only on two components of the three-dimensional $\vec{k}$. This results in flat bands along several lines in the Brillouin zone. Similar features were found for one of the σ and σ^* bands. Such flat regions have significant effects on the density of states and, as will be shown in subsequent chapters, produce characteristics structure in the optical, photoemission spectra.

The p–d mixing of the eigenvectors of the π and π^* bands also vanishes at Γ and also increases with increasing $|\vec{k}|$ away from Γ. Typical mixing ratios at R in the Brillouin zone range from 65% to 85% d and 35% to 15% p for the conduction bands and conversely for the valence bands. There is no d-orbital component for the non-bonding $(\sigma^0$ and $\pi^0)$ bands for the approximation $(pp\sigma) = (pp\pi) = 0$. Even when $(pp\sigma)$ and $(pp\pi)$ are finite there is only a very small d component so that the non-bonding bands may be regarded as having pure p-orbital wavefunction for all practical purposes.

In the next chapter we look at some analytical solutions of the energy bands with $(pp\sigma)$ and $(pp\pi) \neq 0$ and determine in more detail the effect of the oxygen–oxygen interactions.

---

[1] An important exception to this rule occurs for the high-temperature superconductors. For these materials the effective oxygen–oxygen interactions may be comparable to the p–d interactions due to strong electron–electron correlation effects. The d bands of the cuprate HTSCs are discussed in Chapter 11.

Suggested references on "Group theory and ligand fields"

S. Sugano, Y. Tanabe and H. Kamimura, *Multiplets of transition metal ions in crystals*. In *Pure and Applied Physics*, Vol. 33 (New York and London, Academic Press, 1970).

M. Tinkham, *Group theory and quantum mechanics, International Series in Pure and Applied Physics* (New York, McGraw-Hill, 1964).

References

[1] A. H. Kahn and A. J. Leyendecker, *Phys. Rev. A* **135**, 1321 (1964).
[2] J. C. Slater and G. F. Koster, *Phys. Rev.* **94**, 1498 (1954).
[3] T. Wolfram, E. A. Kraut, and F. J. Morin, *Phys. Rev. B* **7**, 1677 (1973).
[4] T. Wolfram, *Phys. Rev. Lett.* **29**, 1383 (1972).
[5] L. F. Mattheiss, *Phys. Rev. B* **6**, 4718 (1972).

Problems for Chapter 4

1. Calculate all of the matrix elements of the forth row of the 14×14 matrix given in Table 4.1.

2. Show by direct matrix multiplication that the eigenvector of (4.42) satisfies (4.33).

3. Calculate the covalent mixing ratio,

$$r_m = \frac{|a_{z^2}|^2 + |a_{x^2}|^2}{|a_z|^2 + |a_x|^2 + |a_y|^2}$$

   for the σ^* band at the symmetry point $M = (\pi/2a)(1, 1, 0)$.

4. Find an analytic expression for the band width of the π^* band and show that it is dependent only on the energy gap $E_g = E_t - E_\perp$ and $(pd\pi)$. (The band width is defined as the difference between the highest and lowest energy of the band.) Assuming a band gap of $3\,\text{eV}$ and $(pd\pi) = 1\,\text{eV}$, calculate the π^* band width.

5. Find an analytic expression for the band width of the σ^* band and show that it is dependent only on $E_{g\sigma} = (E_e - E_\parallel)$ and $(pd\sigma)$. Assuming $E_g = 5\,\mathrm{eV}$ and $(pd\sigma) = 3\,\mathrm{eV}$, calculate the σ^* band width.

6. The average, $\langle \sin^2(k_x a) \rangle$, is defined as

$$\left(\frac{1}{V_{\mathrm{BZ}}}\right) \int_{V_{\mathrm{BZ}}} d\vec{k}\ \sin^2(k_x a)$$

where the integral is over the entire first Brillouin zone and V_{BZ} is the volume of the first Brillouin zone. For the pi bands show that

$$\langle \sin^2(k_x a) \rangle = \frac{1}{2} \frac{(E - E_m)^2 - (E_g/2)^2}{4(pd\pi)^2}.$$

(Hint: Use symmetry arguments.)

7. Using the parameters of Fig. 4.4(b) to calculate the π^*-band energies corresponding to the wavevectors at Γ, X, and M.

5

Analysis of bands at symmetry points

In Chapter 4 the general features of the energy bands of the perovskites were determined for the approximation in which $(pp\sigma) = (pp\pi) = 0$. In this chapter we examine the solutions of the energy bands for $\vec{k}$-vectors at points of high symmetry in the Brillouin zone with $(pp\sigma)$ and $(pp\pi) \neq 0$. From these solutions the role of the oxygen–oxygen interactions in determining the band gap and energy band structure can be assessed.

It is possible to diagonalize the 14×14 energy band matrix exactly at all of the points of high symmetry in the Brillouin zone including Γ, X, M, and R. In what follows we present tables which give the eigenvalues, eigenvectors, and the real-space wavefunctions for each of the 14 energy bands.

5.1 Energy bands at Γ

At Γ in the first Brillouin zone, $\vec{k} = (0,0,0)$ and many of the matrix elements of secular matrix (4.32) vanish. The 14×14 matrix can be block-diagonalized by rearranging rows and columns (equivalent to a unitary transformation).

Table 5.1 summarizes the results. The first column specifies a (arbitrary) numerical label for each of the 14 states. The second column gives the rows and columns of the 14×14 matrix of (4.32) involved in the block which determines energy band states. These numbers also specify the basis orbitals involved according to the notation adopted in Chapter 4. The notation is given by (4.30). By rearranging the columns and rows of (4.32) in the order 1, 3, 6, 9, 12, 2, 11, 14, 4, 7, 10, 5, 8, 13 the secular matrix assumes a block-diagonal form. Rows and columns 1, 3, 6, 9, 12 correspond to 1×1 blocks. Rows and columns 2, 11, 14; 4, 7, 10; and 5, 8, 13 each form 3×3 blocks. The 3×3 blocks are symmetry equivalent. The dimensionality of the blocks can be inferred from column two of Table 5.1.

Column three of Table 5.1 specifies the type of energy band state in terms of the "pi" and "sigma" notation discussed in Chapter 4. An entry such as $\pi + \sigma$

indicates a valence band involving both pi and sigma basis orbitals and $\pi^* + \sigma^*$ would indicate a conduction band involving pi and sigma basis orbitals.

Column four of Table 5.1 gives the energy of the band state, $E_{\Gamma\nu}$. The eigenvector for the state (within the unit cell) is given in column five, where $|n\rangle$ represents the orbital specified by n in (4.30). The total real-space wavefunction can be constructed by summing the local eigenvectors $|n, m\rangle$ over all unit cells, taking into account the phase factors $e^{i\vec{k}\cdot(\vec{R}_m + \vec{\tau}_j)}$ of each orbital, and properly normalizing the total state. For instance, the total real-space wavefunction for the 14th state in the last row of Table 5.1 will read as follows:

$$\frac{1}{\sqrt{NS}} \sum_m e^{i\vec{k}\cdot\vec{R}_m} \left\{ -4c|5, m\rangle \, e^{i\vec{k}\cdot\vec{\tau}_5} + (E_\| - E_{\Gamma 14}) \left[|8, m\rangle \, e^{i\vec{k}\cdot\vec{\tau}_8} + |13, m\rangle \, e^{i\vec{k}\cdot\vec{\tau}_{13}} \right] \right\}.$$

$$(5.1)$$

Since $\vec{k} = 0$ at Γ, the phase factors $e^{i\vec{k}\cdot\vec{R}_m}$, $e^{i\vec{k}\cdot\vec{\tau}_5}$, $e^{i\vec{k}\cdot\vec{\tau}_8}$, and $e^{i\vec{k}\cdot\vec{\tau}_{13}}$ are all unity. The total wavefunction reduces to

$$\frac{1}{\sqrt{NS}} \sum_m \left\{ -4c|5, m\rangle + (E_\| - E_{\Gamma 14}) \left[|8, m\rangle + |13, m\rangle \right] \right\}. \qquad (5.2)$$

Here the notation is

$$
\begin{aligned}
&|1, m\rangle = d_{z^2}(\vec{r} - \vec{R}_m), && |2, m\rangle = p_z(\vec{r} - \vec{R}_m - a\vec{e}_z), \\
&|3, m\rangle = d_{x^2-y^2}(\vec{r} - \vec{R}_m), && |4, m\rangle = p_x(\vec{r} - \vec{R}_m - a\vec{e}_x), \\
& && |5, m\rangle = p_y(\vec{r} - \vec{R}_m - a\vec{e}_y), \\[4pt]
&|6, m\rangle = d_{xy}(\vec{r} - \vec{R}_m), && |7, m\rangle = p_x(\vec{r} - \vec{R}_m - a\vec{e}_y), \\
& && |8, m\rangle = p_y(\vec{r} - \vec{R}_m - a\vec{e}_x), \\[4pt]
&|9, m\rangle = d_{xz}(\vec{r} - \vec{R}_m), && |10, m\rangle = p_x(\vec{r} - \vec{R}_m - a\vec{e}_z), \\
& && |11, m\rangle = p_z(\vec{r} - \vec{R}_m - a\vec{e}_x), \\[4pt]
&|12, m\rangle = d_{yz}(\vec{r} - \vec{R}_m), && |13, m\rangle = p_y(\vec{r} - \vec{R}_m - a\vec{e}_z), \\
& && |14, m\rangle = p_z(\vec{r} - \vec{R}_m - a\vec{e}_y)
\end{aligned}
$$

$$(5.3)$$

where $\vec{R}_m = 2a[n_x(m), n_y(m), n_z(m)]$.

The last column gives the symmetry labels of the irreducible representations of the group of the wavevector in the notation of Bouckaert, Smoluchowski, and Wigner [1].

Table 5.1 shows that the conduction bands, σ^* and π^*, have wavefunctions that involve only d orbitals. There is no mixing between the p and d orbitals at Γ. The energies at Γ of these states are just the diagonal matrix elements which correspond to the ionic model energies including the electrostatic splittings.

The nine valence bands consist of three sets of equivalent states. The π^0 state

Table 5.1. *Energy bands at* Γ, $\vec{k}=(0,0,0)$.

State	Rows	Type	Energy ($E_{\Gamma\nu}$)	Eigenvector	Wavefunction	Sym.
$\Gamma 1$	1	σ^*	E_e	$\lvert 1\rangle$	$\frac{1}{\sqrt{N}}\sum \lvert 1,m\rangle$	Γ_{12}
$\Gamma 2$	3	σ^*	E_e	$\lvert 3\rangle$	$\frac{1}{\sqrt{N}}\sum \lvert 3,m\rangle$	Γ_{12}
$\Gamma 3$	6	π^*	E_t	$\lvert 6\rangle$	$\frac{1}{\sqrt{N}}\sum \lvert 6,m\rangle$	$\Gamma_{25'}$
$\Gamma 4$	9	π^*	E_t	$\lvert 9\rangle$	$\frac{1}{\sqrt{N}}\sum \lvert 9,m\rangle$	$\Gamma_{25'}$
$\Gamma 5$	12	π^*	E_t	$\lvert 12\rangle$	$\frac{1}{\sqrt{N}}\sum \lvert 12,m\rangle$	$\Gamma_{25'}$
$\Gamma 6$	2, 11, 14	π^0	$E_\perp - 4(pp\pi)$	$\frac{1}{\sqrt{2}}\left[\lvert 11\rangle - \lvert 14\rangle\right]$	$\frac{1}{\sqrt{N}}\sum \frac{1}{\sqrt{2}}\left[\lvert 11,m\rangle - \lvert 14,m\rangle\right]$	Γ_{25}
$\Gamma 7$		$\pi+\sigma$	$E_A + E_{Dc}$	$\frac{1}{\sqrt{S}}\left[-4c\lvert 2\rangle + \left(E_\parallel - E_{\Gamma7}\right)\left(\lvert 11\rangle + \lvert 14\rangle\right)\right]$	$\frac{1}{\sqrt{NS}}\sum\left[-4c\lvert 2,m\rangle + \left(E_\parallel - E_{\Gamma7}\right)\left(\lvert 11,m\rangle + \lvert 14,m\rangle\right)\right]$	Γ_{15}
$\Gamma 8$		$\pi+\sigma$	$E_A - E_{Dc}$	$\frac{1}{\sqrt{S}}\left[-4c\lvert 2\rangle + \left(E_\parallel - E_{\Gamma8}\right)\left(\lvert 11\rangle + \lvert 14\rangle\right)\right]$	$\frac{1}{\sqrt{NS}}\sum\left[-4c\lvert 2,m\rangle + \left(E_\parallel - E_{\Gamma8}\right)\left(\lvert 11,m\rangle + \lvert 14,m\rangle\right)\right]$	Γ_{15}
$\Gamma 9$	4, 7, 10	π^0	Same as $\Gamma 6$	$\frac{1}{\sqrt{2}}\left[\lvert 7\rangle - \lvert 10\rangle\right]$	$\frac{1}{\sqrt{N}}\sum \frac{1}{\sqrt{2}}\left[\lvert 7,m\rangle - \lvert 10,m\rangle\right]$	Γ_{25}
$\Gamma 10$		$\pi+\sigma$	Same as $\Gamma 7$	$\frac{1}{\sqrt{S}}\left[-4c\lvert 4\rangle + \left(E_\parallel - E_{\Gamma10}\right)\left[\lvert 7\rangle + \lvert 10\rangle\right]\right]$	$\frac{1}{\sqrt{NS}}\sum\left[-4c\lvert 4,m\rangle + \left(E_\parallel - E_{\Gamma10}\right)\left[\lvert 7,m\rangle + \lvert 10,m\rangle\right]\right]$	Γ_{15}
$\Gamma 11$		$\pi+\sigma$	Same as $\Gamma 8$	$\frac{1}{\sqrt{S}}\left[-4c\lvert 4\rangle + \left(E_\parallel - E_{\Gamma11}\right)\left[\lvert 7\rangle + \lvert 10\rangle\right]\right]$	$\frac{1}{\sqrt{NS}}\sum\left[-4c\lvert 4,m\rangle + \left(E_\parallel - E_{\Gamma11}\right)\left[\lvert 7,m\rangle + \lvert 10,m\rangle\right]\right]$	Γ_{15}
$\Gamma 12$	5, 8, 13	π^0	Same as $\Gamma 6$	$\frac{1}{\sqrt{2}}\left[\lvert 8\rangle - \lvert 13\rangle\right]$	$\frac{1}{\sqrt{N}}\sum \frac{1}{\sqrt{2}}\left[\lvert 8,m\rangle - \lvert 13,m\rangle\right]$	Γ_{25}
$\Gamma 13$		$\pi+\sigma$	Same as $\Gamma 7$	$\frac{1}{\sqrt{S}}\left[-4c\lvert 5\rangle + \left(E_\parallel - E_{\Gamma13}\right)\left[\lvert 8\rangle + \lvert 13\rangle\right]\right]$	$\frac{1}{\sqrt{NS}}\sum\left[-4c\lvert 5,m\rangle + \left(E_\parallel - E_{\Gamma13}\right)\left[\lvert 8,m\rangle + \lvert 13,m\rangle\right]\right]$	Γ_{15}
$\Gamma 14$		$\pi+\sigma$	Same as $\Gamma 8$	$\frac{1}{\sqrt{S}}\left[-4c\lvert 5\rangle + \left(E_\parallel - E_{\Gamma14}\right)\left[\lvert 8\rangle + \lvert 13\rangle\right]\right]$	$\frac{1}{\sqrt{NS}}\sum\left[-4c\lvert 5,m\rangle + \left(E_\parallel - E_{\Gamma14}\right)\left[\lvert 8,m\rangle + \lvert 13,m\rangle\right]\right]$	Γ_{15}

$E_A = \frac{1}{2}\left(E_\parallel + E_\perp\right) + 2(pp\pi),$ $c = (pp\sigma) + (pp\pi),$ $S = 2\left(E_\parallel - E_{\Gamma\nu}\right)^2 + 16c^2.$

$E_D = \frac{1}{2}\left(E_\parallel - E_\perp\right) - 2(pp\pi),$ $E_{Dc} = \sqrt{E_D^2 + 8c^2},$ $R_m = 2a[n_x(m), n_y(m), n_z(m)].$

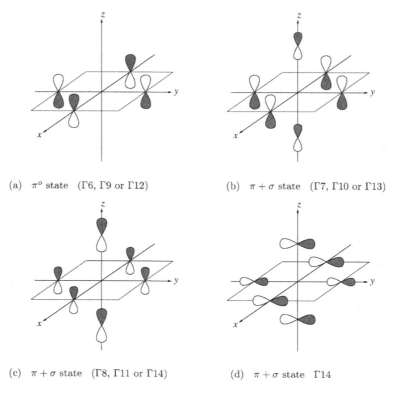

(a) π^o state ($\Gamma 6$, $\Gamma 9$ or $\Gamma 12$)   (b) $\pi + \sigma$ state ($\Gamma 7$, $\Gamma 10$ or $\Gamma 13$)

(c) $\pi + \sigma$ state ($\Gamma 8$, $\Gamma 11$ or $\Gamma 14$)   (d) $\pi + \sigma$ state $\Gamma 14$

Figure 5.1. Schematic representation of the eigenstates at Γ. Symmetry equivalent states can be obtained by switching axes. $\Gamma 8$ in (c) and $\Gamma 14$ in (d) are presented as an example of such an operation. The difference in relative sizes of the planar and apical orbitals stem from the difference in the magnitudes of $(E_{\parallel} - E_{\Gamma 8})$ and $(E_{\parallel} - E_{\Gamma 14})$ as compared to $-4c$ (see Table 5.1 and Fig. 5.3).

is a non-bonding oxygen p-orbital band. Its energy is shifted by the oxygen–oxygen interaction from $E_{\perp}$ to $E_{\perp} - 4(pp\pi)$. There are three degenerate π^0 bands. The wavefunction for a π^0 band is illustrated schematically in Fig. 5.1(a). The wavefunction is repeated throughout the lattice. The $\pi + \sigma$ states, 7, 8; 10, 11; and 13, 14 are admixtures of $p_{\perp}$ and $p_{\parallel}$ orbitals. Schematic representations of these states are given in Figs 5.1(b) and (c).

The effect of the oxygen–oxygen interactions on the band-gap energies at Γ is shown in Fig. 5.2. For the usual case of $|(pp\pi)| < |c|$, the mixed states $\Gamma 7$ (or $\Gamma 10$ or $\Gamma 13$) lie above the non-bonding π^0 ($\Gamma 6$, $\Gamma 9$, or $\Gamma 12$) states when $(pp\pi)$ and $(pp\sigma) \neq 0$, so that the true band gap, $E_g(\Gamma)$, is somewhat smaller than the ionic gap $E_g^0(\Gamma)$. This feature is illustrated in Fig. 5.2. For example, for SrTiO$_3$ the ionic gap is approximately 3.6 eV while $E_g(\Gamma) = 3.24$ eV. Thus the oxygen–oxygen

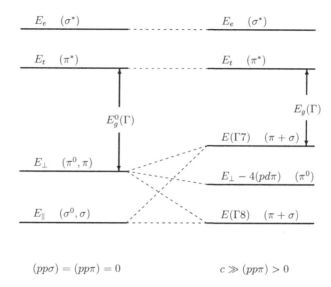

$(pp\sigma) = (pp\pi) = 0$ $\qquad\qquad\qquad c \gg (pp\pi) > 0$

Figure 5.2. Splitting of the states at Γ by the oxygen–oxygen interactions.

interactions cause a 10% reduction in the ionic band gap at Γ. The band-gap energy at Γ can be written in terms of $\Delta(p) = E_\perp - E_\parallel$ as

$$E_g(\Gamma) = E_g^0(\Gamma) - \delta E_g,$$

$$\delta E_g = -\left[\frac{1}{2}\Delta(p) - 2(pp\pi)\right] + \sqrt{\left[\frac{1}{2}\Delta(p) + 2(pp\pi)\right]^2 + 8c^2}. \qquad (5.4)$$

The energies at the various symmetry points are shown in Fig. 5.3. It is noted that there is no mixing between the pi- and sigma-type (t_{2g} and e_g) d orbitals but the pi- and sigma-type p orbitals are mixed by the oxygen–oxygen interactions.

5.2 Energy bands at X

The X points in the Brillouin zone are at $\vec{k} = \frac{\pi}{2a}(\pm 1, 0, 0); \frac{\pi}{2a}(0, \pm 1, 0); \frac{\pi}{2a}(0, 0, \pm 1)$. We consider the X point at $\vec{k} = \frac{\pi}{2a}(1, 0, 0)$ for which $S_x = C_y = C_z = 1$ and $C_x = S_y = S_z = 0$ where $S_\alpha = \sin k_\alpha a$ and $C_\alpha = \cos k_\alpha a$.

The results of the analysis of the secular equation are summarized in Table 5.2. The total real-space wavefunction can be constructed by summing the local eigen-vectors, $|n, m\rangle$, over all unit cells, taking into account the phase factors $e^{i\vec{k}\cdot(\vec{R}_m + \vec{\tau}_j)}$ of each orbital and normalizing the total state. For instance, the 10th state in Table

Table 5.2. Energy bands at X, $\vec{k} = \frac{\pi}{2a}(1, 0, 0)$.

State	Rows	Type	Energy ($E_{X\nu}$)	Eigenvector	Wavefunction	Sym.
X1	12	π^*	E_t	$|12\rangle$	$\frac{1}{\sqrt{N}}\sum_m \eta_m |12, m\rangle$	X_3
X2	7, 10	π^0	$E_\perp + 4(pp\pi)$	$\frac{1}{\sqrt{2}}[|7\rangle - |10\rangle]$	$\frac{1}{\sqrt{2N}}\sum_m \eta_m \,[|7, m\rangle - |10, m\rangle]$	$X_{4'}$
X3	7, 10	π^0	$E_\perp - 4(pp\pi)$	$\frac{1}{\sqrt{2}}[|7\rangle + |10\rangle]$	$\frac{1}{\sqrt{2N}}\sum_m \eta_m \,[|7, m\rangle + |10, m\rangle]$	$X_{3'}$
X4	6, 8	π^*	$E_1 + \sqrt{E_2^2 + 4(pd\pi)^2}$	$\frac{1}{\sqrt{S_2}}[-2i(pd\pi)|6\rangle + (E_t - E_{X4})|8\rangle]$	$\frac{1}{\sqrt{NS_2}}\sum_m \eta_m \,[-2i(pd\pi)|6, m\rangle + (E_t - E_{X4})|8, m\rangle]$	X_5
X5	6, 8	π	$E_1 - \sqrt{E_2^2 + 4(pd\pi)^2}$	$\frac{1}{\sqrt{S_2}}[-2i(pd\pi)|6\rangle + (E_t - E_{X5})|8\rangle]$	$\frac{1}{\sqrt{NS_2}}\sum_m \eta_m \,[-2i(pd\pi)|6, m\rangle + (E_t - E_{X5})|8, m\rangle]$	X_5
X6	9, 11	π^*	Same as X4	$\frac{1}{\sqrt{S_2}}[-2i(pd\pi)|9\rangle + (E_t - E_{X6})|11\rangle]$	$\frac{1}{\sqrt{NS_2}}\sum_m \eta_m \,[-2i(pd\pi)|9, m\rangle + (E_t - E_{X6})|11, m\rangle]$	X_5
X7	9, 11	π	Same as X5	$\frac{1}{\sqrt{S_2}}[-2i(pd\pi)|9\rangle + (E_t - E_{X7})|11\rangle]$	$\frac{1}{\sqrt{NS_2}}\sum_m \eta_m \,[-2i(pd\pi)|9, m\rangle + (E_t - E_{X7})|11, m\rangle]$	X_5
X8	2, 14	$\pi + \sigma$	$E_A + \sqrt{E_D^2 + 4c^2}$	$\frac{1}{\sqrt{S_1}}[-2c|2\rangle + (E_\| - E_{X8})|14\rangle]$	$\frac{1}{\sqrt{NS_1}}\sum_m \eta_m \,[-2c|2, m\rangle + (E_\| - E_{X8})|14, m\rangle]$	$X_{5'}$
X9	2, 14	$\pi + \sigma$	$E_A - \sqrt{E_D^2 + 4c^2}$	$\frac{1}{\sqrt{S_1}}[-2c|2\rangle + (E_\| - E_{X9})|14\rangle]$	$\frac{1}{\sqrt{NS_1}}\sum_m \eta_m \,[-2c|2, m\rangle + (E_\| - E_{X9})|14, m\rangle]$	$X_{5'}$
X10	5, 13	$\pi + \sigma$	Same as X8	$\frac{1}{\sqrt{S_1}}[-2c|5\rangle + (E_\| - E_{X10})|13\rangle]$	$\frac{1}{\sqrt{NS_1}}\sum_m \eta_m \,[-2c|5, m\rangle + (E_\| - E_{X10})|13, m\rangle]$	$X_{5'}$
X11	5, 13	$\pi + \sigma$	Same as X9	$\frac{1}{\sqrt{S_1}}[-2c|5\rangle + (E_\| - E_{X11})|13\rangle]$	$\frac{1}{\sqrt{NS_1}}\sum_m \eta_m \,[-2c|5, m\rangle + (E_\| - E_{X11})|13, m\rangle]$	$X_{5'}$
X12	1, 3, 4	σ^*	E_e	$\frac{1}{\sqrt{2}}\left[\sqrt{3}|1\rangle + |3\rangle\right]$	$\frac{1}{\sqrt{2N}}\sum_m \eta_m \left[\sqrt{3}|1, m\rangle + |3, m\rangle\right]$	X_2
X13	1, 3, 4	σ^*	$E_3 + \sqrt{E_4^2 + 4(pd\sigma)^2}$	$\frac{1}{\sqrt{S_3}}\left[(pd\sigma)\left[|1\rangle - \sqrt{3}|3\rangle\right] - i\left(E_e - E_{X13}\right)|4\rangle\right]$	$\frac{1}{\sqrt{NS_3}}\sum_m \eta_m \left[(pd\sigma)\left[|1\rangle - \sqrt{3}|3, m\rangle\right] + \left(E_e - E_{X13}\right)|4, m\rangle\right]$	X_1
X14	1, 3, 4	σ	$E_3 - \sqrt{E_4^2 + 4(pd\sigma)^2}$	$\frac{1}{\sqrt{S_3}}\left[(pd\sigma)\left[|1\rangle - \sqrt{3}|3\rangle\right] - i\left(E_e - E_{X14}\right)|4\rangle\right]$	$\frac{1}{\sqrt{NS_3}}\sum_m \eta_m \left[(pd\sigma)\left[|1\rangle - \sqrt{3}|3, m\rangle\right] + \left(E_e - E_{X14}\right)|4, m\rangle\right]$	X_1

$$E_A = \tfrac{1}{2}(E_\| + E_\perp), \qquad E_D = \tfrac{1}{2}(E_\| - E_\perp), \qquad c = (pp\sigma) + (pp\pi).$$

$$E_1 = \tfrac{1}{2}(E_t + E_\perp), \qquad E_2 = \tfrac{1}{2}(E_t - E_\perp), \qquad \eta_m = (-1)^{n_x(m)}.$$

$$E_3 = \tfrac{1}{2}(E_e + E_\|), \qquad E_4 = \tfrac{1}{2}(E_e - E_\|), \qquad \vec{R}_m = 2a[n_x(m)\vec{e}_x + n_y(m)\vec{e}_y + n_z(m)\vec{e}_z].$$

$$S_1 = (E_\| - E_{X\nu})^2 + 4c^2, \qquad S_2 = (E_t - E_{X\nu})^2 + 4(pd\pi)^2, \qquad S_3 = (E_e - E_{X\nu})^2 + 4(pd\sigma)^2,$$

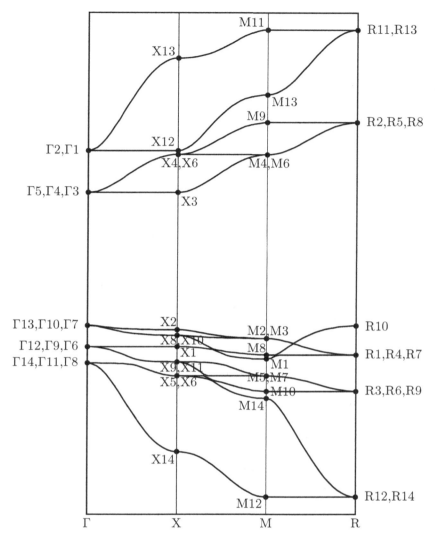

Figure 5.3. The symmetry states listed in Tables 5.1–5.4 indicated on the energy bands diagram calculated from (4.32) with the parameters (all in eV): $E_t = -6.5$, $E_e = -5.5$, $E_\perp = -10.0$, $E_\parallel = -10.5$, $(pd\sigma) = -2.0$, $(pd\pi) = 1.0$, $(pp\sigma) = -0.15$, and $(pp\pi) = 0.05$.

5.2 will read as follows:

$$\frac{1}{\sqrt{NS_1}} \sum_m e^{i\vec{k}\cdot\vec{R}_m} \left[-2c|5,m\rangle\, e^{i\vec{k}\cdot\vec{\tau}_5} + (E_\parallel - E_{X10})|13,m\rangle\, e^{i\vec{k}\cdot\vec{\tau}_{13}} \right] \qquad (5.5)$$

Since $\vec{k}$ at X is equal to $\frac{\pi}{2a}\vec{e}_x$, we have

$$e^{i\vec{k}\cdot\vec{R}_m} = e^{i\frac{\pi}{2a}[2an_x(m)]} = e^{i\pi n_x(m)} = (-1)^{n_x(m)},$$

where $n_x(m)$ is the integer n_x in $\vec{R}_m = 2a(n_x\vec{e}_x + n_y\vec{e}_y + n_z\vec{e}_z)$. Since $\vec{\tau}_5 = a\vec{e}_y$ and $\vec{\tau}_{13} = a\vec{e}_z$ we find that $e^{i\vec{k}\cdot\vec{\tau}_5} = 1$, and $e^{i\vec{k}\cdot\vec{\tau}_{13}} = 1$. The resulting total wavefunction is

$$\frac{1}{\sqrt{NS_1}}\sum_m (-1)^{n_x(m)}\big[-2c|5,m\rangle + (E_\parallel - E_{X10})|13,m\rangle\big]. \tag{5.6}$$

Using the 14th state in Table 5.2 we find the total real-space wavefunction to be

$$\frac{1}{\sqrt{NS_3}}\sum_m (-1)^{n_x(m)}\big\{(pd\sigma)\big[|1,m\rangle - \sqrt{3}|3,m\rangle\big] + (E_e - E_{X14})|4,m\rangle\big\}. \tag{5.7}$$

In obtaining (5.7) we have used the fact that at X, $e^{i\vec{k}\cdot\vec{\tau}_1} = 1$, $e^{i\vec{k}\cdot\vec{\tau}_3} = 1$, and $e^{i\vec{k}\cdot\vec{\tau}_4} = e^{i\frac{\pi}{2a}\vec{e}_x\cdot(a\vec{e}_x)} = e^{i\frac{\pi}{2}} = i$. Thus we have

$$-i(E_e - E_{X14})|4,m\rangle e^{i\vec{k}\cdot\vec{\tau}_4} = +(E_e - E_{X14})|4,m\rangle.$$

The energies at the X point are indicated in Fig. 5.3 for a typical set of energy bands. The highest valence-band state depends upon the values of $(pp\pi)$ and $(pp\sigma)$. Usually X2 is the upper valence band when $|(pp\pi)| \ll |(pp\sigma)|$. In such a case the band gap at X, $E_g(X)$ is $E_t-E(X2)$. From Table 5.1 it can be seen that $E(\Gamma 7) > E(X2)$ and therefore the energy gap at Γ is *smaller* than at X. There is some controversy about whether the minimum band gap is the direct gap between $\Gamma 7$ and $\Gamma 5$ or an indirect gap between $\Gamma 7$ and X3. The model considered here gives the same energy for either gap since $E(\Gamma 5) = E(X3) = E_t$ (because one of the π^* bands is flat along Γ to X). Other effects, such as the spin–orbit interaction and more distant neighbor interactions may alter the equality of $E(\Gamma 5)$ and $E(X3)$. It is noted that there is no mixing between the t_{2g} and e_g orbitals, however, the pi and sigma p orbitals are mixed by the oxygen–oxygen interactions.

5.3 Energy bands at M

The M points in the Brillouin zone are at $\vec{k} = \frac{\pi}{2a}(\pm 1, \pm 1, 0)$; $\frac{\pi}{2a}(\pm 1, 0, \pm 1)$; $\frac{\pi}{2a}(0, \pm 1, \pm 1)$. We consider the M point at $\vec{k} = \frac{\pi}{2a}(1, 1, 0)$.

The secular matrix equation is block-diagonal as follows: three 1×1 blocks corresponding to non-bonding oxygen states, two equivalent 2×2 blocks corresponding to pi conduction and valence bands, one 3×3 which yields a π^0 non-bonding valence band, a π^* conduction band and a π valence band, and one 4×4 involving only the sigma-type orbitals which gives two σ^* conduction bands and two σ valence bands. There is no mixing between pi- and sigma-type orbitals. (See Table 5.3.)

Table 5.3. *Energy bands at* M, $\vec{k} = \frac{\pi}{2a}(1, 1, 0)$.

State	Rows	Type	Energy ($E_{M\nu}$)	Eigenvector	Wavefunction	Sym.
M1	2	σ^0	$E_\parallel$	$\|2\rangle$	$\frac{1}{\sqrt{N}}\sum \xi_m \|2, m\rangle$	$M_{4'}$
M2	10	π^0	$E_\perp$	$\|10\rangle$	$\frac{1}{\sqrt{N}}\sum \xi_m \|10, m\rangle$	$M_{5'}$
M3	13	π^0	$E_\perp$	$\|13\rangle$	$\frac{1}{\sqrt{N}}\sum \xi_m \|13, m\rangle$	$M_{5'}$
M4	9, 11	π^*	$E_1 + \sqrt{E_2^2 + 4(pd\pi)^2}$	$\frac{1}{\sqrt{S_1}}\left[2(pd\pi)\|9\rangle + i\left(E_t - E_{M4}\right)\|11\rangle\right]$	$\frac{1}{\sqrt{NS_1}}\sum \xi_m \left[2(pd\pi)\|9, m\rangle - \left(E_t - E_{M4}\right)\|11, m\rangle\right]$	M_5
M5	9, 11	π	$E_1 - \sqrt{E_2^2 + 4(pd\pi)^2}$	$\frac{1}{\sqrt{S_1}}\left[2(pd\pi)\|9\rangle + i\left(E_t - E_{M5}\right)\|11\rangle\right]$	$\frac{1}{\sqrt{NS_1}}\sum \xi_m \left[2(pd\pi)\|9, m\rangle - \left(E_t - E_{M5}\right)\|11, m\rangle\right]$	M_5
M6	12, 14	π^*	Same as M4	$\frac{1}{\sqrt{S_1}}\left[2(pd\pi)\|12\rangle + i\left(E_t - E_{M6}\right)\|14\rangle\right]$	$\frac{1}{\sqrt{NS_1}}\sum \xi_m \left[2(pd\pi)\|12, m\rangle - \left(E_t - E_{M6}\right)\|14, m\rangle\right]$	M_5
M7	12, 14	π	Same as M5	$\frac{1}{\sqrt{S_1}}\left[2(pd\pi)\|12\rangle + i\left(E_t - E_{M7}\right)\|14\rangle\right]$	$\frac{1}{\sqrt{NS_1}}\sum \xi_m \left[2(pd\pi)\|12, m\rangle - \left(E_t - E_{M7}\right)\|14, m\rangle\right]$	M_5
M8		π^0	$E_\perp + 2b$	$\frac{1}{\sqrt{2}}[\|7\rangle - \|8\rangle]$	$\frac{1}{\sqrt{2N}}\sum \xi_m [\|7, m\rangle - \|8, m\rangle]$	M_4
M9	6, 7, 8	π^*	$(E_1 - b) + \sqrt{(E_2 + b)^2 + 8(pd\pi)^2}$	$\frac{1}{\sqrt{S_2}}\left[4(pd\pi)\|6\rangle + i\left(E_t - E_{M9}\right)[\|7\rangle + \|8\rangle]\right]$	$\frac{1}{\sqrt{NS_2}}\sum \xi_m \left[4(pd\pi)\|6, m\rangle - \left(E_t - E_{M9}\right)[\|7, m\rangle + \|8, m\rangle]\right]$	M_3
M10	6, 7, 8	π	$(E_1 - b) - \sqrt{(E_2 + b)^2 + 8(pd\pi)^2}$	$\frac{1}{\sqrt{S_2}}\left[4(pd\pi)\|6\rangle + i\left(E_t - E_{M10}\right)[\|7\rangle + \|8\rangle]\right]$	$\frac{1}{\sqrt{NS_2}}\sum \xi_m \left[4(pd\pi)\|6, m\rangle - \left(E_t - E_{M10}\right)[\|7, m\rangle + \|8, m\rangle]\right]$	M_3
M11	1, 3, 4, 5	σ^*	$(E_3 + b) + \sqrt{(E_4 - b)^2 + 6(pd\sigma)^2}$	$\frac{1}{\sqrt{S_3}}\left[2\sqrt{3}(pd\sigma)\|3\rangle + i\left(E_e - E_{M11}\right)[\|4\rangle - \|5\rangle]\right]$	$\frac{1}{\sqrt{NS_3}}\sum \xi_m \left[2\sqrt{3}(pd\sigma)\|3, m\rangle - \left(E_e - E_{M11}\right)[\|4, m\rangle - \|5, m\rangle]\right]$	M_2
M12	1, 3, 4, 5	σ	$(E_3 + b) - \sqrt{(E_4 - b)^2 + 6(pd\sigma)^2}$	$\frac{1}{\sqrt{S_3}}\left[2\sqrt{3}(pd\sigma)\|3\rangle + i\left(E_e - E_{M12}\right)[\|4\rangle - \|5\rangle]\right]$	$\frac{1}{\sqrt{NS_3}}\sum \xi_m \left[2\sqrt{3}(pd\sigma)\|3, m\rangle - \left(E_e - E_{M12}\right)[\|4, m\rangle - \|5, m\rangle]\right]$	M_2
M13		σ^*	$(E_3 - b) + \sqrt{(E_4 + b)^2 + 2(pd\sigma)^2}$	$\frac{1}{\sqrt{S_4}}\left[2(pd\sigma)\|1\rangle - i\left(E_e - E_{M13}\right)[\|4\rangle + \|5\rangle]\right]$	$\frac{1}{\sqrt{NS_4}}\sum \xi_m \left[2(pd\sigma)\|1, m\rangle + \left(E_e - E_{M13}\right)[\|4, m\rangle + \|5, m\rangle]\right]$	M_1
M14		σ	$(E_3 - b) - \sqrt{(E_4 + b)^2 + 2(pd\sigma)^2}$	$\frac{1}{\sqrt{S_4}}\left[2(pd\sigma)\|1\rangle - i\left(E_e - E_{M14}\right)[\|4\rangle + \|5\rangle]\right]$	$\frac{1}{\sqrt{NS_4}}\sum \xi_m \left[2(pd\sigma)\|1, m\rangle + \left(E_e - E_{M14}\right)[\|4, m\rangle + \|5, m\rangle]\right]$	M_1

$E_1 = \frac{1}{2}(E_t + E_\perp)$, $E_2 = \frac{1}{2}(E_t - E_\perp)$, $S_3 = 2(E_e - E_{M\nu})^2 + 12(pd\sigma)^2$.

$E_3 = \frac{1}{2}(E_e + E_\parallel)$, $E_4 = \frac{1}{2}(E_e - E_\parallel)$, $S_4 = 2(E_e - E_{M\nu})^2 + 4(pd\sigma)^2$.

$b = (pp\sigma) - (pp\pi)$, $\xi_m = (-1)^{m_x + m_y}$, $R_m = 2a(m_x, m_y, m_z)$.

$S_1 = (E_t - E_{M\nu})^2 + 4(pd\pi)^2$,

$S_2 = 2(E_t - E_{M\nu})^2 + 16(pd\pi)^2$,

The total real-space wavefunction, for example, for the 14th state is

$$\frac{1}{\sqrt{NS_4}} \sum_m (-1)^{n_x(m)+n_y(m)} \left[2(pd\sigma)|1, m\rangle + (E_e - E_{M14})(|4, m\rangle + |5, m\rangle)\right].$$

$$(5.8)$$

In obtaining (5.8) we used the following:

$$e^{i\vec{k}\cdot\vec{R}_m} = e^{i\frac{\pi}{2a}(\vec{e}_x+\vec{e}_y)\cdot 2a(n_x\vec{e}_y+n_y\vec{e}_x+n_z\vec{e}_z)} = e^{i\pi[n_x(m)+n_y(m)]}$$

$$= (-1)^{n_x(m)+n_y(m)},$$

$$e^{i\vec{k}\cdot\vec{\tau}_1} = 1,$$

$$e^{i\vec{k}\cdot\vec{\tau}_4} = e^{i\frac{\pi}{2a}(\vec{e}_x+\vec{e}_y)\cdot a\vec{e}_x} = e^{i\pi/2} = i,$$

$$e^{i\vec{k}\cdot\vec{\tau}_5} = e^{i\frac{\pi}{2a}(\vec{e}_x+\vec{e}_y)\cdot a\vec{e}_y} = e^{i\pi/2} = i.$$

The energies at the M point are shown in Fig. 5.3 for a typical set of energy bands, and the analytical results are given in Table 5.3.

5.4 Energy bands at R

The R points in the Brillouin zone are at $\frac{\pi}{2a}(\pm 1, \pm 1, \pm 1)$. We consider the R point given by $\vec{k} = \frac{\pi}{2a}(1, 1, 1)$. The secular equation is then block-diagonal and consists of a 5×5 and three equivalent 3×3 blocks. The 5×5 block involves only the sigma-type orbitals and the 3×3 blocks involve only the pi-type orbitals. There is no mixing between the pi and sigma orbitals at R.

The energies and eigenvectors at the R point are listed in Table 5.4 and the total real-space wavefunction can be constructed by summing the local eigenvectors $|n, m\rangle$ over all unit cells and properly normalizing. The total real-space wavefunction for the 14th state in Table 5.4 is, for example,

$$\frac{1}{\sqrt{NS_3}} \sum (-1)^{n_x(m)+n_y(m)+n_z(m)}$$

$$\left[-2\sqrt{3}(pd\sigma)|3, m\rangle + (E_e - E_{R14})(|4, m\rangle - |5, m\rangle)\right]. \qquad (5.9)$$

In obtaining (5.9) we used the following results:

$$e^{i\vec{k}\cdot\vec{R}_m} = (-1)^{n_x(m)+n_y(m)+n_z(m)},$$

$$e^{i\vec{k}\cdot\vec{\tau}_3} = 1,$$

$$e^{i\vec{k}\cdot\vec{\tau}_4} = e^{i\vec{k}\cdot\vec{\tau}_5} = e^{i\pi/2} = i.$$

Schematic representations of the wavefunctions are shown in Fig. 5.4(a)–5.4(g).

The splitting of the d bands at R is of interest since this quantity is the band-

Table 5.4. *Energy bands at* R, $\vec{k} = \frac{\pi}{2a}(1,1,1)$.

State	Rows	Type	Energy ($E_{R\nu}$)	Eigenvector	Wavefunction	Sym.
R1		π^0	$E_\perp + 2b$	$\frac{1}{\sqrt{2}}\left[|7\rangle - |8\rangle\right]$	$\frac{1}{\sqrt{2N}}\sum \gamma_m\left[|7,m\rangle - |8,m\rangle\right]$	R$_{15'}$
R2	6, 7, 8	π^*	$E_1 + E_5$	$\frac{1}{S_1}\left[4(pd\pi)|6\rangle + i\left(E_t - E_{R2}\right)[|7\rangle + |8\rangle]\right]$	$\frac{1}{\sqrt{NS_1}}\sum \gamma_m\left[4(pd\pi)|6,m\rangle + \left(E_t - E_{R2}\right)[|7,m\rangle + |8,m\rangle]\right]$	R$_{25'}$
R3		π	$E_1 - E_5$	$\frac{1}{S_1}\left[4(pd\pi)|6\rangle + i\left(E_t - E_{R3}\right)[|7\rangle + |8\rangle]\right]$	$\frac{1}{\sqrt{NS_1}}\sum \gamma_m\left[4(pd\pi)|6,m\rangle + \left(E_t - E_{R3}\right)[|7,m\rangle + |8,m\rangle]\right]$	R$_{25'}$
R4		π^0	Same as R1	$\frac{1}{\sqrt{2}}\left[|10\rangle - |11\rangle\right]$	$\frac{1}{\sqrt{2N}}\sum \gamma_m\left[|10,m\rangle - |11,m\rangle\right]$	R$_{15'}$
R5	9, 10, 11	π^*	Same as R2	$\frac{1}{S_1}\left[4(pd\pi)|9\rangle + i\left(E_t - E_{R5}\right)[|10\rangle + |11\rangle]\right]$	$\frac{1}{\sqrt{NS_1}}\sum \gamma_m\left[4(pd\pi)|9,m\rangle + \left(E_t - E_{R5}\right)[|10,m\rangle + |11,m\rangle]\right]$	R$_{25'}$
R6		π	Same as R3	$\frac{1}{S_1}\left[4(pd\pi)|9\rangle + i\left(E_t - E_{R6}\right)[|10\rangle + |11\rangle]\right]$	$\frac{1}{\sqrt{NS_1}}\sum \gamma_m\left[4(pd\pi)|9,m\rangle + \left(E_t - E_{R6}\right)[|10,m\rangle + |11,m\rangle]\right]$	R$_{25'}$
R7		π^0	Same as R1	$\frac{1}{\sqrt{2}}\left[|13\rangle - |14\rangle\right]$	$\frac{1}{\sqrt{2N}}\sum \gamma_m\left[|13,m\rangle - |14,m\rangle\right]$	R$_{15'}$
R8	12, 13, 14	π^*	Same as R2	$\frac{1}{S_1}\left[4(pd\pi)|12\rangle + i\left(E_t - E_{R8}\right)[|13\rangle + |14\rangle]\right]$	$\frac{1}{\sqrt{NS_1}}\sum \gamma_m\left[4(pd\pi)|12,m\rangle + \left(E_t - E_{R8}\right)[|13,m\rangle + |14,m\rangle]\right]$	R$_{25'}$
R9		π	Same as R3	$\frac{1}{S_1}\left[4(pd\pi)|12\rangle + i\left(E_t - E_{R9}\right)[|13\rangle + |14\rangle]\right]$	$\frac{1}{\sqrt{NS_1}}\sum \gamma_m\left[4(pd\pi)|12,m\rangle + \left(E_t - E_{R9}\right)[|13,m\rangle + |14,m\rangle]\right]$	R$_{25'}$
R10		σ^0	$E_\parallel - 4b$	$\frac{1}{\sqrt{3}}\left[|2\rangle + |4\rangle + |5\rangle\right]$	$\frac{1}{\sqrt{3N}}\sum \gamma_m\left[|2,m\rangle + |4,m\rangle + |5,m\rangle\right]$	R$_1$
R11	1, 2, 3, 4, 5	σ^*	$E_3 + E_6$	$\frac{1}{3S_2}\left[6(pd\sigma)|1\rangle + i\left(E_e - E_{R11}\right)[2|2\rangle - |4\rangle - |5\rangle]\right]$	$\frac{1}{\sqrt{3NS_2}}\sum \gamma_m\left[6(pd\sigma)|1,m\rangle - \left(E_e - E_{R11}\right)[2|2,m\rangle - |4,m\rangle - |5,m\rangle]\right]$	R$_{12}$
R12		σ	$E_3 - E_6$	$\frac{1}{3S_2}\left[6(pd\sigma)|1\rangle + i\left(E_e - E_{R12}\right)[2|2\rangle - |4\rangle - |5\rangle]\right]$	$\frac{1}{\sqrt{3NS_2}}\sum \gamma_m\left[6(pd\sigma)|1,m\rangle - \left(E_e - E_{R12}\right)[2|2,m\rangle - |4,m\rangle - |5,m\rangle]\right]$	R$_{12}$
R13		σ^*	Same as R11	$\frac{1}{S_2}\left[-2\sqrt{3}(pd\sigma)|3\rangle - i\left(E_e - E_{R13}\right)[|4\rangle - |5\rangle]\right]$	$\frac{1}{\sqrt{NS_2}}\sum \gamma_m\left[-2\sqrt{3}(pd\sigma)|3,m\rangle + \left(E_e - E_{R13}\right)[|4,m\rangle - |5,m\rangle]\right]$	R$_{12}$
R14		σ	Same as R12	$\frac{1}{S_2}\left[-2\sqrt{3}(pd\sigma)|3\rangle - i\left(E_e - E_{R14}\right)[|4\rangle - |5\rangle]\right]$	$\frac{1}{\sqrt{NS_2}}\sum \gamma_m\left[-2\sqrt{3}(pd\sigma)|3,m\rangle + \left(E_e - E_{R14}\right)[|4,m\rangle - |5,m\rangle]\right]$	R$_{12}$

$E_1 = \frac{1}{2}(E_t + E_\perp) - b$, $\quad E_2 = \frac{1}{2}(E_t - E_\perp) + b$, $\quad S_1 = 2(E_t - E_{R\nu})^2 + 16(pd\pi)^2$, $\quad \gamma_m = (-1)^{n_x(m) + n_y(m) + n_z(m)}$.

$E_3 = \frac{1}{2}(E_e + E_\parallel) + b$, $\quad E_4 = \frac{1}{2}(E_e - E_\parallel) - b$, $\quad S_2 = 2(E_e - E_{R\nu})^2 + 12(pd\sigma)^2$, $\quad R_m = 2a[n_x(m) + n_y(m) + n_z(m)]$.

$E_5^2 = E_2^2 + 8(pd\pi)^2$, $\quad E_6^2 = E_4^2 + 6(pd\sigma)^2$, $\quad b = (pp\sigma) - (pp\pi)$.

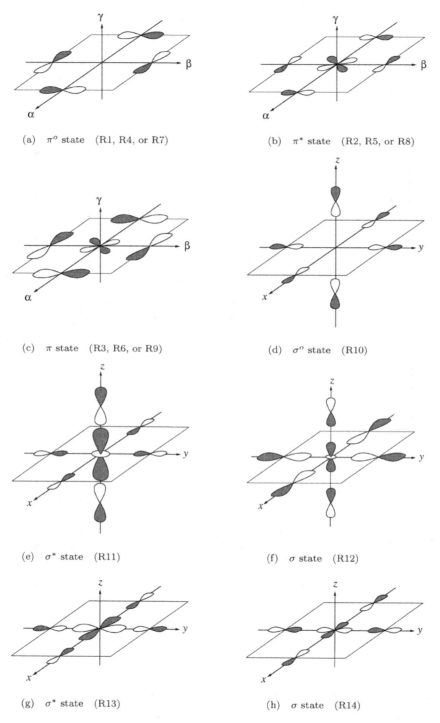

Figure 5.4. Schematic representation of the real-space wavefunctions of the energy band states at R in the Brillouin zone. To display the local symmetry, parts of the wavefunction in neighboring unit cells are shown as well.

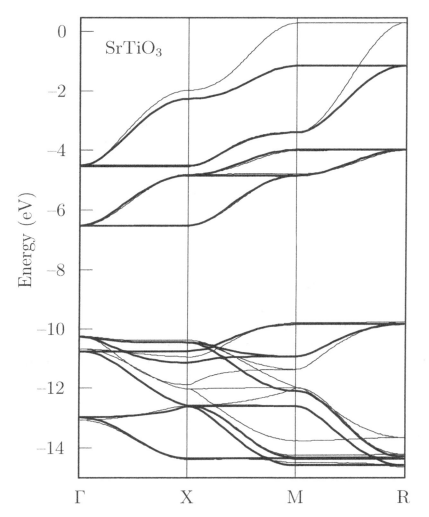

Figure 5.5. Comparison of LCAO energy bands (thick lines) with LDA calculations [2] (thin lines) for SrTiO$_3$. The LCAO parameters used for the fit are $E_t = -6.52$, $E_e = -4.52$, $E_\perp = -10.95$, $E_\parallel = -12.10$, $(pd\sigma) = -2.35$, $(pd\pi) = 1.60$, $(pp\sigma) = -0.05$, and $(pp\pi) = 0.50$ (all in eV) .

theory analog of the ligand-field splitting of d ions. The splitting between the R11 and R2 energies is given by

$$\Delta_R(d) = \frac{1}{2}\left(E_e - E_t + E_\parallel - E_\perp + 4b\right) + \sqrt{\left[\frac{1}{2}\left(E_e - E_\parallel - 2b\right)\right]^2 + 6(pd\sigma)^2}$$

$$+ \sqrt{\left[\frac{1}{2}\left(E_t - E_\perp + 2b\right)\right]^2 + 8(pd\pi)^2} \, . \tag{5.10}$$

It is readily seen that $\Delta_R(d)$ depends on the ionic splittings, $\Delta(d)$ and $\Delta(p)$, and on the two-center LCAO integrals $(pd\sigma)$ and $(pd\pi)$, which lead to covalent mixing between the cation and the oxygen ions. As mentioned earlier, the splitting at R is larger than at Γ. For $SrTiO_3$, $\Delta(d)$ is about 2.4 eV and $\Delta_R(d)$ is about 3.5 eV. For $KTaO_3$, $\Delta(d) \simeq 3.4$ eV and $\Delta_R(d) \simeq 6$ eV, and for ReO_3, $\Delta(d) \simeq 4$ eV and $\Delta_R(d) \simeq 7$ eV.

In Fig. 5.5, the energy bands for $SrTiO_3$ obtained from LCAO model (in thick lines) are shown. The band parameters are determined by fitting the important bands to the results [2] of *ab initio* density functional calculations in local density approximation (in thin lines). The π^* conduction band, lower part of σ^* conduction band, valence-band width are in good agreement with the local density approximation (LDA) results.

5.5 Cluster electronic states

Calculations for a small cluster of atoms are used to explore the properties of small particles. They are also sometimes used to infer the electronic and optical properties of a solid with similar constituents. It is therefore useful to understand the electronic states of a cluster and how they relate to the energy band states discussed in the previous sections.

The local characters of band wavefunctions, when extended to the neighboring unit cells (Figs 5.1 and 5.4), have a close resemblance to the localized wavefunctions of a cluster of atoms composed of a B ion surrounded by an octahedron of oxygen ions.

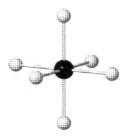

Figure 5.6. BO_6 cluster.

In order to analyze the cluster states, we consider the BO_6 cluster shown in Fig. 5.6. There are 23 basis orbitals for the cluster: three p orbitals on each of the six oxygen ions and five d orbitals on the central B ion. The orbitals are labeled as follows:

$$
\begin{aligned}
&|1\rangle = d_{z^2}(\vec{r}), && |2\rangle = p_z(\vec{r} - a\vec{e}_z), && |2'\rangle = p_z(\vec{r} + a\vec{e}_z), \\
&|3\rangle = d_{x^2-y^2}(\vec{r}), && |4\rangle = p_x(\vec{r} - a\vec{e}_x), && |4'\rangle = p_x(\vec{r} + a\vec{e}_x), \\
& && |5\rangle = p_y(\vec{r} - a\vec{e}_y), && |5'\rangle = p_y(\vec{r} + a\vec{e}_y), \\
&|6\rangle = d_{xy}(\vec{r}), && |7\rangle = p_x(\vec{r} - a\vec{e}_y), && |7'\rangle = p_x(\vec{r} + a\vec{e}_y), \\
& && |8\rangle = p_y(\vec{r} - a\vec{e}_x), && |8'\rangle = p_y(\vec{r} + a\vec{e}_x), \\
&|9\rangle = d_{xz}(\vec{r}), && |10\rangle = p_x(\vec{r} - a\vec{e}_z), && |10'\rangle = p_x(\vec{r} + a\vec{e}_z), \\
& && |11\rangle = p_z(\vec{r} - a\vec{e}_x), && |11'\rangle = p_z(\vec{r} + a\vec{e}_x), \\
&|12\rangle = d_{yz}(\vec{r}), && |13\rangle = p_y(\vec{r} - a\vec{e}_z), && |13'\rangle = p_y(\vec{r} + a\vec{e}_z), \\
& && |14\rangle = p_z(\vec{r} - a\vec{e}_y), && |14'\rangle = p_z(\vec{r} + a\vec{e}_y).
\end{aligned}
\tag{5.11}
$$

The LCAO matrix elements may be calculated as in the preceding section with the neglect of interactions beyond second nearest neighbors. The resulting Hamiltonian, shown in Table 5.5, is a 23×23 matrix.

The matrix of Table 5.5 possesses many zero elements, but straightforward diagonalization is extremely tedious. The complexity of the problem can be greatly reduced using well-known group theory methods [3, 4]. By transforming the Hamiltonian to a symmetry-coordinates representation the matrix can be reduced to a block-diagonal form in which the largest blocks are 2×2s. We will not go into all the details of the group theory method here, but a brief outline of the procedure is described in the next subsection. Readers unfamiliar with group theory may wish to skip the next subsection and go directly to Subsection (b).

A more detailed analysis of the BO_6 cluster states that includes overlap integrals as well as the B-ion s orbitals may be found in [5]. The results are not qualitatively different from those presented here.

(a) Block-diagonalizing the Hamiltonian using symmetry coordinates

In this section we give a brief description of how group theory can be used to block-diagonalize the matrix of Table 5.5. A central result of group representation theory is that the matrix elements of a Hamiltonian vanish between functions that transform according to different IRs (irreducible representations) of the symmetry group or between functions which transform according to different rows of the same IR. Thus, if we transform the Hamiltonian matrix to a representation labeled by basis functions for the IRs of the point group of the cluster, non-zero matrix element will occur only between the functions which transform according to the same row of the same IR.

The BO_6 cluster possesses O_h symmetry. To determine the nature of the block-

Table 5.5. *Hamiltonian matrix for the BO_6 cluster.*

	d_{z2}	d_{x2}	d_{xy}	d_{xz}	d_{yz}	2	2'	4	4'	5	5'	7	7'	8	8'	10	10'	11	11'	13	13'	14	14'
d_{z2}	E_e					$2s$	$-2s$	$-s$	s	$-s$	s												
d_{y2}		E_e						t	$-t$	$-t$	t												
d_{xy}			E_t									d	$-d$	d	$-d$								
d_{xz}				E_t												d	$-d$	d	$-d$				
d_{yz}					E_t															d	$-d$	d	$-d$
2	$2s$					E_s		$-v$	v	$-v$	v							u	u			u	u
2'	$-2s$						E_s	v	$-v$	v	$-v$							u	u			u	u
4	$-s$	t				$-v$	v	E_s		$-v$	v	u	u			u	u						
4'	s	$-t$				v	$-v$		E_s	v	$-v$	u	u			u	u						
5	$-s$	$-t$				$-v$	v	$-v$	v	E_s				u	u					u	u		
5'	s	t				v	$-v$	v	$-v$		E_s			u	u					u	u		
7		d						u	u			E_p		$-v$	v	p	p						
7'		$-d$						u	u				E_p	v	$-v$	p	p						
8		d								u	u	$-v$	v	E_p						p	p		
8'		$-d$								u	u	v	$-v$		E_p					p	p		
10			d					u	u			p	p			E_p		$-v$	v				
10'			$-d$					u	u			p	p				E_p	v	$-v$				
11			d			u	u									$-v$	v	E_p				p	p
11'			$-d$			u	u									v	$-v$		E_p			p	p
13				d						u	u			p	p					E_p		$-v$	v
13'				$-d$						u	u			p	p						E_p	v	$-v$
14				d		u	u											p	p	$-v$	v	E_p	
14'				$-d$		u	u											p	p	v	$-v$		E_p

$E_p = E_\perp$, $d = (pd\pi)$, $s = (pd\sigma)/2$, $u = [(pp\sigma) + (pp\pi)]/2 = c/2$.
$E_s = E_\parallel$, $p = (pp\pi)$, $t = \sqrt{3}(pd\sigma)/2$, $v = [(pp\sigma) - (pp\pi)]/2 = b/2$.

diagonal form of the Hamiltonian we need only know the numbers and types of IRs that will occur in the solution of the cluster problem. That information can be determined by decomposing the representation based on the 23 cluster orbitals (5 d orbitals plus 18 p orbitals) into the IRs of O_h point group. Using the character table for the O_h group and the calculated characters for the cluster–orbital representation we arrive at the results shown in (5.12a-e) and in the table that follows:

$$\Gamma(\text{orbital}) = \Gamma(B) + \Gamma(\text{oxygens}) \tag{5.12a}$$

$$\Gamma(B) = e_g + t_{2g} \tag{5.12b}$$

$$\Gamma(\text{oxygens}) = a_{1g} + e_g + t_{1g} + t_{2g} + t_{2u} + 2t_{1u} \tag{5.12c}$$

$$\Gamma(\text{orbital}) = a_{1g} + 2e_g + t_{1g} + 2t_{2g} + t_{2u} + 2t_{1u} . \tag{5.12d}$$

In the equations (5.12), $\Gamma(B)$ is the representation based on the five d orbitals of the B ion, $\Gamma(\text{oxygens})$ is the representation based on the 18 p orbitals of the six

	a_{1g}	e_g	t_{1g}	t_{2g}	t_{2u}	t_{1u}
Number of states that transform as the IR	1	4	3	6	3	6
Dimensions of the block-diagonalized Hamiltonian	(1×1)	$2(2\times2)$	$3(1\times1)$	$3(2\times2)$	$3(1\times1)$	$3(2\times2)$

$$(5.12\text{e})$$

oxygens. The O_h irreducible representations (IRs) have the following dimensions: a_{1g} is one-dimensional, e_g is two-dimensional, and all others in (5.12c) are three-dimensional representations. The subscript 'g' (gerade) indicates basis functions that are invariant under inversion through the origin (the center of the cluster) and 'u' (ungerade) indicates functions that are antisymmetric. Since there can be no non-zero matrix elements of the Hamiltonian between different rows of the same IRs, the dimensionalities of the blocks in the block-diagonalized Hamiltonian will be determined solely by the decomposition coefficients (i.e., the number of times an IR is contained in the orbital representation). The number and size of the blocks in the transformed Hamiltonian are shown in (5.12e). For example, e_g is a two-dimensional representation. It occurs twice in the decomposition formula, once in $\Gamma(B)$ and once in Γ(oxygens). Therefore, the block-diagonalized matrix will have a 2×2 which involves the first-row e_g d orbitals with the first-row e_g p orbital combination. The second 2×2 involves the second-row IR basis functions for the p and the d orbitals. Each of the 2×2 blocks will yield the same two eigenvalues, E and E', even though the eigenvectors of the 2×2 blocks are different. Thus, the e_g solutions consist of two pairs of doubly degenerate states. That is, two states with eigenvalues E and two states with eigenvalues E'.

To calculate the block-diagonalized Hamiltonian we need to find combinations of the cluster orbitals that form bases for the IRs of O_h that appear in the decomposition. Group theory provides a method for doing this using that are called projection operators [3, 4]. A brief description of the process is as follows. A set of cluster orbitals is selected as a "trial" function. To generate a function which can be used as the basis for the nth row of the jth IR, the trial function is subjected to an operation of the group (a rotation, for example) and the resulting function multiplied by the (n, n) diagonal element of the jth IR matrix representing that operation. The sum obtained by repeating this procedure for all of the operations of the group is either a null function or a function that transforms according to the nth row of the jth IR. (Although it is usually not necessary, the trial function can be taken as the sum of all of the cluster orbitals and that guarantees that a null function will not be generated for any IR contained in the orbital representation.) The orbital basis functions generated in this way are often referred to as symmetry

coordinates or symmetry functions. A particular eigenfunction of the Hamiltonian
will always be a linear combination of symmetry coordinates which belong to the
same row of the same IR. The collection of symmetry coordinates, when normalized,
can be used to form a unitary matrix, $\mathbb{U}$, which will block-diagonalize the cluster
Hamiltonian matrix; that is,

$$\mathbb{U}^{\dagger}\mathbb{H}\mathbb{U} = \mathbb{H}' \qquad \text{(a block diagonalized matrix)}.$$

The decomposition (5.12b) shows that the d orbitals are invariant under inversion. Thus, the antisymmetric u states (ungerade) are composed entirely of p
orbitals. This means that the u states are non-bonding. That is, states that do not
contribute to the B–O chemical bond. Also, it is clear that the bonding and antibonding states will have e_g or t_{2g} symmetries because these are the only symmetries
that can produce an eigenstate with both p and d orbitals. Furthermore, the basis
functions of the IRs (a_{1g}, t_{1g}, and t_{2u}) which occur with a coefficient of unity in the
decomposition equation of (5.12c) must be cluster *eigenfunctions* since they can
have no matrix elements with any other symmetry coordinate. Thus, group theory
gives the eigenfunctions for seven of the 23 cluster states. These eigenfunctions are
independent on the values of the parameters that enter into the Hamiltonian matrix
elements appearing in Table 5.5.

The 23 symmetry coordinates for the BO_6 cluster are given in Table 5.6.

(b) Calculation of cluster states

The Hamiltonian matrix in Table 5.5 can be block-diagonalized by transforming it
to the representation based on the symmetry coordinates listed in Table 5.6. The
matrix, $\mathbb{U}$, which accomplishes this is given in Table 5.7.

(c) Types of cluster states

The block-diagonalized Hamiltonian is shown in Tables 5.8 and 5.9. The types
of states for the BO_6 cluster that result are briefly discussed in this section. For
convenience we denote a cluster-state eigenfunction by CN ($N = 1, 2, \ldots, 23$) and
the corresponding eigenvalue by E_{CN}.

The a_{1g} state

The a_{1g} state is a totally symmetric combination of $p_{\parallel}$ orbitals that form a non-
bonding cluster state. The symmetry coordinate, $S(1)$, is the eigenfunction of the

Table 5.6. *Symmetry coordinates for the BO_6 cluster.*

Name	IR	Row	Symmetry coordinates
$S(1)$	a_{1g}	1	$\frac{1}{\sqrt{6}}[(|2\rangle - |2'\rangle) + (|4\rangle - |4'\rangle) + (|5\rangle - |5'\rangle)]^*$
$S(2)$	e_g	1	$|1\rangle$
$S(3)$	e_g	1	$\frac{1}{2\sqrt{3}}[2(|2\rangle - |2'\rangle) - (|4\rangle - |4'\rangle) - (|5\rangle - |5'\rangle)]$
$S(4)$	e_g	2	$|3\rangle$
$S(5)$	e_g	2	$\frac{1}{2}[(|4\rangle - |4'\rangle) - (|5\rangle - |5'\rangle)]$
$S(6)$	t_{2g}	1	$|6\rangle$
$S(7)$	t_{2g}	1	$\frac{1}{2}[(|7\rangle - |7'\rangle) + (|8\rangle - |8'\rangle)]$
$S(8)$	t_{2g}	2	$|9\rangle$
$S(9)$	t_{2g}	2	$\frac{1}{2}[(|10\rangle - |10'\rangle) + (|11\rangle - |11'\rangle)]$
$S(10)$	t_{2g}	3	$|12\rangle$
$S(11)$	t_{2g}	3	$\frac{1}{2}[(|13\rangle - |13'\rangle) + (|14\rangle - |14'\rangle)]$
$S(12)$	t_{1g}	1	$\frac{1}{2}[(|11\rangle - |11'\rangle) - (|10\rangle - |10'\rangle)]^*$
$S(13)$	t_{1g}	2	$\frac{1}{2}[(|14\rangle - |14'\rangle) - (|13\rangle - |13'\rangle)]^*$
$S(14)$	t_{1g}	3	$\frac{1}{2}[(|8\rangle - 8'\rangle) - (|7\rangle - |7'\rangle)]^*$
$S(15)$	t_{1u}	1	$\frac{1}{\sqrt{2}}[|2\rangle + |2'\rangle]$
$S(16)$	t_{1u}	1	$\frac{1}{2}[(|11\rangle + |11'\rangle) + (|14\rangle + |14'\rangle)]$
$S(17)$	t_{1u}	2	$\frac{1}{\sqrt{2}}[(|4\rangle + |4'\rangle)]$
$S(18)$	t_{1u}	2	$\frac{1}{2}[(|7\rangle + |7'\rangle) + (|10\rangle + |10'\rangle)]$
$S(19)$	t_{1u}	3	$\frac{1}{2}[|5\rangle + |5'\rangle]$
$S(20)$	t_{1u}	3	$\frac{1}{2}[(|8\rangle + |8'\rangle) + (|13\rangle + |13'\rangle)]$
$S(21)$	t_{2u}	1	$\frac{1}{2}[(|7\rangle + |7'\rangle) - (|10\rangle + |10'\rangle)]^*$
$S(22)$	t_{2u}	2	$\frac{1}{2}[(|13\rangle + |13'\rangle) - (|8\rangle + |8'\rangle)]^*$
$S(23)$	t_{2u}	3	$\frac{1}{2}[(|14\rangle + |14'\rangle) - (|11\rangle + |11'\rangle)]^*$

* Indicates that the symmetry function is also an eigenfunction.

Table 5.7. *Symmetry coordinate transformation matrix,* $\mathbb{U}$.

	d_{z^2}	d_{x^2}	d_{xy}	d_{xz}	d_{yz}	2	2'	4	4'	5	5'	7	7'	8	8'	10	10'	11	11'	13	13'	14	14'
d_{z^2}	1																						
d_{y^2}		1																					
d_{xy}			1																				
d_{xz}				1																			
d_{yz}					1																		
2	q	$2r$												s									
2'	$-q$	$-2r$												s									
4	q	$-r$	t												s								
4'	$-q$	r	$-t$												s								
5	q	$-r$	$-t$													s							
5'	$-q$	r	t														s						
7									t				$-t$					t		t			
7'									$-t$				t					t		t			
8									t				t							t		$-t$	
8'									$-t$				$-t$							t		$-t$	
10										t	$-t$							t		$-t$			
10'										$-t$	t							t		$-t$			
11										t	t					t							$-t$
11'										$-t$	$-t$					t							$-t$
13												t	$-t$							t		t	
13'												$-t$	t							t		t	
14												t	t			t							t
14'												$-t$	$-t$			t							t

$q = \frac{1}{\sqrt{6}}, \qquad r = \frac{1}{2\sqrt{3}}, \qquad s = \frac{1}{\sqrt{2}}, \qquad t = \frac{1}{2}.$

Hamiltonian with energy, E_{C1}:

$$E_{C1} = E_{\parallel} - 2b,$$
$$C1 = S(1). \tag{5.13}$$

The e_g states

The e_g states are bonding and antibonding states involving admixtures of e_g-type d orbitals with $p_{\parallel}$ orbitals. The two 2×2 blocks in Table 5.8 yield identical pairs of energies, E_{C2} and E_{C3} where

$$E_{C2} = \frac{1}{2}(E_e + E_{\parallel} + b) + \sqrt{\left[\frac{1}{2}(E_e - E_{\parallel} - b)\right]^2 + 3(pd\sigma)^2},$$

$$E_{C3} = \frac{1}{2}(E_e + E_{\parallel} + b) - \sqrt{\left[\frac{1}{2}(E_e - E_{\parallel} - b)\right]^2 + 3(pd\sigma)^2}, \tag{5.14}$$

$$E_{C4,5} = E_{C2,3}.$$

Table 5.8. *Block-diagonalized g-cluster states.*

$$a_{1g} \quad \boxed{E_\parallel - 2b}$$

$$e_g \quad \left\{ \begin{array}{cc} E_e & \sqrt{3}(pp\sigma) \\ \sqrt{3}(pp\sigma) & E_\parallel + b \end{array} \right. \quad \begin{array}{cc} E_e & \sqrt{3}(pp\sigma) \\ \sqrt{3}(pp\sigma) & E_\parallel + b \end{array}$$

$$t_{2g} \quad \left\{ \begin{array}{cc} E_t & 2(pd\pi) \\ 2(pd\pi) & E_\perp + b \end{array} \quad \begin{array}{cc} E_t & 2(pd\pi) \\ 2(pd\pi) & E_\perp + b \end{array} \quad \begin{array}{cc} E_t & 2(pd\pi) \\ 2(pd\pi) & E_\perp + b \end{array} \right.$$

$$t_{1g} \quad \left\{ E_\perp + b \quad E_\perp + b \quad E_\perp + b \right.$$

Table 5.9. *Block-diagonalized u-cluster states.*

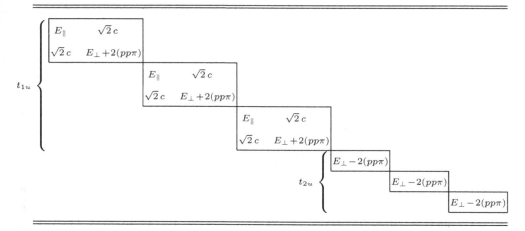

$C2$ and $C4$ are antibonding states and $C3$ and $C5$ are bonding states. The eigenfunctions are given below:

(row 1 eigenfunctions):

$$C2,3 = \left\{ \sqrt{3}(pd\sigma)S(2) - (E_e - E_{C2,3})S(3) \right\}/N_{C2,3} \tag{5.15}$$

$$N_{C2,3} = \sqrt{(E_e - E_{C2,3})^2 + 3(pd\sigma)^2} \tag{5.16}$$

(row 2 eigenfunctions):

$$C4,5 = \left\{ \sqrt{3}(pd\sigma)S(4) - (E_e - E_{C2,3})S(5) \right\}/N_{C2,3}. \tag{5.17}$$

The t_{2g} states

The t_{2g} states are bonding and antibonding states involving admixtures of t_{2g}-type d orbitals with $p_\perp$ orbitals. The three identical 2×2 matrices in Table 5.8 yield a pair of triply degenerate energies, E_{C6} and E_{C7}, where

$$E_{C6,7} = \frac{1}{2}(E_t + E_\perp - b)\pm\sqrt{\left[\frac{1}{2}(E_t - E_\perp + b)\right]^2 + 4(pd\pi)^2}, \tag{5.18}$$

$$E_{C8,9} = E_{C10,11} = E_{C6,7}. \tag{5.19}$$

The six eigenfunctions are given by

$$C6, 7 = \left\{ 2(pd\pi)S(6) + \frac{1}{2}(E_t - E_{C6,7})S(7) \right\} / N_{C6,7} \tag{5.20}$$

$$C8, 9 = \left\{ 2(pd\pi)S(8) + \frac{1}{2}(E_t - E_{C6,7})S(9) \right\} / N_{C6,7} \tag{5.21}$$

$$C10, 11 = \left\{ 2(pd\pi)S(10) + \frac{1}{2}(E_t - E_{C6,7})S(11) \right\} / N_{C6,7} \tag{5.22}$$

$$N_{C6,7} = \sqrt{\left[\frac{1}{2}(E_t - E_{C6,7}) \right]^2 + 4(pd\pi)^2} \, . \tag{5.23}$$

The t_{1g} states

The t_{1g} states are non-bonding combinations of $p_\perp$ orbitals. They form a triply degenerate set with energy,

$$E_{C12,13,14} = E_\perp + b, \tag{5.24}$$
$$C12 = S(12), \tag{5.25}$$
$$C13 = S(13), \tag{5.26}$$
$$C14 = S(14). \tag{5.27}$$

The eigenfunctions are the same as the symmetry coordinates.

The t_{1u} states

The t_{1u} states are non-bonding combinations of $p_\parallel$ orbitals, and $p_\perp$ orbitals. They form a pair of triply degenerate states (Table 5.9). The energies are given by

$$E_{C15,16} = \frac{1}{2} \left[E_\parallel + E_\perp + 2(pp\pi) \right] \pm \sqrt{\left[E_\parallel - E_\perp - 2(pp\pi) \right]^2 + 2c^2} \, . \tag{5.28}$$

The eigenfunctions are:

$$C15, 16 = \left\{ \sqrt{2}cS(15) - (E_\parallel - E_{C15,16})S(16) \right\} / N_{C15,16} \tag{5.29}$$

$$C17, 18 = \left\{ \sqrt{2}cS(17) - (E_\parallel - E_{15,16})S(18) \right\} / N_{C15,16} \tag{5.30}$$

$$C19, 20 = \left\{ \sqrt{2}cS(19) - (E_\parallel - E_{15,16})S(20) \right\} / N_{C15,16} \tag{5.31}$$

$$N_{C15,16} = \sqrt{2c^2 + (E_\parallel - E_{C15,16})^2} \, . \tag{5.32}$$

The t_{2u} states

The t_{2u} states are non-bonding combinations of $p_\perp$ orbitals. They form a triply degenerate set with energy,

$$E_{C21,22,23} = E_\perp - 2(pp\pi), \tag{5.33}$$
$$C21 = S(21), \tag{5.34}$$
$$C22 = S(22), \tag{5.35}$$
$$C23 = S(23). \tag{5.36}$$

The t_{2u} eigenfunctions are the same as the symmetry coordinates.

A schematic diagram of the cluster states is shown in Fig. 5.7. The ordering of the states is dependent upon the values of the LCAO parameters.

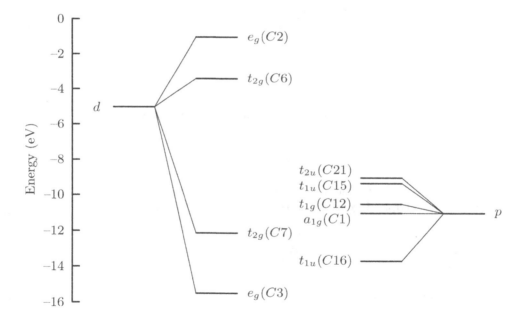

Figure 5.7. Cluster states for the following parameters (in eV): $E_e = -4.0$, $E_t = -6.0$, $E_\parallel = -12.0$, $E_\perp = -10.0$, $(pd\pi) = -2.0$, $(pd\sigma) = -4.0$, $(pp\pi) = -0.5$, $(pp\sigma) = -1.0$. The levels are at: $a_{1g} = -11.0$, $e_g = -1.02$ and -15.48, $t_{2g} = -3.38$ and -12.12, $t_{1g} = -10.5$, $t_{1u} = -9.32$ and -13.68, $t_{2u} = -9.0$, all in eV.

In some clusters or analogs of perovskite metals the states are completely occupied below the t_{2g} levels, but have only one or two electrons in the triply degenerate t_{2g} levels. In this case the cluster ground state is electronically degenerate since several arrangements of the electrons in the t_{2g} levels have the same energy. According to the Jahn–Teller theorem [6] if a nonlinear molecule's ground state is electronically degenerate the molecule must distort to lower its symmetry and energy. For example, a compression along the z-axis of the BO_3 cluster would lower the symmetry from cubic to tetragonal. It would also raise the energies of the xz and yz levels, but lower the xy level. The xy level could then accommodate one or two electrons. The total energy of the completely filled states below t_{2g} is often nearly unchanged by the perturbation. Some levels move up, and some down in such a way that "energy center of gravity" is nearly unchanged. Therefore, the new ground state has a lower total energy than that for the cubic case.

(d) Comparison of cluster states with energy band states

A reasonable correspondence between an LCAO energy band states and the BO_6 cluster states can be established by inspecting the symmetry and composition of the eigenfunctions. In Fig. 5.4 the local symmetries of the band wavefunctions are made evident by showing the orbitals in a unit cell and also those in neighboring cells which would belong to a BO_6 cluster. We shall refer to this combination of orbitals as the *local band function*. In what follows we shall show that there is a one-to-one correspondence between the cluster eigenfunctions and the local band functions.

The energy-band wavefunctions possess full O_h symmetry at Γ and R and therefore we look for correlations with the cluster states at these points in the Brillouin zone. In general, the u-cluster states will be correlated with band states at Γ and the g-cluster states correlated with band states at R. To understand this assignment consider a p orbital, $p(\vec{r}-a\vec{e}_\alpha)$, involved in an energy-band wavefunction in a particular unit cell. Its partner orbital, $p(\vec{r}+a\vec{e}_\alpha)$, will be located in an adjacent unit cell and therefore the amplitude will differ by a phase factor of $e^{i(2k_\alpha a)}$. This phase factor is $+1$ at Γ and -1 at R. For the u-cluster states $p(\vec{r}-a\vec{e}_\alpha)$ is always combined with $+p(\vec{r}+a\vec{e}_\alpha)$ to form a function that is antisymmetric under inversion. For the g states, $p(\vec{r}-a\vec{e}_\alpha)$ is always combined with $-p(\vec{r}+a\vec{e}_\alpha)$ to form a function that is symmetric under inversion. Therefore the u states will correlate with energy-band states at Γ and the g-cluster states will correlate with energy-band states at R.

The eigenfunctions of the seven cluster states, $C1$, $C12$, $C13$, $C14$, $C21$, $C22$, and $C23$ are the symmetry coordinates $S(1)$, $S(12)$, $S(13)$, $S(14)$, $S(21)$, $S(22)$, and $S(23)$, respectively. To illustrate how a correlation with a band state can be established consider the energy band state in Table 5.4 labeled R10. The wavefunction for the unit cell is proportional to $[|2\rangle + |4\rangle + |5\rangle]$. The partner p orbitals (the p' orbitals), which lie in adjacent unit cells, will have the negative of these amplitudes because the phase factor $e^{i(2k_\alpha a)} = -1$ at R. Combining unit-cell orbitals with their partner orbitals yields the local band function for R10 that is given by,

$$[|2\rangle - |2'\rangle + |4\rangle - |4'\rangle + |5\rangle - |5'\rangle] \quad \text{(R10 local band function)}. \tag{5.37}$$

The function in (5.37) is proportional to the symmetry coordinate $S(1)$. According to (5.14) the eigenfunction for the $C1$ cluster state is $S(1)$ (a_{1g} symmetry). Therefore, except for a constant normalization factor, the $C1$ cluster-state eigenfunction is identical to the local band function for the R10 energy band state. In a similar fashion we can show that the local band functions for R4, R7, and R1 are proportional to the symmetry coordinates $S(12)$, $S(13)$, and $S(14)$ and hence are, within a constant factor, the same as the eigenfunctions for the cluster states (t_{1g} symmetry) $C12$, $C13$, and $C14$, respectively.

Next, consider the band state, Γ_9. The unit-cell eigenfunction is proportional to $|7\rangle - |10\rangle$ so the local band function is

$$|7\rangle + |7'\rangle - |10\rangle - |10'\rangle, \tag{5.38}$$

since the phase factor $e^{i(2k_\alpha a)} = +1$ at Γ. The function in (5.38) is proportional to the t_{2u} symmetry coordinate, $S(21)$. Thus, the local band function for the energy band state Γ_9 is, apart from a constant, the same as the eigenfunction for the cluster state $C21$. Similarly, Γ_{12}, and Γ_6 have local band functions that are proportional to the symmetry coordinates $S(22)$ and $S(23)$, respectively. That is, the local band functions for Γ_{12}, and Γ_6 are the same as the eigenfunctions for the cluster states $C22$ and $C23$ (t_{2u} symmetry).

The remaining cluster states have wavefunctions that are admixtures of two different functions that transform according to the same row of the IR representation. As a last example, consider the band states in Table 5.4 labeled R11 and R12. The local band functions are constructed as before except that we need to include an internal phase factor $e^{i(2k_\alpha a)} = i$ for the p orbitals relative to the d orbital. The orbitals of the eigenfunction in the unit cell take the form

$$6(pd\sigma)|1\rangle - (E_e - E_{R11,12})[2|2\rangle - |4\rangle - |5\rangle],$$

so the local band function is

$$6(pd\sigma)|1\rangle - 2\sqrt{3}(E_e - E_{R11,12})[2|2\rangle - 2|2'\rangle - |4\rangle + |4'\rangle - |5\rangle + |5'\rangle]$$

$$\propto \sqrt{3}(pd\sigma)S(2) - (E_e - E_{R11,12})S(3). \tag{5.39}$$

Comparison of this result with the e_g cluster eigenvectors in (5.15) and (5.17) shows that the band states R11 and R12 are identical in form to the cluster states $C2$

and $C3$. They differ only by the eigenvalues that enter into the coefficient of $S(3)$. The band state R11 (R12) involves E_{R11} (E_{R12}) and the cluster state $C2$ ($C3$) involves the energy E_{C2} (E_{C3}). In general, the band energies are not the same as the cluster energies so the eigenfunctions are different. However, the wavefunctions are identical with respect to their symmetry properties. We discuss the energy differences between the cluster states and the energy band states at the end of this section.

The correlations between the energy-band states and the cluster states is summarized as follows:

$$
\begin{aligned}
C1\,(a_{1g}) &\rightarrow R10 & &(\sigma^0 \text{ valence band at R}) \\
C2, 4\,(e_g) &\rightarrow R11, 13 & &(\sigma^* \text{ conduction band at R}) \\
C3, 5\,(e_g) &\rightarrow R12, 14 & &(\sigma \text{ valence band at R}) \\
C6, 8, 10\,(t_{2g}) &\rightarrow R2, 5, 8 & &(\pi^* \text{ conduction band at R}) \\
C7, 9, 11\,(t_{2g}) &\rightarrow R3, 6, 9 & &(\pi \text{ valence band at R}) \quad (5.40) \\
C12, 13, 14\,(t_{1g}) &\rightarrow R1, 4, 7 & &(\pi^0 \text{ valence band at R}) \\
C15, 17, 19\,(t_{1u}) &\rightarrow \Gamma 7, 10, 13 & &(\pi + \sigma \text{ valence band at } \Gamma) \\
C16, 18, 20\,(t_{1u}) &\rightarrow \Gamma 8, 11, 14 & &(\pi + \sigma \text{ valence band at } \Gamma) \\
C21, 22, 23\,(t_{2u}) &\rightarrow \Gamma 6, 9, 12 & &(\pi^0 \text{ valence band at } \Gamma)
\end{aligned}
$$

where the arrow indicates that the states have identical symmetry properties.

Figure 5.8 shows the locations of the energy-band states which correlate with the cluster state symmetries.

As we have seen, the cluster eigenstates can be correlated with energy-band wavefunctions at Γ or R in the Brillouin zone. The eigenvalues of the cluster states are also closely related to the energy band eigenvalues, but they are clearly different. For example, the eigenvalue of the cluster state, $C1$, is $E_\| - 2b$, but the energy of the correlated band state, R10, is $E_\| - 4b$. Similarly, the energy of E_{R11} (see Table 5.4) involves a term $6(pp\sigma)^2$, whereas for the correlated cluster state, $C2$, the corresponding factor is $3(pp\sigma)^2$. These differences are due to the fact that an oxygen ion in the cluster interacts with only one B ion while in the solid it interacts with two B ions. Also each oxygen interacts with twice as many other neighboring oxygen ions in the solid as in the cluster. If the following replacements

$$
\begin{aligned}
(pp\pi) &\rightarrow 2(pp\pi), \\
(pp\sigma) &\rightarrow 2(pp\sigma), \\
(pd\sigma) &\rightarrow \sqrt{2}(pd\sigma), \\
(pd\pi) &\rightarrow \sqrt{2}(pd\pi)
\end{aligned}
$$

are made in the expressions for the cluster energies to correct for the difference in the

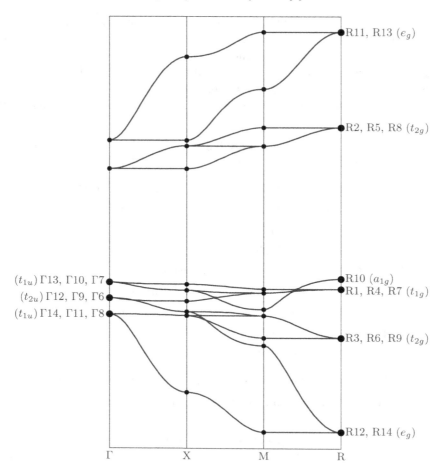

Figure 5.8. Symmetries of energy band states and of corresponding cluster states (in parentheses).

numbers of interactions, the cluster energies become identical to the corresponding band energies.

   In summary, there are remarkable similarities between the electronic states of the BO_6 cluster and those of the corresponding perovskite solid. However, there are also significant differences that must be borne in mind. For an insulating perovskite (unoccupied conduction bands) the usual, direct band gap is the energy difference between the top of the valence band at Γ (t_{1u} in the Fig. 5.8 or t_{2u} in Fig. 5.7) and the bottom of the π^* conduction band at Γ. However, the cluster model does not have a state corresponding to the bottom of the π^* (or σ^*) conduction band at Γ. The closest analog to the band gap for the cluster is the energy difference between the t_{2g} state a R and the t_{1u} state at Γ. Consequently, the analogs of the electronic transitions involved to the onset of optical absorption are not represented in the cluster model. Furthermore, the π, π^*, σ, and σ^* band states at Γ are composed entirely of d orbitals. There are no analogs of these states in the cluster model. It is sometimes argued that the cluster levels should occur roughly at the "center of gravity" of the corresponding energy bands of the solid. This suggestion is certainly not valid for the perovskites. As will be discussed in Chapter 6, the energy bands of the perovskites possess critical singularities in the electronic density of states that are responsible for structure in the optical and electronic properties. These important characteristics can not be accounted for with a cluster model. One conclusion of the above analysis is the cluster model can not give even a rough idea of the electronic properties of the solid. Conversely, the band model can not give even a rough idea of the electronic properties of an actual cluster particle. Clearly, such particles can possess electronic and optical properties that are quite different from those of the corresponding solid.

References

[1] L. B. Bouckaert, R. Smoluchowski, and E. P. Wigner, *Phys. Rev.* **50**, 65 (1936).

[2] E. Mete, R. Shaltaf, and Ş. Ellialtıoğlu, *Phys. Rev.* B **68**, 035119 (2003).

[3] E. P. Wigner, *Group theory and its application to the quantum mechanics of atomic spectra* (New York and London, Academic Press, 1959).

[4] D. C. Harris and M. D. Bertolucci, *Symmetry and spectroscopy* (New York, Dover, 1989).

[5] T. Wolfram, R. Hurst, and F. J. Morin, *Phys. Rev.* B **15**, 1151 (1977).

[6] H. A. Jahn and E. Teller, *Proc. Roy. Soc.* A **161**, 220 (1937).

Problems for Chapter 5

1. Derive the energies and wavefunctions for the π and π^* bands at Γ and R given in Tables 5.1 and 5.4.

   Hints: (i) Rearrange rows and columns of the eigenvalue matrix to achieve a block-diagonal form. (ii) Use unit-cell symmetry coordinates to transform the 5×5 to a 1×1 and two 2×2s.

2. Calculate the orbital amplitudes for the eigenvector of the state R_{11}, using the LCAO parameters given in Fig. 5.7.

3. (a) Explain why there are no BO_6 cluster states corresponding to the pure d-orbital energy band states at Γ. (Hint: refer to Problem 7 in Chapter 1.)

   (b) What symmetry does the infinite cubic array have that the BO_6 cluster lacks?

4. Using the symmetry coordinates of Table 5.6 derive the block-diagonals shown in Table 5.8 for the a_{1g} and t_{2g} states.

5. (a) Calculate the energies of the $t_{2g}(C6)$ and $t_{2u}(C21)$ states using the parameters of Fig. 5.7.

   (b) Compare the energy difference $[E(C6) - E(C1)]$ with $(E_{\pi^*} - E_\pi)$ at Γ.

6

Density of states

The density of states (DOS) of a solid is of fundamental importance in determining its electronic, optical, and transport properties. The DOS of an ideal perovskite is unusual in that it possesses peculiar structures called van Hove singularities which are associated with the two-dimensional character of the pi bands and portions of the Brillouin zone where the sigma bands are flat. The effects of these singularities show up in the optical reflectivity and absorption and in the photoelectron energy distributions observed for perovskites.

In this chapter we determine the DOS functions and use them to calculate the Fermi surface and effective masses for the d-band perovskites.

6.1 Definitions

A solid consisting of N unit cells will have $n_s \times N$ electronic states, where n_s is the number of basis orbitals per unit cell. The electronic states are characterized by a wavevector $\vec{k}_i$ ($i = 1, 2, \ldots, N$) and a band index ν ($\nu = 1, 2, \ldots, n_s$). For an infinite crystal ($N \to \infty$) the $\vec{k}_i$'s form a dense set. One simple way of choosing the set of $\vec{k}_i$ is to divide the Brillouin zone into N cells of equal volume and take the $\vec{k}$-vectors which lie at the center of each of the N cells. As $N \to \infty$, each cell becomes a differential volume element in $\vec{k}$-space with volume $d\vec{k} = \Omega/N$ where Ω is the volume of the first Brillouin zone. In this limit we may use the relation

$$\frac{1}{N} \sum_{i=1}^{N} f_{\vec{k}_i \nu} \underset{N \to \infty}{\Longrightarrow} \frac{1}{\Omega} \int_{\text{BZ}} d\vec{k} \, f_\nu(\vec{k})$$

where $f_\nu(\vec{k})$ is a continuous function with value $f_{\vec{k}_i \nu}$ at $\vec{k} = \vec{k}_i$ and the integral is over the first Brillouin zone.

We define the density of (electronic) states, $\rho(E)$ by the condition that the quantity $\rho(E)\Delta E$ is the number of states per unit cell in the energy range between

E and $E + \Delta E$. If we consider one of the N cells of the first Brillouin zone with wavevector $\vec{k}_i$ then the number of energy band eigenvalues corresponding to $\vec{k}_i$ with energy less than $E + \Delta E$ is given by

$$\sum_{\nu=1}^{n_s} \Theta\left(E + \Delta E - E_{\vec{k}_i\nu}\right), \tag{6.1}$$

where $\Theta(x)$ is the unit step function; $\Theta(x) = 1$ for $x > 0$ and is zero otherwise. Similarly, the number of eigenvalues with energy less than E is

$$\sum_{\nu=1}^{n_s} \Theta\left(E - E_{\vec{k}_i\nu}\right). \tag{6.2}$$

It is then obvious that the number of eigenvalues corresponding to $\vec{k}_i$ which lie in the energy range between E and $E + \Delta E$ is just the difference between (6.1) and (6.2). If we sum this difference over all of the $\vec{k}_i$-vectors we will have the total number of states with energies between E and $E + \Delta E$. Thus,

$$\rho(E)\,\Delta E = \frac{1}{N} \sum_{\nu=1}^{n_s} \sum_i \left\{ \Theta\left(E + \Delta E - E_{\vec{k}_i\nu}\right) - \Theta\left(E - E_{\vec{k}_i\nu}\right) \right\}, \tag{6.3}$$

where the $1/N$ factor is introduced to give the number of states per unit cell.

In the limit as $\Delta E \to 0$ the difference term on the right-hand side of (6.3) can be replaced by

$$\frac{d}{dE}\left\{\Theta(E - E_{\vec{k}_i\nu})\right\}\Delta E, \tag{6.4}$$

so that

$$\rho(E) = \frac{1}{N} \sum_{\nu=1}^{n_s} \sum_i \frac{d}{dE}\Theta(E - E_{\vec{k}_i\nu}). \tag{6.5}$$

The step function Θ may be written as

$$\Theta(E - E_{\vec{k}_i\nu}) = \int_{-\infty}^{E} \delta(t - E_{\vec{k}_i\nu})\,dt \tag{6.6}$$

where $\delta(t - E_{\vec{k}_i\nu})$ is the delta function with its singularity at $t = E_{\vec{k}_i\nu}$. (A detailed discussion of the delta function and properties of the DOS functions is given in Appendix B.) From (6.6) it is apparent that

$$\frac{d}{dE}\Theta(E - E_{\vec{k}_i\nu}) = \delta(E - E_{\vec{k}_i\nu}). \tag{6.7}$$

Thus, we arrive at the result that

$$\rho(E) = \frac{1}{N} \sum_{\nu=1}^{n_s} \sum_i \delta(E - E_{\vec{k}_i\nu}). \tag{6.8}$$

Next, we need to pass to the limit as $N \to \infty$ and convert the sum over $\vec{k}_i$ into an integral. This requires that we weigh each term by the volume it represents in $\vec{k}$-space. Each $\vec{k}_i$ represents a volume $\Omega/N = \Delta\vec{k}$, where $\Delta\vec{k}$ is the differential volume element in $\vec{k}$-space. Thus,

$$\rho(E) = \frac{1}{\Omega} \sum_{\nu=1}^{n_s} \int d\vec{k} \; \delta(E - E_{\vec{k}\nu}), \tag{6.9}$$

where the integral is over the volume of the first Brillouin zone. It is clear from (6.9) that

$$\int_{-\infty}^{\infty} \rho(E) \, dE = n_s. \tag{6.10}$$

It is often convenient to work with other density functions. For example, if $f(E)$ is a single-valued function of the energy, then

$$\rho(f)_\nu = \frac{1}{\Omega} \int d\vec{k} \; \delta[f(E) - f(E_{\vec{k}\nu})] \tag{6.11}$$

$$\rho(f) = \sum_{\nu=1}^{n_s} \rho(f)_\nu \; .$$

The relation between these densities is

$$\rho(E) = \rho(f) \left| \frac{df}{dE} \right| . \tag{6.12}$$

In utilizing (6.9) or (6.11) the δ function can be represented as follows:

$$\delta(x - x_0) = -\frac{1}{\pi} \, \Im \left\{ \frac{1}{x - x_0 + i0^+} \right\}$$

where 0^+ is a positive infinitesimal and $\Im\{\cdots\}$ indicates the imaginary part of the expression within the brackets. Thus, we have that

$$\rho(f)_\nu = -\frac{1}{\pi} \, \Im \, \frac{1}{\Omega} \int d\vec{k} \left\{ \frac{1}{f(E) - f(E_{\vec{k}\nu}) + i0^+} \right\}, \tag{6.13a}$$

$$\rho(E)_\nu = -\frac{1}{\pi} \, \Im \, \frac{|df/dE|}{\Omega} \int d\vec{k} \left\{ \frac{1}{f(E) - f(E_{\vec{k}\nu}) + i0^+} \right\} . \tag{6.13b}$$

6.2 DOS for the pi bands

In Chapter 4 we derived simple analytical expressions for the 14 energy bands of a d-band perovskite for the approximation $(pp\sigma) = (pp\pi) = 0$. In this section we make use of these formulae for the energy bands to derive the DOS functions.

The energy bands are classified as pi and sigma bands and DOS expressions can be derived for each type separately. We begin by considering the pi bands. There are nine pi bands described by

$$E_{\vec{k}\pi^0(\alpha\beta)} = E_\perp \quad (\text{non-bonding bands}), \tag{6.14}$$

$$E_{\vec{k}\pi^*(\alpha\beta)} = \frac{1}{2}(E_t + E_\perp) + \sqrt{\left[\frac{1}{2}(E_t - E_\perp)\right]^2 + 4(pd\pi)^2\left(S_\alpha^2 + S_\beta^2\right)} \tag{6.15}$$

$$E_{\vec{k}\pi(\alpha\beta)} = \frac{1}{2}(E_t + E_\perp) - \sqrt{\left[\frac{1}{2}(E_t - E_\perp)\right]^2 + 4(pd\pi)^2\left(S_\alpha^2 + S_\beta^2\right)} \tag{6.16}$$

$$(\alpha\beta = xy, \ xz, \text{ or } yz).$$

We can express both equations (6.15) and (6.16) in the dimensionless form:

$$\frac{\left[E_{\vec{k}\nu} - \frac{1}{2}(E_t + E_\perp)\right]^2 - \left[\frac{1}{2}(E_t - E_\perp)\right]^2}{2(pd\pi)^2} - 2 = -(C_{2\alpha} + C_{2\beta}) \tag{6.17}$$

where $C_{2\alpha} \equiv \cos(2k_\alpha a)$ and $C_{2\beta} \equiv \cos(2k_\beta a)$. If (6.17) is solved for $E_{\vec{k}\nu}$, the two solutions are $E_{\vec{k}\pi^*}$ and $E_{\vec{k}\pi}$. This strange form turns out to be a mathematically convenient function for investigating the density of states of the π^* and π bands. Rather than working with the actual energies, $E_{\vec{k}\pi^*}$ and $E_{\vec{k}\pi}$, it is much easier to calculate the density of state of the function on the left-hand side of (6.17). To do this we define the dimensionless function

$$\varepsilon_\pi(\vec{k}) \equiv \frac{\left[E_{\vec{k}\nu} - \frac{1}{2}(E_t + E_\perp)\right]^2 - \left[\frac{1}{2}(E_t - E_\perp)\right]^2}{2(pd\pi)^2} - 2 \tag{6.18}$$

so that (6.17) is given by

$$\varepsilon_\pi(\vec{k}) = -(C_{2\alpha} + C_{2\beta}). \tag{6.19}$$

Now using (6.13a) we calculate the density, $\rho(\varepsilon_\pi)$, defined by

$$\begin{aligned}
\rho(\varepsilon_\pi) &= -\frac{1}{\pi}\,\Im m\,\frac{1}{\Omega}\int \frac{d\vec{k}}{\varepsilon_\pi - \varepsilon_\pi(\vec{k}) + i0^+} \\
&= -\frac{1}{\pi}\,\Im m\,\left(\frac{2a}{2\pi}\right)^3 \int\!\!\int\!\!\int_{-\pi/2a}^{\pi/2a} \frac{dk_\alpha\,dk_\beta\,dk_\gamma}{\varepsilon_\pi + (C_{2\alpha} + C_{2\beta}) + i0^+}
\end{aligned} \tag{6.20}$$

where we have used the fact that $\Omega = (2\pi/2a)^3$ for the cubic perovskites, with a being the B–O distance.

The integration over dk_γ may be performed immediately and then introducing the variables $x = 2k_\alpha a$ and $y = 2k_\beta a$ we find

$$\rho(\varepsilon_\pi) = -\Im m\,\frac{4}{\pi}\left(\frac{1}{2\pi}\right)^2 \int_0^\pi dx \int_0^\pi dy\,\frac{1}{\varepsilon_\pi + C_x + C_y + i0^+}. \tag{6.21}$$

We proceed as follows:

$$\Im \int_0^\pi \frac{dy}{\varepsilon_\pi + C_x + C_y + i0+} = -\pi \int_0^\pi dy \, \delta[(\varepsilon_\pi + C_x) + C_y] \tag{6.22}$$

$$= -\pi \int_0^\pi \frac{d(-C_y)}{S_y} \, \delta[(\varepsilon_\pi + C_x) - (-C_y)]$$

$$= -\pi \frac{\Theta[1 - (\varepsilon_\pi + C_x)^2]}{\sqrt{1 - (\varepsilon_\pi + C_x)^2}}. \tag{6.23}$$

The Θ function ensures that the value of ε_π is such that the δ function is satisfied in the integration range. Thus, we find that

$$\rho(\varepsilon_\pi) = \frac{1}{\pi^2} \int_0^\pi \frac{\Theta[1 - (\varepsilon_\pi + C_x)^2]}{\sqrt{1 - (\varepsilon_\pi + C_x)^2}} \, dx. \tag{6.24}$$

The Θ function is non-zero only when ε_π lies within the pi bands; that is when $|\varepsilon_\pi(k)| \leq 2$. From (6.24) it is apparent that when $|\varepsilon_\pi| > 2$, then $\Theta[1 - (\varepsilon_\pi + C_x)^2] = 0$ at all points in the range of the integration. Thus, the density of states vanishes for ε_π outside of the range of the π or π^* bands, as of course it must. We may then confine our attention to the range $|\varepsilon_\pi| \leq 2$. We make the substitution $z = \varepsilon_\pi + C_x$ and obtain

$$\rho(\varepsilon_\pi) = \frac{1}{\pi^2} \int_c^b \frac{dz}{\sqrt{(z-a)(z-b)(z-c)(z-d)}}, \tag{6.25}$$

where $a > b > c > d$. For $-2 \leq \varepsilon_\pi < 0$, we have $a = 1, b = 1 + \varepsilon_\pi, c = -1$, and $d = \varepsilon_\pi - 1$. For $0 \leq \varepsilon_\pi < 2$, we have $a = 1 + \varepsilon_\pi, b = 1, c = \varepsilon_\pi - 1$, and $d = -1$. The required integral is given by Gradshteyn and Ryzhik [1] (see 3.147.4) with the result that

$$\rho(\varepsilon_\pi) = \frac{1}{\pi^2} K\left(\sqrt{1 - (\varepsilon_\pi/2)^2}\right) \Theta[1 - (\varepsilon_\pi/2)^2] \tag{6.26}$$

where $K(k)$ is the complete elliptic integral of the first kind:

$$K(k) = \int_0^{\pi/2} \frac{d\theta}{\sqrt{1 - k^2 \sin^2 \theta}}. \tag{6.27}$$

A graph of $\rho(\varepsilon_\pi)$ is shown in Fig. 6.1(a). It is seen that $\rho(\varepsilon_\pi)$ has jump discontinuities (designated as P_0 and P_2 critical points) at the band edges ($\varepsilon_\pi = \pm 2$). The peak at $\varepsilon_\pi = 0$ is a logarithmic singularity (a P_1 critical point). The discontinuities arise from the fact that the π^* and π bands depend only on two components k_α and k_β, of the three-dimensional k-vector. They therefore behave as two-dimensional energy bands and have constant energy with variation of k_γ. The logarithmic singularity occurs at $\varepsilon_\pi = 0$ and arises from the saddle points in the energy band dispersion near the X-points in the Brillouin zone.

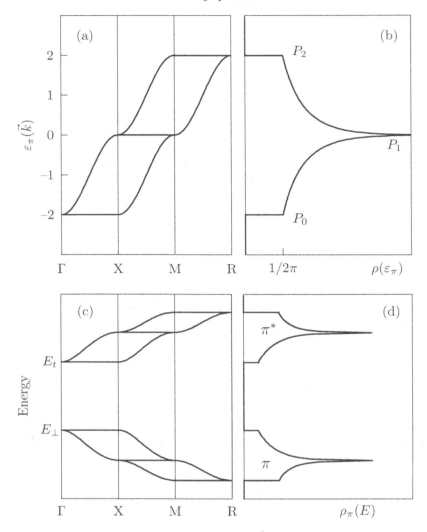

Figure 6.1. Density of states functions for the pi bands. (a) The three reduced bands $\varepsilon_\pi(\vec{k})$ for α, β, $\gamma = x$, y or, z; $\alpha \neq \beta$, and (b) the corresponding reduced density of states $\rho(\varepsilon_\pi)$ with P_0, P_1, and P_2 being van Hove singularities related to the critical points in 2D. (c) Energy bands $E_\pi(\vec{k})$ and (d) the corresponding density of states $\rho_\pi(E)$ in E-space.

The function $\varepsilon_\pi(\vec{k})$ defined by (6.17) and (6.18) describe both the π^* conduction bands and the π valence bands.

To convert to the density of states in energy space, $\rho(E)$, we employ (6.13b) and (6.18) to find that

$$\rho_\pi(E) = \frac{1}{\pi^2} \left| \frac{E - \frac{1}{2}(E_t + E_\perp)}{(pd\pi)^2} \right| K\left(\sqrt{1 - (\varepsilon_\pi(E)/2)^2}\right) \Theta[1 - (\varepsilon_\pi(E)/2)^2].$$

(6.28)

The function $\rho(E)$ has two regions where it is non-vanishing as illustrated in Fig. 6.1(b). The upper region is the DOS for a π^* conduction band and the lower region is the DOS for a π valence band.

The DOS for a π^0 non-bonding band is simply a δ function since these bands are flat. The broadening of the bands due to the oxygen–oxygen interactions can be included approximately by replacing the δ function with a Gaussian shape of the form:

$$\rho_{\pi^0}(E) \cong \sqrt{\frac{\lambda_{\pi^0}}{\pi}} \, e^{-\lambda_{\pi^0}(E-E_{\pi^0})^2},$$

(6.29)

where λ is a parameter that determines the band width and E_{π^0} is the position of the center of the π^0 band. These parameters can be estimated from the results of Chapter 5. An interpolation formula for the π^0 bands is

$$E_{\vec{k}\pi^0} \simeq E_\perp - 4(pp\pi)C_\alpha^2 C_\beta^2 + 2bS_\alpha^2 S_\beta^2,$$

(6.30)

where $b = (pp\sigma) - (pp\pi)$. At Γ, (6.30) gives a triply degenerate energy, $E_{\pi^0}(\Gamma) = E_\perp - 4(pp\pi)$ and at M it gives $E_{\pi^0}(M) = E_\perp + 2b$, and a doubly degenerate energy $E_\perp$. At R (6.30) gives a triply degenerate energy $E_{\pi^0}(R) = E_\perp + 2b$.

Thus (6.30) give exact energies at the Γ, M and R points. However, at X it gives two energies at $E_\perp$, and one at $E_\perp - 4(pp\pi)$. The exact solution at X gives $E = E_\perp + 4(pp\pi)$ and $E_\perp - 4(pp\pi)$ for the two π^0 energies. Despite this shortcoming, (6.30) is a reasonably good approximation from which the band width and DOS of the π^0 bands can be estimated. If $(pp\pi)$ and b have the same sign then the π^0 bands extend from $E_\perp - 4(pp\pi)$ to $E_\perp + 2b$ and the band width is $|2b + 4(pp\pi)| = |2(pp\sigma) + 2(pp\pi)| = 2|c|$. If $(pp\pi)$ and b differ in sign then the band extends from $E_\perp$ to $E_\perp + 2b$, and the band width is $2|b|$. Since $(pp\pi)$ is generally smaller than $(pp\sigma)$ the band width will be approximately equal to $2|(pp\sigma)|$.

If we define the band width to be the full-width at half-maximum of the Gaussian of (6.29) then $\lambda_{\pi^0} = \ln 2/(pp\sigma)^2$. For SrTiO$_3$, $(pp\sigma) \simeq 0.2$ eV so one has $\lambda \simeq 17$ eV^{-2}.

An analytical formula [2] for the DOS for π^0 bands corresponding to the interpolation formula (6.30) can be obtained in terms of complete elliptic functions of the first kind. The resultant DOS contains jump discontinuities and a logarithmic singularity, shown in Fig. 6.2.

The total density of states arising from the nine pi bands can finally be written

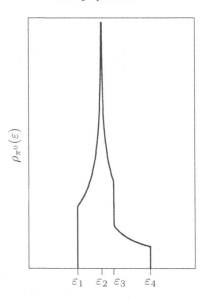

Figure 6.2. Density of states [2] for the π^0 bands given by (6.30) with parameters b and $(pp\pi)$ opposite in sign. $\varepsilon_1 = E_\perp$, $\varepsilon_2 = E_\perp - 4(pp\pi)$ and $\varepsilon_4 = E_\perp + 2b$.

as

$$\rho_{\text{pi}}(E) = 3\rho_{\pi^0}(E) + 3\rho_\pi(E).\tag{6.31}$$

The units of ρ_{pi} are electron states per unit cell per spin per unit energy or simply states/(spin-cell-energy).

$$\int_{-\infty}^{\infty} \rho_{\text{pi}}(E)\,dE = 9 \quad \text{and} \quad \int_{-\infty}^{E_\perp} \rho_{\text{pi}}(E)\,dE = 6.\tag{6.32}$$

Including spin there are 18 states. The 12 valence-band states are occupied in a perovskite.

6.3 DOS for the sigma bands

According to (4.57) the eigenvalue equation for the sigma bands is

$$(E_e - E)(E_\parallel - E) = 2(pd\sigma)^2 \Big\{ (S_x^2 + S_y^2 + S_z^2)$$

$$\pm \sqrt{(S_x^4 + S_y^4 + S_z^4) - (S_x^2 S_y^2 + S_y^2 S_z^2 + S_z^2 S_x^2)} \Big\}.\tag{6.33}$$

It is convenient to introduce the dimensionless variable, $\varepsilon_\sigma(E)$, defined by

$$\varepsilon_\sigma \equiv [(E_e - E)(E_\parallel - E)/(pd\sigma)^2] - 3.\tag{6.34}$$

Equation (6.33) can be expressed as, $\varepsilon_\sigma(E) - \varepsilon_\sigma^\pm(\vec{k}) = 0$ where

$$\varepsilon_\sigma^\pm(\vec{k}) = -(C_{2x} + C_{2y} + C_{2z})$$
$$\pm \sqrt{C_{2x}^2 + C_{2y}^2 + C_{2z}^2 - C_{2x}C_{2y} - C_{2y}C_{2z} - C_{2z}C_{2x}} \qquad (6.35)$$

with $C_{2\alpha} = \cos 2k_\alpha a$. The function $\varepsilon_\sigma(\vec{k})$ describes two branches $\varepsilon_\sigma^+(\vec{k})$ and $\varepsilon_\sigma^-(\vec{k})$ as shown in Fig. 6.3.

Along the line $\overline{\Gamma X}$, $C_{2y} = C_{2z} = 1$ and

$$\varepsilon_\sigma^\pm(\vec{k}) = -(C_{2x} + 2) \pm \sqrt{(1 - C_{2x})^2}, \qquad (6.36)$$
$$\varepsilon_\sigma^+(\vec{k}) = -1 - 2C_{2x}, \qquad (6.37)$$
$$\varepsilon_\sigma^-(\vec{k}) = -3. \qquad (6.38)$$

Along the line $\overline{XM}$, $C_{2x} = -1$ and $C_{2z} = +1$ so

$$\varepsilon_\sigma^\pm(\vec{k}) = -C_{2y} \pm \sqrt{3 + C_{2y}^2}, \qquad (6.39)$$

$$\varepsilon_\sigma^+(\vec{k}) = -C_{2y} + \sqrt{3 + C_{2y}^2}, \qquad (6.40)$$

$$\varepsilon_\sigma^-(\vec{k}) = -C_{2y} - \sqrt{3 + C_{2y}^2}. \qquad (6.41)$$

Along the line $\overline{MR}$, $C_{2x} = C_{2y} = -1$ and

$$\varepsilon_\sigma^\pm(\vec{k}) = 2 - C_{2z} \pm \sqrt{(C_{2z} + 1)^2}, \qquad (6.42)$$
$$\varepsilon_\sigma^+(\vec{k}) = 3, \qquad (6.43)$$
$$\varepsilon_\sigma^-(\vec{k}) = 1 - 2C_{2z}. \qquad (6.44)$$

According to (6.38) $\varepsilon_\sigma^-(\vec{k})$ is flat along $\overline{\Gamma X}$ and according to (6.43) $\varepsilon_\sigma^+(\vec{k})$ is flat along $\overline{MR}$. As we shall show shortly, these lines of constant energy produce jump discontinuities in the DOS at the bottom ($\varepsilon_\sigma = -3$) and top ($\varepsilon_\sigma = +3$) of the σ^* and σ bands as shown in Fig. 6.3. Such jumps are designated as P_0 and P_2 critical points [3].

In order to determine the nature of the critical points we investigate the analytic properties of the sigma energy bands near particular energies for which the derivative of the density of states is infinite or discontinuous. The form of $\rho(\varepsilon_\sigma)$ near $\varepsilon_\sigma = \pm 3$ and ± 1 can be determined by using a power-series expansion of the band energies near these special points. We consider first the branch $\varepsilon_\sigma^-(k)$ near $\varepsilon_\sigma = -3$. Along $\overline{\Gamma X}$ the band is flat. Consider a cylinder with axis along $\overline{\Gamma X}$. The contribution to $\rho(\varepsilon_\sigma)$ from this cylinder can be obtained by expanding the energy $\varepsilon_\sigma^-(\vec{k})$ in powers of k_y and k_z near the $\overline{\Gamma X}$ line along the k_x-axis.

$$\varepsilon_\sigma^- = -3 + \frac{3}{4}(2k_y a)^2 + \frac{3}{4}(2k_z a)^2 = -3 + \frac{3}{4}r^2,$$

$$r^2 \equiv (2k_y a)^2 + (2k_z a)^2 . \tag{6.45}$$

Then the contribution $\Delta\rho(\varepsilon_\sigma)$ to ρ is

$$\Delta\rho(\varepsilon_\sigma) = -\frac{1}{\pi} \Im \left(\frac{2a}{2\pi}\right)^3 \int_{-\pi/2a}^{\pi/2a} dk_x \int\int_{\text{cylinder}} \frac{dk_y\, dk_z}{\varepsilon_\sigma + 3 - \frac{3}{4}r^2 + i0+}$$

$$= -\frac{1}{\pi} \Im \frac{1}{(2\pi)^2} \int_0^{2\pi} d\phi \int_0^{r_0} \frac{r\, dr}{(\varepsilon_\sigma + 3) - \frac{3}{4}r^2 + i0+}$$

$$= -\frac{1}{\pi} \Im \frac{4/3}{4\pi} \int_0^{r_0^2} \frac{dx}{\frac{4}{3}\xi - x + i0+} \quad \text{with} \quad \xi \equiv (\varepsilon_\sigma + 3)$$

$$= \frac{1}{3\pi} \int_0^{r_0^2} dx\, \delta\left(\frac{4}{3}\xi - x\right) = \frac{1}{3\pi} . \tag{6.46}$$

The result of $1/(3\pi)$ is valid provided the cylinder radius $r_0 > \sqrt{4\xi/3}$, where $\xi = (\varepsilon_\sigma + 3)$. We are interested in the result as $\varepsilon_\sigma \to -3$ so that $\xi \to 0$, therefore the result is valid and independent of r_0 so long as r_0 is not identically zero. Thus, as ε_σ varies from $-3+0^-$ to $-3+0^+$ there is a jump in the density equal to $1/(3\pi)$. Since there are three equivalent $\overline{\Gamma X}$ lines the total jump is $1/\pi$. Symmetry requires that the same result holds for the ε_σ^+ behavior of the density near $\varepsilon_\sigma = +3$ along the $\overline{MR}$ lines along the edges of the first Brillouin zone. Each $\overline{MR}$ line is shared by four zones so that the weight of each line is $1/4$. Thus the 12 $\overline{MR}$ lines have a total weight of 3 and the total jump in $\rho(\varepsilon_\sigma)$ is again $1/\pi$.

The ε_σ^+ branch is quadratic near $\varepsilon_\sigma = -3$ and contributes nothing to the jump in $\rho(\varepsilon_\sigma)$ as $\varepsilon_\sigma \to -3$. Similarly, ε_σ^- does not contribute to the jump in $\rho(\varepsilon_\sigma)$ as $\varepsilon_\sigma \to +3$.

Next, we consider the behavior of the branches near $\varepsilon_\sigma = \pm 1$, where the ε_σ^+ branch has the expansion

$$\varepsilon_\sigma^+ = 1 - \alpha^2 + \frac{1}{4}r^2 \tag{6.47}$$

with $\alpha = (\pi - 2k_x a)$ and $r^2 = (2k_y a)^2 + (2k_z a)^2$. Equation (6.47) describes a saddle point region of the energy dispersion and the associated singular points are designated as M_1 and M_2 [3].

Consider a cylinder with radius r_0 extending from $\alpha = 0$ to α_0. The contribution to the DOS from this cylinder is

$$\Delta\rho(\varepsilon_\sigma) = -\frac{1}{\pi} \Im \left(\frac{1}{2\pi}\right)^3 \int_0^{\alpha_0} d\alpha \int_0^{r_0} \frac{r\, dr}{\varepsilon_\sigma - 1 + \alpha^2 - \frac{1}{4}r^2} \int_0^{2\pi} d\phi$$

$$= -\frac{1}{\pi} \Im \frac{2(2\pi)}{(2\pi)^3} \int_0^{\alpha_0} d\alpha \int_0^{r_0^2/4} \frac{dx}{\varepsilon_\sigma - 1 + \alpha^2 - x}$$

$$= \frac{1}{2\pi^2} \int_0^{\alpha_0} d\alpha \int_0^{r_0^2/4} \delta(\varepsilon_\sigma - 1 + \alpha^2 - x)\, dx$$

$$= \frac{1}{2\pi^2} \int_{L_1}^{\sqrt{r_0^2/4 - (\varepsilon_\sigma - 1)}} d\alpha, \tag{6.48}$$

where

$$L_1 = \begin{cases} 0 & \text{if } \varepsilon_\sigma - 1 > 0, \\ \sqrt{1 - \varepsilon_\sigma} & \text{if } \varepsilon_\sigma - 1 < 0. \end{cases} \tag{6.49}$$

Thus, apart from the numerical factor $1/(2\pi^2)$

$$\Delta\rho(\varepsilon_\sigma) \rightarrow \begin{cases} \sqrt{r_0^2/4 - (\varepsilon_\sigma - 1)} & \text{if } \varepsilon_\sigma > 1, \\ \sqrt{r_0^2/4 + (1 - \varepsilon_\sigma)} - \sqrt{(1 - \varepsilon_\sigma)} & \text{if } \varepsilon_\sigma < 1. \end{cases} \tag{6.50}$$

In the limit of $\varepsilon_\sigma \rightarrow 1^+$,

$$\Delta\rho(\varepsilon_\sigma) \rightarrow \frac{1}{2} r_0 - \frac{(\varepsilon_\sigma - 1)}{r_0} \tag{6.51}$$

which shows that $\Delta\rho(\varepsilon_\sigma)$ contains a contribution that *decreases* linearly as ε_σ increases from 1. On the other hand, as $\varepsilon_\sigma \rightarrow 1^-$

$$\Delta\rho(\varepsilon_\sigma) \rightarrow \frac{1}{2} r_0 + \frac{(1 - \varepsilon_\sigma)}{r_0} - \sqrt{(1 - \varepsilon_\sigma)}. \tag{6.52}$$

In this limit as ε_σ approaches 1 from below, the second term on the right-hand side of (6.52) is negligible compared to the square root term. Thus, we see that $\Delta\rho(\varepsilon_\sigma)$ *increases* as ε_σ increases towards 1 from below. More significantly, the derivative $d(\Delta\rho)/d\varepsilon_\sigma$ becomes infinite (tends to $-\infty$) as $\varepsilon_\sigma \rightarrow 1^-$. At $\varepsilon_\sigma = 1^+$ or 1^-, $\Delta\rho = r_0/(4\pi^2)$ and therefore the DOS near $\varepsilon_\sigma = -1$ is the mirror reflection of the analytic behavior near $\varepsilon_\sigma = +1$.

A similar analysis of the ε_σ^- branch near $\varepsilon_\sigma = -1$ yields the result that:

$$\Delta\rho(\varepsilon_\sigma) \rightarrow \begin{cases} \frac{1}{2} r_0 + (\varepsilon_\sigma + 1)/r_0 & \text{if } \varepsilon_\sigma < -1, \\ \frac{1}{2} r_0 + (\varepsilon_\sigma + 1)/r_0 - \sqrt{1 + \varepsilon_\sigma} & \text{if } \varepsilon_\sigma > -1. \end{cases} \tag{6.53}$$

A schematic of $\rho(\varepsilon_\sigma)$ is shown in Fig. 6.3(b). The DOS in ε_σ-space is symmetric about $\varepsilon_\sigma = 0$.

An exact result for $\rho(\varepsilon_\sigma)$ has not yet been obtained, but knowing the analytical behavior at the critical points suggests a simple approximation. We express $\rho(\varepsilon_\sigma)$

Density of states

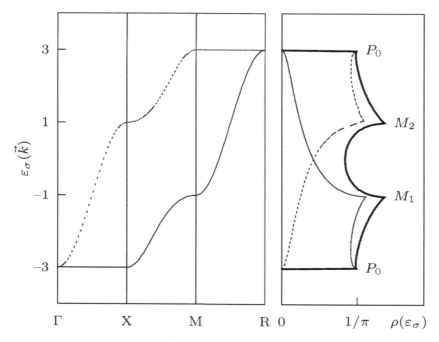

Figure 6.3. Schematic of $\rho(\varepsilon_\sigma)$.

as a linear combination of terms which have the correct analytical behavior:

$$\rho(\varepsilon_\sigma) = \begin{cases} \rho_1(\varepsilon_\sigma) & \text{for } |\varepsilon_\sigma| \leq 1, \\ \\ \rho_2(\varepsilon_\sigma) & \text{for } 1 \leq |\varepsilon_\sigma| \leq 3, \end{cases} \tag{6.54}$$

with

$$\rho_1(\varepsilon_\sigma) = A + B\sqrt{1 - \varepsilon_\sigma^2} + F(1 - |\varepsilon_\sigma|)\,|\varepsilon_\sigma|\,, \tag{6.55}$$

$$\rho_2(\varepsilon_\sigma) = C + Dx^2 + F(1 - x)\sqrt{x}\,, \tag{6.56}$$

$$x \equiv (3 - |\varepsilon_\sigma|)/2.$$

To determine the constants A, B, C, D, and F, we use the following conditions:

$$\rho_2(\pm 3) = 1/\pi \qquad \text{(jump at the band edges)}, \tag{6.57}$$

$$\rho_1(\pm 1) = \rho_2(\pm 1) \qquad \text{(continuity)}, \tag{6.58}$$

$$\int_{-3}^{3} \rho(\varepsilon_\sigma)\,d\varepsilon_\sigma = 2 \qquad \text{(two sigma bands are represented)}, \tag{6.59}$$

$$\rho_1(\pm 1) = \rho_{\text{Num}}(\pm 1), \tag{6.60}$$

where $\rho_{\text{Num}}(1) \approx 0.432$ is the value of $\rho_1(1)$ obtained by numerical evaluation of

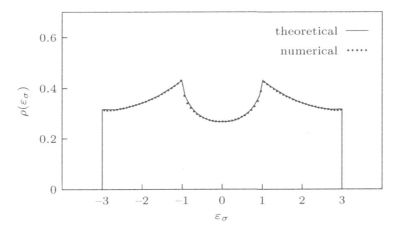

Figure 6.4. The function $\rho(\varepsilon_\sigma)$ of (6.54) compared with numerical calculations of the exact DOS.

the density of states. Imposing equations (6.57)–(6.60) yields the following results:

$$A = \rho_{\text{Num}}(\pm 1) \approx 0.432$$
$$B = -0.1646$$
$$C = \frac{1}{\pi} = 0.3183 \tag{6.61}$$
$$D = A - C \approx 0.1136$$
$$F = -0.0151.$$

Figure 6.4 shows the sigma density of states computed from (6.54) and (6.61) compared with results obtained by numerical integration. The approximation for $\rho(\varepsilon_\sigma)$ given by (6.55) and (6.56) is remarkably good [4], being within 1% of the numerical value for all values of ε_σ. A comparison of the approximate analytic expression for $\rho(\varepsilon)$ with numerical calculations is shown in Fig. 6.4.

To calculate the DOS in E-space we need only multiply the function in (6.54) by $|\partial \varepsilon_\sigma / \partial E|$. This gives

$$\rho_\sigma(E) = \left| \frac{E - \frac{1}{2}(E_e + E_\parallel)}{(pd\sigma)^2} \right| \rho(\varepsilon_\sigma(E)). \tag{6.62}$$

The factor $\frac{1}{2}(E_e + E_\parallel)$ is the mid-gap energy between the σ and σ^* band edges at Γ. Thus, the conversion factor $|E - \frac{1}{2}(E_e + E_\parallel)|/(pd\sigma)^2$ introduces a distortion of the DOS which enhances the DOS at the bottom of the σ valence band and the top of the σ^* conduction bands. A sketch of $\rho_\sigma(E)$ is shown in Fig. 6.5.

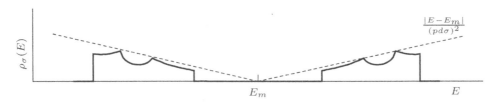

Figure 6.5. Sigma-band DOS as a function of E.

The non-bonding σ^0 band may be treated in the same manner as the π^0 bands. Thus

$$\rho_{\sigma^0}(E) \cong \sqrt{\frac{\lambda_{\sigma^0}}{\pi}}\, e^{-\lambda_{\sigma^0}(E-E_{\sigma^0})^2} \tag{6.63}$$

where λ_{σ^0} is a parameter that determines the band width of the σ^0 band and E_{σ^0} is the center of the band.

The width of the σ^0 band can be calculated approximately using first-order perturbation theory to include the effects of the $(pp\pi)$ and the $(pp\sigma)$ interactions. We can write the Hamiltonian, H (discussed in Chapter 4), as $H = H^0 + H'$, where H^0 is H with $(pp\pi)$ and $(pp\sigma)$ set equal to zero, and H' is the part of H which contains matrix elements involving $(pp\pi)$ and/or $(pp\sigma)$. Using the matrix elements described in Chapter 4 we find that

$$E_{\vec{k}\sigma^0} \cong E_{\parallel} + \langle\sigma^0|H'|\sigma^0\rangle = E_{\parallel} - \frac{12\,b\,S_x^2\,S_y^2\,S_z^2}{(S_x^2\,S_y^2 + S_y^2\,S_z^2 + S_x^2\,S_z^2)}. \tag{6.64}$$

The approximate band width is therefore $4b$, which is about twice the band width of the π^0 bands. If we equate $4b$ to the full-width at half-maximum of the function in (6.63) then we find that

$$\lambda_{\sigma^0} \simeq \frac{\ln 2}{4b^2}. \tag{6.65}$$

An analytical formula [2] for the σ^0 DOS corresponding to the approximate expression (6.64) gives the structure shown in Fig. 6.6 which has a base width of $4b$. With the analytic forms of the "pi" and "sigma" DOS functions we are now able to give a simple expression for the total density of states including the contributions from all of the 14 primary energy bands,

$$\rho_{\text{total}}(E) = 2\left[3\Big(\rho_{\pi^0}(E) + \rho_{\text{pi}}(E)\Big) + \Big(\rho_{\sigma^0}(E) + \rho_{\text{sigma}}(E)\Big)\right], \tag{6.66}$$

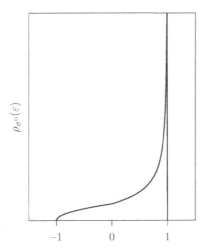

Figure 6.6. Density of states [2] for the σ^0 band given by (6.64) with parameters $\varepsilon = 1$ corresponding to $E = E_\parallel$, and $\varepsilon = -1$ to $E = E_\parallel - 4b$.

where the factor of 2 accounts for the two spin states. We note that

$$\int_{-\infty}^{\infty} \rho_{\text{total}}(E)\, dE = 28\,,\tag{6.67}$$

$$\int_{-\infty}^{E_\perp} \rho_{\text{total}}(E)\, dE = 18\,.\tag{6.68}$$

Equation (6.68) expresses the fact that there are 18 electrons per unit cell in the occupied valence bands. The valence-band wavefunctions, however, are admixtures of p and d orbitals so that d orbitals are partially occupied as a result of the filled valence-band states.

Mattheiss [5] has carried out augmented plane wave (APW) calculations of the energy bands of several of the perovskites. Using the numerical results he has constructed histograms of the DOS for SrTiO$_3$, KTaO$_3$, and ReO$_3$. In Figs 6.7, comparisons of the results of Mattheiss with the analytical result of (6.66) are presented. Also shown is a similar comparison for NaWO$_3$ with the result of Kopp et al. [6]. The model parameters have been selected to produce the prominent structures in the same locations as in the histograms. It is seen that the analytical density of states $\rho_{\text{total}}(E)$ reproduces the DOS with considerable success.

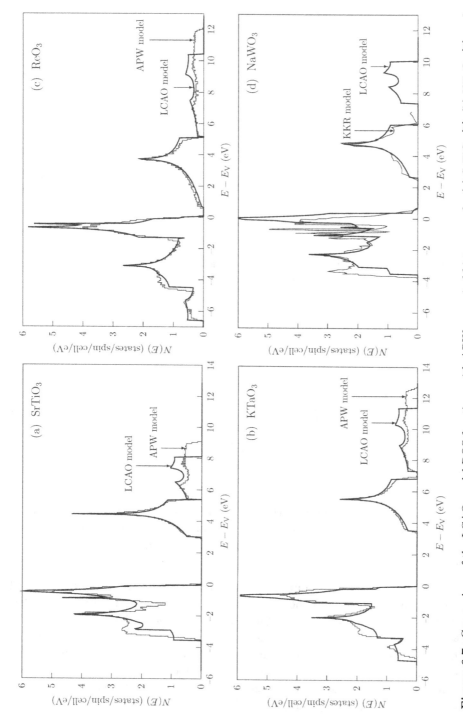

Figure 6.7. Comparison of the LCAO model DOS functions with APW numerical histograms for (a) $SrTiO_3$ [5], (b) $KTaO_3$ [5], and (c) ReO_3 [7]. (d) Comparison of the LCAO model DOS function with KKR numerical histogram for Na_xWO_3 [6].

6.4 The Fermi surface and effective mass

(a) Semiconducting perovskites

The Fermi surface is an important feature of a metallic or semiconducting solid that plays a vital role in determining the physical and chemical properties of the material. The Fermi surface of a crystal is defined as the surface of constant energy in $\vec{k}$-space that separates the occupied states from the unoccupied states. At absolute zero $(T=0\,\mathrm{K})$ the lowest energy states are occupied up to a particular energy called the Fermi energy, E_F, and the states lying above E_F are unoccupied. Hence at $T=0$, the surface is sharply defined. At finite temperatures the surface will be fuzzy with a few "holes" below E_F and a few electrons in states above E_F. The fuzzyness is described by the Fermi distribution function, $f(E,T) = [1 + e^{(E-E_F)/k_B T}]^{-1}$, where k_B is Boltzmann's constant. In this section we discuss the shape of the Fermi surface, the density of states at the Fermi energy, and the effective mass of electrons in the conduction bands (π^* bands) of a semiconducting cubic perovskite.

Let $E_{\alpha\beta}(\vec{k})$ ($\alpha\beta = xy, xz,$ or yz) denote a particular π^* band of an insulating perovskite. The crystal can be made conducting or semiconducting by any of a number of "doping" methods. It can be converted to an n-type semiconductor by heating in a hydrogen atmosphere, introducing oxygen vacancies, or adding cation impurities. For example, $SrTiO_3$ becomes n-type when doped with Nb atoms. For an n-type material, the donated electrons occupy the lowest energy states of the π^* bands. Typical n-type materials usually have electron concentration of the order of 10^{18}–10^{20} electrons/cm^3. For a lattice parameter of 3.9 Å this gives a concentration of 6×10^{-5} to 6×10^{-3} electrons per unit cell. Therefore, the occupied states will be confined to a very small volume in $\vec{k}$-space near Γ.

According to (6.28), at the bottom of the π^* bands ($E = E_t$) the total density of states is

$$\rho_{total}(E_t) = 2(\text{spins}) \times 3(\text{equivalent bands}) \times \frac{1}{4\pi} \frac{E_g^0}{(pd\pi)^2}. \tag{6.69}$$

Assuming a typical band gap, E_g^0, of 3.2 eV, and $(pd\pi)$ of 1.2 eV, we find that

$$\rho_{total}(E_t) = 1.06 \text{ states}/(\text{unit cell eV}). \tag{6.70}$$

For typical donor concentrations $\rho_\pi(E)$ can be taken to be constant for the small range of occupied states near E_t with $[\rho_{\pi^*}(E_F)]_{total} = [\rho_{\pi^*}(E_t)]_{total} \equiv \rho^0$. We have that $\Delta_F \rho^0 = n_{cell}$ where $\Delta_F = (E_F - E_t)$ and n_{cell} is the donor concentration per unit cell. Thus,

$$\Delta_F = 0.943 \, n_{cell} \quad (\text{in eV}). \tag{6.71}$$

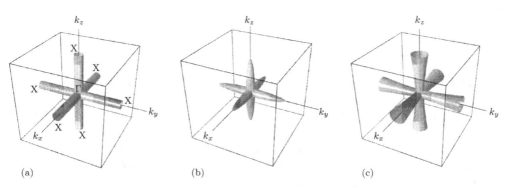

Figure 6.8. Surfaces of constant energy for E near E_t. For $E = E_F$ any wavevector on the surface of the "jack" corresponds to a state of energy E_F. $\vec{k}$-vectors inside and on the surface of the "jack" correspond to filled states while those outside the "jack" are unoccupied at $T = 0\,\text{K}$. (a) For $\delta = 0$: the surfaces are those of three circular cylinders aligned along the three principal axes. Each cylinder extends between equivalent X points of symmetry on the opposite faces of the Brillouin zone. The axes of the three cylinders intersect at Γ. (b) For $\delta > 0$: the arms of the "jack" become slender "cigars". (c) For $\delta < 0$: the arms of the "jack" flare out, forming six "trumpets".

For a concentration of 10^{20} electrons/cm^3 ($n_{\text{cell}} = 6 \times 10^{-3}$ electrons/cell) we find that $\Delta_F = 5.6 \times 10^{-3}\,\text{eV}$.

For E near E_t we may write the energy of the $\alpha\beta$-type π^* band as

$$E^{\pi^*}_{\alpha\beta}(\vec{k}) - E_t \approx \frac{4(pd\pi)^2}{E^0_g}a^2[(k_\alpha)^2 + (k_\beta)^2] \quad (\alpha\beta = xy, xz, \text{or } yz) \tag{6.72}$$

From (6.72) it follows that for the $\pi^*(xy)$ band the surface of constant energy in $\vec{k}$-space is that of a circular cylinder with its axis along k_z. Similarly, for $\pi^*(xz)$ and $\pi^*(yz)$ the surfaces of constant energy are those of circular cylinders oriented along the k_y and k_x axes, respectively. The three cylinders form an object that resembles a "jack" as shown in Fig. 6.8(a). The shape of the Fermi surface for a free-electron model and for the energy bands of many materials is the surface of a sphere or spheroid in $\vec{k}$-space. Clearly, for the cubic perovskite the Fermi surface is quite different.

The transport properties of the electrons (holes) in a crystal can often be expressed in terms of the effective mass of the carriers. It can be shown that in an external electric field an electron acquires an acceleration, $\vec{a}$ given by

$$\vec{a} = \frac{1}{\hbar^2}\frac{d^2E(\vec{k})}{dk^2}\,q\vec{\mathcal{E}}, \tag{6.73}$$

where $\hbar$ is Planck's constant divided by 2π, q is the electron charge, and $\mathcal{E}$ is the external electric field strength. The effective mass, m^*, is defined so that a Newton-like equation, $(1/m^*)\vec{F} = \vec{a}$, applies to the motion of an electron in a solid. For electronic bands with the energy dependent only on the square of the magnitude of the wavevector, $E(\vec{k}) = E(k^2)$, the surfaces of constant energy are spherical in $\vec{k}$-space and one needs only a single number, m^*, given by

$$\frac{1}{m^*} = \frac{1}{\hbar^2}\frac{\partial^2 E(\vec{k})}{\partial k^2}.$$

For more complex energy bands a tensor is required to describe the electron's dynamics in a magnetic or electric field. The inverse mass tensor is defined so that

$$\mathbf{a}_\alpha = \sum \left[\frac{1}{\mathbf{m}^*}\right]_{\alpha\beta}\mathbf{F}_\beta,$$

where

$$\left[\frac{1}{\mathbf{m}^*}\right]_{\alpha\beta} = \frac{1}{\hbar^2}\frac{\partial^2 E(\vec{k})}{\partial k_\alpha \partial k_\beta}. \tag{6.74}$$

Using (6.74) we find that for an n-type perovskite,

$$\left[\frac{1}{\mathbf{m}^*}\right]_{\alpha\alpha} = \left[\frac{1}{\mathbf{m}^*}\right]_{\beta\beta} = \frac{8(pd\pi)^2 a^2}{\hbar^2 E_g^0} \quad \text{for } \pi^*(\alpha\beta) \text{ band,} \tag{6.75}$$

$$\left[\frac{1}{\mathbf{m}^*}\right]_{\gamma\gamma} = \left[\frac{1}{\mathbf{m}^*}\right]_{\alpha\beta} = \left[\frac{1}{\mathbf{m}^*}\right]_{\alpha\gamma} = 0 \quad \text{for } \pi^*(\alpha\beta) \text{ band,} \tag{6.76}$$

where $E_g^0 = (E_t - E_\perp)$ is the energy gap at Γ and "a" is half the lattice constant. With $E_g^0 = 3.2\,\text{eV}$, $(pd\pi) = 1.2\,\text{eV}$, and $a = 1.95\,\text{Å}$, numbers appropriate for SrTiO$_3$, we find $m^* = 0.56\,m_0$, where m_0 is the free-electron mass, 9.11×10^{-28} g. This result means, for example, that an electron in the bottom of the $\pi^*(\alpha\beta)$ band moves under the influence of an electric field oriented along the α- or β-axis as if it had only 56% of its free-electron mass. Conversely, in a field oriented along the γ-direction the electron behaves as if it had an infinite mass.

Of course, there are three symmetry-equivalent $\pi^*(\alpha\beta)$ bands so the effective mass is isotropic for cubic perovskites. It should also be mentioned that different average effective masses are employed in the analysis of different types of experiments. In the case of conductivity, for example, the quantity $\langle m^*\rangle_{\text{cond}}$ is employed, where

$$\left\langle\frac{1}{m^*}\right\rangle_{\text{cond}} = \frac{1}{3}\left(\frac{1}{m^*_{\alpha\alpha}} + \frac{1}{m^*_{\beta\beta}} + \frac{1}{m^*_{\gamma\gamma}}\right). \tag{6.77}$$

For the example above, this would yield $\langle m^*\rangle_{\text{cond}} = \frac{3}{2}m^*_{\alpha\alpha}$.

As can be seen from (6.75), the effective mass increases linearly with increasing energy gap, E_g^0, but decreases quadratically with $(pd\pi)$. Therefore the result is strongly dependent on the LCAO parameters employed. This is not true for all quantities. For example, $\Delta_F/[1/\mathbf{m}^*]_{\alpha\alpha} = \Delta_F m_{\alpha\alpha}^* = \pi\hbar^2 n_{\text{cell}}/(12a^2)$ involves only physical constants and the electron concentration and is therefore independent of the LCAO parameters, $(pd\pi)$, and E_g^0.

The interesting shape of the Fermi surface (Fig. 6.8) is a direct result of the two-dimensional character of the π and π^* bands. In a more elaborate theory (with more distant interactions, additional orbitals, or spin-orbit effects), the $\pi^*(\alpha\beta)$ and $\pi(\alpha\beta)$ bands would have at least a weak dependence on k_γ. The dispersion would be expected to take the form

$$E_{\pi^*}^{\alpha\beta}(\vec{k}) - E_t \approx \frac{4(pd\pi)^2}{E_g^0}a^2\left[(k_\alpha)^2 + (k_\beta)^2 + \delta(k_\gamma)^2\right] \tag{6.78}$$

where δ is a dimensionless number that is small compared to 1. The effect of the δ-term would be to convert the circular cylinders in Fig. 6.8 into thin, cigar-shaped rods (Fig. 6.8(b) for $\delta > 0$) or flare the cylinders into trumpet-shaped objects (Fig. 6.8(c) for $\delta < 0$). Using (6.75) we find for n-type $SrTiO_3$ that

$$\left[\frac{1}{\mathbf{m}^*}\right]_{\gamma\gamma} = \delta\left[\frac{1}{\mathbf{m}^*}\right]_{\alpha\alpha} = \delta\left(\frac{1}{0.56}\right) \quad \text{or} \quad m_{\gamma\gamma}^* = \frac{0.56}{\delta}. \tag{6.79}$$

Recent numerical calculations for $SrTiO_3$ reported by Marques et al. [8] give $m_{\alpha\alpha}^* = 0.408\,m_0$ (light mass) and $m_{\gamma\gamma}^* = 7.357\,m_0$ (heavy mass),[1] while older calculations by Kahn [9] give "bare"[2] values of $m_{\alpha\alpha}^* = 0.96\,m_0$ and $m_{\gamma\gamma}^* = 4.7\,m_0$ and Frederikse et al. [10] deduced values of $1.5\,m_0$ and $6.0\,m_0$ from magnetoresistance measurements. The scatter of these results is great, but they have in common the existence of a "light" and "heavy" effective mass which is predicted by the simple LCAO model to result from the two-dimensional character of the π^* bands. Referring to (6.79) it is seen that a small value of δ, for example $\delta = 0.1$, would bring the heavy mass into the range of the numbers quoted above.

It should be mentioned that $m_{\alpha\alpha}^*$ $(m_{\gamma\gamma}^*)$ is often referred to as the *transverse* (*longitudinal*) effective mass, m_t^* (m_l^*) because it applies to motion transverse (longitudinal) to the long axis of the Fermi surface.

Since the π valence bands are the mirror reflection of the π^* conduction bands, the results for the effective mass tensor of a hole at the top of the π valence band are the same as for the π^* band (but opposite in sign because the hole acts as a

---

[1] Marques et al. [8] reports both relativistic and non-relativistic values for the effective masses. The results quoted above are their non-relativistic masses. Relativistic effects have only a very small effect on the light mass, but greatly reduce the heavy mass according to these authors.

[2] The effective mass tensor components derived here are called the "bare" effective masses because they do not include the effects of phonons or polarons. Such lattice effects can in some cases "clothe" an electron, greatly increasing its effective mass. This is particularly important when the crystal has a "soft" phonon mode as is the case for crystals that become ferroelectric at lower temperatures.

positive charge). However, the non-bonding bands usually lie near to or above the top of the π valence bands. Consequently, the effective mass at the Fermi surface will likely be determined by the curvature of the non-bonding bands. Since the non-bonding bands have little curvature, heavy hole masses are expected.

(b) Metallic perovskites

Next, we consider a metallic perovskite with a significant density of electrons in the π^* conduction band. In this case the Fermi energy is well above the conduction-band edge and the two-dimensional cross-section of the Fermi surface will depart from circularity. Figure 6.9 shows the cross-section of the constant energy surfaces in $\vec{k}$-space for any one of the three π^* conduction bands. The contours would correspond to the Fermi surface cross-section as the number of electrons per band increases from

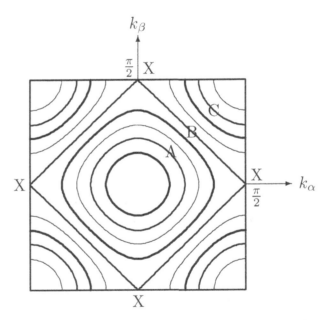

Figure 6.9. Cross-section of the constant energy surfaces for the $\pi^*(\alpha\beta)$ band for a cubic perovskite. The figure is a plot of the curves $S_\alpha^2 + S_\beta^2 = \Omega$ for various values of Ω. For small Ω (curves inside A) the cross-section is nearly circular. The square cross-section (B) occurs for $\Omega = 1$ and corresponds to the cross-section of the Fermi surface for a half-filled band. For Ω increasing beyond 1 the surfaces curve around the four X points in the Brillouin zone. For a nearly filled band, (C) the cross-section in the first Brillouin zone consists of four, quarter-round rods. In an extended Brillouin-zone view the Fermi surface would consist of four circular rods each centered on one of the X points and extending perpendicular to the plane of the diagram.

0 up to 1 (2 including spin). Using the definition (6.74) we find that the effective mass is given by

$$\left[\frac{1}{m^*}\right]_{\alpha\alpha} = \frac{4(pd\pi)^2 a^2}{\hbar^2(E - E_m)} \left[\cos(2k_\alpha a) - \frac{(pd\pi)^2 \sin^2(2k_\alpha a)}{(E - E_m)^2} \right], \tag{6.80}$$

for the $\pi(\alpha\beta)$ band. An equivalent expression holds for $[1/m^*]_{\beta\beta}$ with k_β replacing k_α in (6.80) and as before $[1/m^*]_{\gamma\gamma} = 0$. In contrast to the results of the previous section for low electron densities, it is clear that the effective mass components depend upon the magnitude of the wavevector and will therefore vary on the Fermi surface. In addition, it can be seen from (6.80) that the sign of $[1/m^*]_{\alpha\alpha}$ may change from positive to negative for a value of $k_\alpha a$ sufficiently large (but still $< \pi/2$). In such a case an electron would be accelerated by an electric field as if it had a positive charge (hole-like behavior). The effective mass will change sign at the point for which $[1/m^*]_{\alpha\alpha} = 0$ on the Fermi surface. The condition for this is

$$\cos^2(2k_\alpha a) + 2\beta \cos(2k_\alpha a) - 1 = 0, \text{ where } 2\beta = \frac{(E - E_m)^2}{(pd\pi)^2}. \tag{6.81}$$

The solution to (6.81) is

$$k_c a = \frac{1}{2} \arccos\left[-\beta + \sqrt{\beta^2 + 1}\,\right], \tag{6.82}$$

where k_c is the "critical" value of k_α for which $[1/m^*]_{\alpha\alpha} = 0$.

As a concrete example consider the compound $NaWO_3$, a metallic, cubic perovskite at room temperature. For this compound we have W^{6+}, Na^{1+}, and O^{2-} ions so that there is one electron in the π^* conduction bands (1/6 of an electron in each of the six degenerate π^* bands including the two spin states). To determine the Fermi energy, E_F, we make use of the DOS function and write,

$$\int_{-2}^{\varepsilon_F} \rho_{\pi^*}(\varepsilon) \, d\varepsilon = \frac{1}{6}, \tag{6.83}$$

where $\rho_{\pi^*}(\varepsilon)$ is given by equation (6.26) and ε is defined by (6.19). For the case of $NaWO_3$ the relevant parameters are, $2a = 3.87\,\text{Å}$, $|(pd\pi)| = 1.54\,\text{eV}$ and $E_g = 3.72\,\text{eV}$. Using these values we find with the help of (6.26) that $\varepsilon_F = -1.086$. Figure 6.10(a) shows the density of states and the number of electrons as a function of ε. The position of ε_F is shown in the figure. Using the definition in (6.19) we can also write,

$$E_F - E_m = \sqrt{(E_g/2)^2 + 2(pd\pi)^2(\varepsilon_F + 2)} = 2.79 \text{ eV}. \tag{6.84}$$

For the $\pi(xy)$ band, we define Ω_F to be the projection of the Fermi surface on

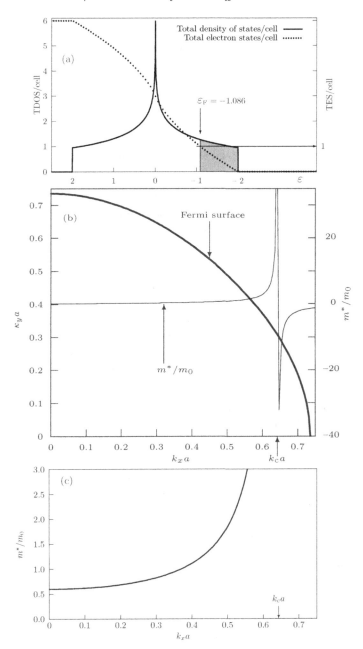

Figure 6.10. (a) Total density of states/cell for the π^* bands and the total number of electron states/cell as a function of ε. The arrow indicates the value of the Fermi energy, ε_F, for one electron occupying the bands. (b) The upper curve is the projection of the Fermi surface on the k_x–k_y plane, Ω_F. The lower curve is the effective mass, m^*, as a function of k_x for wavevectors on Ω_F. k_c is indicated. (c) Expanded plot of $m^* > 0$.

the k_x-k_y plane. For NaWO$_3$, ε_F is defined by the equation

$$S_x^2 + S_y^2 = \frac{(\varepsilon_F + 2)}{2} = 0.457, \qquad (6.85)$$

which gives an upper limit of 0.742 radians for $k_a a$ and (6.82) gives $k_c a = 0.643$ radians. The pairs of wavevectors (k_x and k_y) lying on Ω_F can easily be found. For example, one can pick a value for k_x then use (6.85) to calculate k_y with the requirement that k_y must be real. Once the values of the two-dimensional vectors, (k_x, k_y) lying on Ω_F are determined the effective mass can be calculated from (6.80) as a function of k_x or k_y.

Figure 6.10(b) shows the right-hand, upper quadrant of the Brillouin zone with the values of m^* (the inverse of $[1/m^*]_{xx}$) as a function of the value of $k_x a$ on Ω_F for the $\pi^*(xy)$ conduction band of NaWO$_3$.

The singularity in m^* occurs when $k_x a = k_c a = 0.643$. At that point $[1/m^*]_{xx} = 0$ so m^* diverges to plus infinity on one side and minus infinity on the other. Both m^* and Ω_F are independent of k_z and do not depend upon the sign of k_x or k_y, so the results for the entire k_x-k_y plane of the Brillouin zone can be obtained by reflecting the results in the quadrant through the k_x and k_y axes. From Fig. 6.10(b) it can be seen that the effective mass is positive for $k_x a$ less than $k_c a$ and negative in the range $k_c a < k_x a < 0.742$. The upper limit is the largest value of $k_x a$ which lies on Ω_F. The average effective mass, $\langle 1/m^* \rangle^{-1}$, averaged over the Fermi surface is 1.66 m_0 for NaWO$_3$.

It should be noted that as the number of electrons in the π^* bands is decreased the critical value of $k_x a$ given by (6.82) will eventually be larger than the maximum value of $k_x a$ on Ω_F. When this happens, the electron effective mass will be positive everywhere on Ω_F.

(c) Electronic properties of Na$_x$ WO$_3$

In this section we discuss some of the electronic properties of the sodium–tungsten bronze alloys, Na$_x$WO$_3$, as a function of x. In particular, we use our LCAO model to calculate the x dependence of the effective mass, the electronic specific heat, and the magnetic susceptibility.

The properties of Na$_x$WO$_3$ depend strongly upon the value of x. To a very good approximation each Na atom contributes one electron to the π^* conduction bands. Thus, the conduction-band electron concentration is equal to x. For $x < 0.24$, Na$_x$WO$_3$ is believed to be a Mott-type insulator [11–13]. For $0.24 < x < 0.49$, it is a superconductor with a tetragonal structure [14–16]. In the range $0.5 \leq x \leq 1$,

$Na_x WO_3$ is metallic and has the aristotype, cubic structure [17, 18]. We confine our discussions to the latter range of x values where the material is cubic.

Specific heat

The specific heat due to the conduction electrons, C_{el}, is defined as the rate of change of the average total electronic energy with temperature per unit cell at constant volume:

$$C_{el} = \frac{d}{dT} \int E\, \rho(E)\, f(E,T)\, dE = \int E\, \rho(E)\, \frac{df(E,T)}{dT}\, dE \qquad (6.86)$$

where T is the temperature, E is the electronic energy, $\rho(E)$ is the electronic density of states, and $f(E,T)$ is the Fermi distribution function. Since $df(E,T)/dT$ is negligible except in a small range of thermal energies about the Fermi energy, E_F, the limits of the integral can be taken arbitrarily as E_1 and E_2 as long as they are far from the Fermi energy. In the small thermal range where df/dT is significant, the density of states is nearly constant and hence it may replaced by $\rho(E_F)$. It is convenient to measure all of the energies from the Fermi energy so that

$$C_{el} = k_B^2 \rho(0) \int_{-(E_1 - E_F)}^{(E_2 - E_F)} \frac{[(E - E_F)/k_B T]^2 \exp[(E - E_F)/k_B T] d[(E - E_F)/k_B T]}{[1 + \exp[(E - E_F)/k_B T]^2}$$

$$= k_B^2 T \rho(0) \int_{-L_1}^{L_2} \frac{\alpha^2 \exp(\alpha)\, d\alpha}{[1 + \exp(\alpha)]^2} \equiv \gamma T, \qquad (6.87)$$

$$\gamma = \frac{1}{3}\, \pi^2 k_B^2 \rho(0) = 2.36\, \rho(E_F) \qquad \text{in} \quad \text{mJ/mole K}^2. \qquad (6.88)$$

In (6.87) $\alpha = (E - E_F)/k_B T$, $\rho(0) = \rho(E_F)$, $L_1 = (E_1 - E_F)/k_B T$, and $L_2 = (E_2 - E_F)/k_B T$. The specific heat coefficient, γ, given in (6.88) is the well-known result for metals, but care must be used when employing it. The integrand in (6.87) vanishes as $\alpha^2 e^{-|\alpha|}$ when $\alpha \to \pm\infty$. Therefore, if L_1 and L_2 are sufficiently large the actual limits become unimportant provided there are no energy gaps near E_F; that is, provided $|E_F - E_{edge}| \gtrsim 10 k_B T$, where E_{edge} is the edge of an energy band beyond which there is an energy gap. At room temperature $k_B T \approx 0.026$ eV so that the criterion would require $|E_F - E_{edge}| \gtrsim 0.25$ eV. At higher temperatures the requirement is more demanding. In the previous section on doped insulators we found that $|E_F - E_{edge}| \approx 0.006$ eV and therefore (6.86) can not be used for the electronic specific heat of these materials.

Specific heat of $Na_x WO_3$ as a function of x

The Fermi energy of $Na_x WO_3$ lies in the π^* conduction bands and for $x \simeq 0.5$, $E_F - E_t \simeq 0.55$ eV so the above criterion, $|E_F - E_{edge}| = |E_F - E_t| \gtrsim 10 k_B T$ is satisfied for $T \leq 600$ K. The application of (6.88) requires $\rho(E_F)$ as a function of x. One approximation often made is the so-called "rigid-band model" (RBM). According to this model the band structure itself is independent of x. That is, the energy bands are "rigid" and as x increases the added electrons donated by the Na atoms simply occupy the fixed (in energy) π^* states. This is *not* a good approximation for $Na_x WO_3$ [19]. Optical experiments [20] show that the band gap at Γ for $x = 0.5$ is about $0.5\,\mathrm{eV}$ smaller than for $x = 1.0$. In addition, the lattice constant $2a$ increases by $0.04\,\text{Å}$ [18] over this same range.

To calculate γ we make use of the results for the density of states for the π^* bands,

$$\rho_\pi(\varepsilon) = \frac{6}{\pi^2}\, K\left(\sqrt{1 - (\varepsilon/2)^2}\right), \tag{6.89}$$

$$\text{number of conduction electrons} = x = \int_{-2}^{\varepsilon_F(x)} \rho_\pi(\varepsilon)\, d\varepsilon, \tag{6.90}$$

$$\rho_\pi(E_F) = \frac{\sqrt{(E_g/2)^2 + 2(pd\pi)^2(\varepsilon_F + 2)}}{(pd\pi)^2}\, \rho_\pi(\varepsilon_F). \tag{6.91}$$

Equation (6.89) gives the total DOS in terms of the dimensionless variable, ε. The prefactor 6 accounts for the three degenerate bands and two spin states. Equation (6.90) expresses the number of conduction electrons, x, in terms of the integral over the dimensionless density of states. This expression determines the dimensionless Fermi energy, $\varepsilon_F(x)$. The DOS, $\rho_\pi(E_F)$ (in real energy) is then obtained by use of (6.91). It should be noted that the result for $\rho_\pi(E_F)$ (and hence γ) depends only on two energies, E_g and $(pd\pi)$. Specific energies such as E_t, and $E_\perp$ are not required. For the specific heat coefficient in units of $mJ/(mole\ K^2)$ we have:

$$\gamma(x) = 2.36\, \frac{\sqrt{(E_g(x)/2)^2 + 2(pd\pi(x))^2(\varepsilon_F(x) + 2)}}{(pd\pi(x))^2}\, \rho_\pi(\varepsilon_F(x)), \tag{6.92}$$

The calculation can be simplified by making use of approximate formulae for the x dependence of the quantities. The following approximate relationships have been shown to agree well with a wide range of experiments [19]:

$$E_g(x) = 2.85 + 0.84\,x, \qquad\qquad \text{(in eV)} \tag{6.93a}$$

$$(pd\pi(x)) = 2.44 - 0.86\,x, \qquad\quad \text{(in eV)} \tag{6.93b}$$

$$a(x) = \frac{1}{2}(3.785 + 0.0818\,x), \qquad \text{(W–O distance in Å)} \tag{6.93c}$$

$$\varepsilon_F(x) = -2 + (2 - 1.086)\,x, \qquad \text{(in eV)} \tag{6.93d}$$

$$\rho_\pi(x) = \frac{6}{2\pi}(1 + 0.326\,x), \qquad\quad \text{(total density of states/cell).} \tag{6.93e}$$

The results obtained from equations (6.92) and (6.93) for $\gamma(x)$ are shown in Fig. 6.11 where they are compared with experimental data [21–23]. The dashed lines

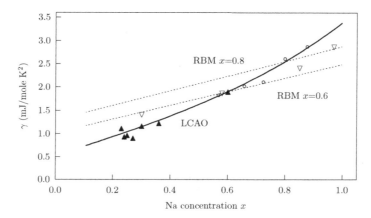

Figure 6.11. Electronic specific heat coefficient, γ, as a function of x for Na_xWO_3. The solid line gives the result of (6.92). The dashed lines show the results for the rigid-band model fits at $x = 0.6$ and 0.8. The experimental data are from Höchst *et al.* [21] ($\triangledown$), Vest *et al.* [22] ($\circ$), and Zumsteg [23] ($\blacktriangle$).

marked 'RBM' result from applying the rigid-band approximation [21] where the band structure is assumed to be independent of x. The x-dependent model clearly fits the experimental data better than the RBM.

Magnetic susceptibility and effective mass of Na_xWO_3 as functions of x

The magnetic susceptibility consists of a paramagnetic contribution from the electron spins (Pauli paramagnetic term) and a diamagnetic contribution due to the orbital motion of the electrons. An approximate expression that combines these two terms (in units of emu/mole), is

$$\chi = 4.0424 \times 10^{-6} \, a^2 \left(1 - \frac{1}{3}\left\langle \frac{m_0}{m^*} \right\rangle^2\right) N_\pi(E_F), \tag{6.94}$$

where $N_\pi(E_F)$ is the total density of states at E_F and m_0 is the electron rest mass. To calculate χ the inverse effective mass averaged over the Fermi surface, $\langle m_0/m^* \rangle$, is required. This quantity can be obtained from the expression given for the inverse of the effective mass in (6.80). The averages, $\langle\cos(2k_\alpha a)\rangle$, and $\langle\sin^2(2k_\alpha a)\rangle$ are given by line integrals around the curve that is the projection of the Fermi surface on the $\alpha\beta$-plane:

$$\langle\cos(2k_\alpha a)\rangle = \frac{1}{\ell} \int_0^\ell \cos[2k_\alpha(\ell)a] \, d\ell \,, \tag{6.95}$$

$$\langle\sin^2(2k_\alpha a)\rangle = \frac{1}{\ell} \int_0^\ell \sin^2[2k_\alpha(\ell)a] \, d\ell \,, \tag{6.96}$$

where ℓ is the length of the Fermi-surface curve in the $\alpha\beta$-plane. The average, $\langle\cos(2k_\alpha a)\rangle$, can be obtained by symmetry arguments. The Fermi surface for our LCAO energy bands is defined by the equation, $S_\alpha^2 + S_\beta^2 = \Omega_F(\varepsilon_F) = \frac{1}{2}(\varepsilon_F + 2)$, a constant for $k_\alpha a$ and $k_\beta a$ on the Fermi surface. Since $\cos(2k_\alpha a) = 1 - 2\sin^2(k_\alpha a)$ we have that

$$\langle\cos(2k_\alpha a)\rangle = 1 - \langle 2\sin^2(k_\alpha a)\rangle. \tag{6.97}$$

Whatever shape the Fermi-surface curve has on the $\alpha\beta$-plane it must be symmetric in the variables along α and β because these "directions" are physically equivalent and indistinguishable. Thus, the average $\langle\sin^2(k_\alpha a)\rangle$ must be equal to the average $\langle\sin^2(k_\beta a)\rangle$. That being the case, (6.97) may be written as

$$\langle\cos(2k_\alpha a)\rangle = 1 - \langle\sin^2(k_\alpha a) + \sin^2(k_\beta a)\rangle = 1 - \Omega_F(\varepsilon_F) = -\frac{1}{2}\,\varepsilon_F. \tag{6.98}$$

The other average, $\langle\sin^2(2k_\alpha a)\rangle$, can not be obtained analytically, but a simple interpolation formula that errors by less than 0.3% over the cubic range is given by

$$\langle\sin^2(2k_\alpha a)\rangle \approx \Omega_F(x)[2 - 1.49\,\Omega_F(x)]. \tag{6.99}$$

Using (6.99), (6.80), and (6.84) yields

$$\left\langle\frac{m_0}{m^*(x)}\right\rangle = \frac{4m_0(pd\pi(x))^2 a(x)^2}{\hbar^2[(E_g(x)/2)^2 + 4(pd\pi(x))^2\Omega_F(x)]^{1/2}}$$

$$\times\left\{[1 - \Omega_F(x)] - \frac{(pd\pi(x))^2\Omega_F(x)[2 - 1.49\,\Omega_F(x)]}{[(E_g(x)/2)^2 + 4(pd\pi(x))^2\Omega_F(x)]}\right\}. \tag{6.100}$$

The x-dependent quantities in (6.100) are given by (6.93) and (6.99). The results for $\langle m^*(x)/m_0\rangle$, shown in Fig. 6.12, agree well with the data of Camagni *et al.* [24] but lie well above the data of Owen *et al.* [20].

Use of equation (6.100) in (6.94) yields an expression for the magnetic susceptibility as a function of x:

$$\chi(x) = 4.0424 \times 10^{-6}\,a(x)^2\left(1 - \frac{1}{3}\left\langle\frac{m_0}{m^*(x)}\right\rangle^2\right)N_\pi(x), \tag{6.101a}$$

$$N_\pi(x) = \frac{(E_F - E_m)}{(pd\pi(x))^2}\,\rho_\pi(\varepsilon_F(x)) \tag{6.101b}$$

$$= \frac{\sqrt{(E_g(x)/2)^2 + 2(pd\pi(x))^2(\varepsilon_F(x) + 2)}}{(pd\pi(x))^2}\,\rho_\pi(x) \tag{6.101c}$$

where the x dependence of the parameters on the right-hand side of (6.102) is given by equations (6.93a-e). The results of (6.101) are shown in Fig. 6.13 and compared

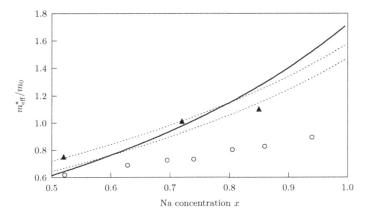

Figure 6.12. Effective mass averaged over the Fermi surface as a function of x for $Na_x WO_3$. The solid curve is obtained from the results of equation (6.100). The Dashed curves show the RBM results fixed at $x=0.6$ and 0.8. The experimental data are from Camagni *et al.* [24] (▲) and Owen *et al.* [20] (○).

with experimental data. The theoretical curve is somewhat higher than the experimental data but follows closely the general shape. Also shown in Fig. 6.13 is the susceptibility calculated from the effective mass data of Owen *et al.* [20]. The data of Greiner *et al.* lie about equally between the LCAO model results and those from Owen *et al.* The above results show that the simple empirical, nearest neighbor, LCAO model is able to predict reasonable values for the electronic properties of the cubic perovskites including the effective mass, specific heat, and magnetic suscepti-

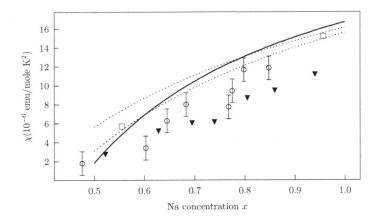

Figure 6.13. Contribution of the conduction electrons to the magnetic susceptibility, χ, as a function of x for $Na_x WO_3$. The solid curve gives the result of (6.101). The dashed curve is the result obtained using the data for the effective mass from Owen *et al.* [20] (▼). The other data are from Kupka *et al.* [25] (□) and Greiner *et al.* [26] (○).

bility. In addition, the model shows that these electronic properties are principally determined by two parameters, the $(pd\pi)$ integral and the band gap E_g. The density of states, $\rho(\varepsilon)$, is a dimensionless function determined entirely by symmetry. It is a universal function in that it applies to all cubic perovskites and is independent of the values of the empirical parameters.

References

[1] I. S. Gradshteyn and I. M. Ryzhik, *Table of Integrals, Series, and Products*, 4th edn (New York, Academic Press, 1965).

[2] T. Wolfram and Ş. Ellialtıoğlu, *Phys. Rev.* B **25**, 2697 (1982).

[3] L. van Hove, *Phys. Rev.* **89**, 1189 (1953); J. C. Phillips, *Phys. Rev.* **104**, 1263 (1956).

[4] Ş. Ellialtıoğlu and T. Wolfram, *Phys. Rev.* B **15**, 5909 (1977).

[5] L. F. Mattheiss, *Phys. Rev.* B **6**, 4718 (1972).

[6] L. Kopp, B. N. Harmon, and S. H. Liu, *Solid State Commun.* **22**, 677 (1977).

[7] L. F. Mattheiss, *Phys. Rev.* **181**, 987 (1969).

[8] M. Marques, L. K. Teles, V. Anjos, L. M. R. Scolfaro, J. R. Leite, V. N. Freire, G. A. Farias, and E. F. da Silva, Jr., *Appl. Phys. Lett.* **82**, 3074 (2003).

[9] A. H. Kahn, *Phys. Rev.* **172**, 813 (1968).

[10] H. P. R. Frederikse, W. R. Hosler, W. R. Thurber, J. Babiskin, and P. G. Siebenmann, *Phys. Rev.* **159**, 775 (1967); H. P. R. Frederikse, W. R. Hosler, and W. R. Thurber, *Phys. Rev.* **143**, 648 (1966); H. P. R. Frederikse and G. A. Candela, *Phys. Rev.* **147**, 583 (1966).

[11] J. B. Goodenough, *Prog. Solid State Chem.* **5**, 195 (1971).

[12] A. Ferreti, D. B. Rogers, and J. B. Goodenough, *J. Phys. Chem. Solids* **26**, 2007 (1965).

[13] N. F. Mott, *Philos. Mag.* **35**, 111 (1977).

[14] C. J. Raub, A. R. Sweedler, M. A. Jensen, S. Broadston, and B. T. Matthias, *Phys. Rev. Lett.* **13**, 746 (1964).

[15] H. R. Shanks, *Solid State Commun.* **15**, 753 (1974).

[16] K. L. Ngai and R. Silberglitt, *Phys. Rev.* B **13**, 1032 (1976).

[17] G. Hagg, *Z. Phys. Chem. Abt.* B **29**, 192 (1935).

[18] B. W. Brown and E. Banks, *J. Am. Chem. Soc.* **76**, 963 (1954).

[19] T. Wolfram and L. Sutcu, *Phys. Rev.* B **31**, 7680, (1985).

[20] J. F. Owen, K. J. Teegarden, and H. R. Shanks, *Phys. Rev.* B **18**, 3827 (1978).

[21] H. Höchst, R. D. Bringans, and H. R. Shanks, *Phys. Rev.* B **26**, 1702 (1982).

[22] R. W. Vest, M. Griffel, and J. F. Smith, *J. Chem. Phys.* **28**, 293 (1958).

[23] C. Zumsteg, *Phys. Rev.* B **14**, 1401 (1976).

[24] P. Camagni, A. Manara, G. Campagnoli, A. Gustinetti, and A. Stella, *Phys. Rev.* B **15**, 4623 (1977).

[25] F. Kupka and M. J. Sienko, *J. Chem. Phys.* **18**, 1296 (1950).

[26] J. D. Greiner, H. R. Shanks, and D. C. Wallace, *J. Chem. Phys.* **36**, 772 (1962).

Problems for Chapter 6

1. Find expressions for the DOS, $\rho(E)$, for the following energy bands whose dispersion is given by:

   (a) $E = E_0 + \sqrt{E_1^2 + E_2^2 S_x^2}$, where $S_x = \sin(k_x a)$, and $-\frac{\pi}{2} < k_x a < \frac{\pi}{2}$.

   (b) $E = E_0 + E_1 S_x$.

   (Hint: Make use of the expression $\rho(E) = \rho(f(E))|df(E)/dE|$.)

   (c) What is the energy dependence of $\rho(E)$ for
   $$E = E_0 + E_1[\sin^2(k_x a) + \sin^2(k_y a) + \sin^2(k_z a)]$$
   when E is near E_0?

2. Discuss the nature of the singularities (if any) of the DOS for (a), (b), and (c) in Problem 1.

3. A saddle point is said to exist at E_0 if the dispersion takes the form $E = E_0 + \alpha k_x^2 - \beta k_y^2$ near E_0, where α and β are real positive numbers. Show that this form leads to a logarithmic singularity at as $E \to E_0$.

4. Show that the π and π^* energy bands have saddle points at the X points in the Brillouin zone. (For simplicity ignore the oxygen–oxygen interactions.)

5. Show that for $\Omega = S_x^2 + S_y^2 \le 1$, the minimum value of the dimensionless Fermi energy, ϵ_F, for which $1/m^*$ has a zero, is determined by the condition $-\beta + \sqrt{\beta^2 + 1} = -(\epsilon_F + 1)$, where $\beta = [(w_g/2)^2 + 2(\epsilon_F + 2)]/2$, and $w_g = E_g/(pd\pi)$.

6. For $Na_x WO_3$ the total number of electrons in the π^* bands, n, is described accurately by the expression $n \approx (\epsilon_F + 2)/(1.828)$ for $0 < n < 1$. Use this result and the result of Problem 5 find an expression for the *minimum* total number of electrons per unit cell, n_{min}, for which $1/m^*$ has a zero. Using $E_g = 3.72 \, \text{eV}$ and $(pd\pi) = 1.54 \, \text{eV}$ find the value of n_{min}.

7

Optical properties of the d-band perovskites

Light interacts with the electrons of a solid through the electromagnetic field associated with the light wave. The electric field exerts an oscillating force on the electrons and ions which produces electronic transitions and other excitations in the solid.

There are several different types of optical adsorption mechanisms for ionic solids such as the perovskites. In the infrared region the electromagnetic field of the photons is strongly coupled to the polarization field of the vibrating ions and "optical" phonons can be created. If the solid is magnetic then adsorption of light can occur due to the excitation of spin waves or magnons. absorption by free electrons (or holes) is also important in the infrared optical region for metallic or semiconducting perovskites. Another important source of absorption is the excitation of plasmons. In doped semiconducting perovskites, plasmon absorption may occur in the infrared region while for metallic materials it is in the visible to ultraviolet region.

Photons with energy greater than the electronic band gap between the highest occupied and the lowest unoccupied bands can cause *interband* transitions. An interband transition involves the excitation of an electron from a filled valence band to an unoccupied state in another band. For an insulating material interband transitions can occur for photon energy $\hbar\omega > E_g$, where E_g is the fundamental band gap. The optical properties of insulating perovskites in the visible and ultraviolet regions are mainly determined by such interband transitions. This chapter deals principally with the nature of interband transitions in the insulating perovskites.

There are many other mechanisms of importance to the optical properties such as absorption by excitons and indirect interband transitions, which involve an electronic transition accompanied by the simultaneous creation or absorption of a phonon or magnon. These topics will not be discussed.

The optical response of a solid to higher-energy photons ($\hbar\omega$ of the order of

10^3 eV) such as are employed in x-ray photoelectron spectroscopy (XPS) will be discussed in Chapter 8.

7.1 Review of semiclassical theory

In this section we briefly review the semiclassical theory of the optical properties of solids. More detailed discussions may be found in the references given at the end of this chapter.

In the presence of an electromagnetic field the kinetic energy operator for an electron, $p^2/2m$, must be replaced by the new operator

$$\frac{1}{2m}\left[\vec{p} - \frac{e}{c}\vec{A}(\vec{r},t)\right]^2 , \tag{7.1}$$

where $\vec{p} = -i\hbar\vec{\nabla}$, $\vec{A}(\vec{r},t)$ is the vector potential of the electromagnetic field, e is the magnitude of the electron charge, and c is the velocity of light. The electromagnetic field may be described in the Coulomb gauge where $\vec{\nabla} \cdot \vec{A} = 0$.

The vector potential is usually a small perturbation on the electrons of a solid so that the term in $\vec{A}(\vec{r},t)^2$ in (7.1) may be neglected and the term linear in $\vec{A}(\vec{r},t)$ may be treated by means of first-order time-dependent perturbation theory.

The one-electron Hamiltonian for the solid is written as

$$H = H^0 + H'(\vec{r},t) \tag{7.2}$$

$$H^0 = \frac{p^2}{2m} + V(\vec{r})$$

$$H'(\vec{r},t) = \left(\frac{e}{mc}\right)\vec{A}(\vec{r},t) \cdot \vec{p}.$$

The vector potential for a field of angular frequency ω is

$$\vec{A}(\vec{r},t) = A\,\vec{a}_0\,e^{[i(\vec{q}\cdot\vec{r}-\omega t)]} + \text{c.c.} \tag{7.3}$$

where A is a scalar amplitude, $\vec{a}_0$ is a unit vector (the polarization vector) perpendicular to the propagation vector $\vec{q}$, and "c.c." means the complex conjugate. The electric field strength, $\vec{\mathcal{E}}(\vec{r},t)$ is related to $\vec{A}(\vec{r},t)$ by the equation

$$\vec{\mathcal{E}}(\vec{r},t) = -\frac{1}{c}\frac{d}{dt}\vec{A}(\vec{r},t)$$

$$= \left(\frac{i\omega}{c}\right)A\,\vec{a}_0\,e^{i(\vec{q}\cdot\vec{r}-\omega t)} + \text{c.c.} \tag{7.4}$$

It is useful to write $\vec{\mathcal{E}}(\vec{r},t) = \vec{\mathcal{E}}_1(\vec{r},t) + \vec{\mathcal{E}}_1(\vec{r},t)^*$. The electric displacement field

$\vec{\mathcal{D}}(\vec{r}, t)$ can then be written as

$$\vec{\mathcal{D}}(\vec{r}, t) = \varepsilon(\vec{q}, \omega)\, \vec{\mathcal{E}}_1(\vec{r}, t) + \varepsilon(\vec{q}, \omega)^*\, \vec{\mathcal{E}}_1(\vec{r}, t)^* \tag{7.5}$$

where $\varepsilon(\vec{q}, \omega)$, the frequency- and wavevector-dependent dielectric function, is a 3×3 tensor whose elements are complex:

$$\varepsilon(\vec{q}, \omega)_{\alpha\beta} = \varepsilon_1(\vec{q}, \omega)_{\alpha\beta} + i\, \varepsilon_2(\vec{q}, \omega)_{\alpha\beta}. \tag{7.6}$$

Since we are concerned here with cubic crystals we may assume that ε is a multiple of the unit tensor and consequently it may be treated as a scalar quantity.

According to electromagnetic theory, the average rate of loss of energy density (energy/volume/s) from an electromagnetic field in a medium with a dielectric function $\varepsilon(\vec{q}, \omega)$ is

$$\frac{1}{4\pi} \left\langle \vec{\mathcal{E}} \cdot \frac{d\vec{\mathcal{D}}}{dt} \right\rangle, \tag{7.7}$$

where $\langle \cdots \rangle$ means the time average over a period of oscillation, $T = 2\pi/\omega$. Using (7.4) and (7.5) one finds the total energy loss per second

$$\frac{1}{4\pi} \left\langle \vec{\mathcal{E}} \cdot \frac{d\vec{\mathcal{D}}}{dt} \right\rangle = \frac{\omega}{2\pi} \int_0^{2\pi/\omega} \left(\vec{\mathcal{E}} \cdot \frac{d\vec{\mathcal{D}}}{dt} \right) dt$$

$$= \frac{1}{2\pi} \frac{|A|^2 \omega^3}{c^2}\, \varepsilon_2(\vec{q}, \omega). \tag{7.8}$$

Equation (7.8) provides a means of relating *macroscopic* electromagnetic theory to *microscopic* quantum theory. The approach is to find the rate of electronic transitions, dW/dt, due to the perturbation H' of (7.2). Each transition corresponds to the absorption of a photon of energy $\hbar\omega$ and therefore to a loss rate, $(dW/dt)\hbar\omega$, of energy from the electromagnetic field. By equating this loss to that given by (7.8) one can find an expression for ε_2 in terms of the quantum states of the solid.

According to first-order time-dependent perturbation theory, an oscillating perturbation of the form $H'(\vec{r})e^{-i\omega t}$, where $H'(\vec{r})$ is a constant or operator function of $\vec{r}$, will cause transitions between the unperturbed states of the system at a rate (transitions per second) given by

$$\frac{dW_{if}(\hbar\omega)}{dt} = \frac{2\pi}{\hbar} |\langle f|H'(\vec{r})|i\rangle|^2\, \delta(E_f - E_i - \hbar\omega)\, \delta_{S_i, S_f}. \tag{7.9}$$

In (7.9), $|i\rangle$ and $\langle f|$ designate the initial and final states, respectively, E_i and E_f are the corresponding energies and S_i and S_f indicate the spin states. Equation (7.9) shows that the strength of the transition is proportional to the absolute square of the matrix element and the δ function ensures conservation of energy in the process.

Using (7.2), the rate of energy loss per unit volume is

$$\frac{1}{N(2a)^3} \hbar\omega \left(\frac{dW}{dt}\right) = \frac{1}{N(2a)^3} \sum_{if} \sum_{S_i S_f} \hbar\omega \left(\frac{dW_{if}}{dt}\right) f(E_i) \left[1 - f(E_f)\right]$$

$$= \frac{\hbar\omega}{N(2a)^3} \frac{4\pi}{\hbar} |A|^2 \left(\frac{e}{mc}\right)^2 \sum_{if} |\langle f| e^{i\vec{q}\cdot\vec{r}} \vec{a}_0 \cdot \vec{p}|i\rangle|^2$$

$$\times \delta(E_f - E_i - \hbar\omega) f(E_i) \left[1 - f(E_f)\right], \quad (7.10)$$

where $f(E)$ is the Fermi distribution function. The term $f(E_i)$ is the probability that the initial state was occupied and the term $[1 - f(E_f)]$, is the probability that the final state is empty.

Then equating (7.10) to the result of (7.8) and solving for $\varepsilon_2(\vec{q}, \omega)$ one has

$$\varepsilon_2(\vec{q}, \omega) = \frac{8\pi^2}{N(2a)^3} \left(\frac{e}{m\omega}\right)^2 \sum_{if} |\langle f| e^{i\vec{q}\cdot\vec{r}} \vec{a}_0 \cdot \vec{p}|i\rangle|^2$$

$$\times \delta(E_f - E_i - \hbar\omega) f(E_i) \left[1 - f(E_f)\right]. \quad (7.11)$$

Equation (7.11) is the fundamental relation between macroscopic electromagnetic theory and the energy band states of quantum theory. Once ε_2 is known, all of the optical constants can be derived from it.

The real part of the dielectric function ε_1 is calculated from ε_2 by means of Kramers–Kronig dispersion relation:

$$\varepsilon_1(\vec{q}, \omega) = 1 + \frac{2}{\pi} P \int_0^\infty \frac{\omega' \varepsilon_2(\vec{q}, \omega')}{\omega'^2 - \omega^2} d\omega', \quad (7.12)$$

where P indicates the principal value.

The complex index of refraction $N(\vec{q}, \omega)$ is related to $\varepsilon = \varepsilon_1 + i\varepsilon_2$ by

$$N^2(\vec{q}, \omega) = \varepsilon(\vec{q}, \omega) \quad (7.13)$$

$$N(\vec{q}, \omega) = n(\vec{q}, \omega) + i\kappa(\vec{q}, \omega) \quad (7.14)$$

where n and κ being the index of refraction and the extinction coefficient, respectively. The absorption coefficient, α, is given by

$$\alpha(\vec{q}, \omega) = \frac{(\omega/c)\,\varepsilon_2(\vec{q}, \omega)}{n(\vec{q}, \omega)}. \quad (7.15)$$

Finally, we note that for radiation incident normal to the solid surface the reflectance is

$$R(\vec{q}, \omega) = \left|\frac{\sqrt{\varepsilon(\vec{q}, \omega)} - 1}{\sqrt{\varepsilon(\vec{q}, \omega)} + 1}\right|^2. \quad (7.16)$$

7.2 Qualitative theory of $\varepsilon_2(\omega)$

In the preceding section we saw that the dielectric function depends upon the transition matrix elements M_{if} where

$$M_{if} = \langle f | e^{i\vec{q}\cdot\vec{r}} \, \vec{a}_0 \cdot \vec{p} | i \rangle. \tag{7.17}$$

The initial and final states are to be energy band states which are of the form

$$\psi_{\vec{k}\nu}(\vec{r}) = \frac{1}{\sqrt{N}} \sum_{mj\alpha} a_{j\alpha}(\vec{k}, \nu) \, e^{i\vec{k}\cdot\vec{R}_{mj}} \, \xi_\alpha(\vec{r} - \vec{R}_{mj}) \tag{7.18}$$

where $a_{j\alpha}(\vec{k}, \nu)$ are the eigenvector components of the band state with wavevector $\vec{k}$ and band index ν. The quantity $\vec{R}_{mj} = \vec{R}_m + \vec{\tau}_j$, where $\vec{R}_m$ is the position vector of the mth unit cell and $\vec{\tau}_j$ locates the jth atom relative to the origin of the mth unit cell. The function $\xi_\alpha(\vec{r} - \vec{R}_{mj})$ is a Löwdin orbital with symmetry index α, centered at $\vec{R}_{mj}$.

The matrix element $M_{if} \equiv M_{\vec{k}\nu, \vec{k}'\nu'}$ is

$$M_{\vec{k}\nu, \vec{k}'\nu'} = \int \psi_{\vec{k}'\nu'}(\vec{r})^* \, e^{i\vec{q}\cdot\vec{r}} \, (\vec{a}_0 \cdot \vec{p}) \, \psi_{\vec{k}\nu}(\vec{r}) \, d\vec{r}. \tag{7.19}$$

Using (7.18) it is easily shown that $M_{\vec{k}\nu, \vec{k}'\nu'}$ involves a factor

$$\sum_{\vec{R}} e^{i\frac{1}{2}[\vec{k} - \vec{k}' + \vec{q}]\cdot\vec{R}} = N \sum_{\vec{G}} \delta_{\vec{k} + \vec{q} - \vec{k}', \vec{G}} \tag{7.20}$$

where $\vec{G}$ is any reciprocal lattice vector including zero. Thus, $M_{\vec{k}\nu, \vec{k}'\nu'}$ vanishes unless $\vec{k}' = \vec{k} + \vec{q} \pm \vec{G}$. The wavevector of the light, $\vec{q}$, is in most cases small compared with any $\vec{G} \neq 0$. For example, the $|\vec{q}|$ for a 1-eV photon is about $5 \times 10^4 \, \text{cm}^{-1}$, while $|\vec{G}| = 2\pi/a \simeq 2 \times 10^8 \, \text{cm}^{-1}$. Thus $M_{\vec{k}\nu, \vec{k}'\nu'}$ will be small unless $\vec{k}' \cong \vec{k} \pm \vec{G}$. However, if both $\vec{k}'$ and $\vec{k}$ are confined to the interior of the first Brillouin zone, then their vector difference or sum can not be equal to a (non-zero) reciprocal lattice vector. Thus the only non-zero matrix elements are those for which $\vec{k}' \cong \vec{k}$ and $\vec{G} = 0$ in (7.20). A schematic of the situation is shown in Fig. 7.1. It is apparent that the *interband* transitions are essentially "vertical" on an energy band diagram such as that of Fig. 7.1. On the other hand, the *intraband* transition is nearly "horizontal". If $\hbar\omega = (E_{\vec{k}'\nu'} - E_{\vec{k}\nu})$ and is large compared with $(E_{\vec{k}'\nu} - E_{\vec{k}\nu})$ or $(E_{\vec{k}'\nu'} - E_{\vec{k}\nu'})$ then the intraband transition is far from "resonance" and can not occur. In the case of insulating perovskites the energy gap between the filled valence bands and the empty conduction bands is about $3 \, \text{eV}$. Interband transitions can occur then only for $\hbar\omega \gtrsim 3 \, \text{eV}$. Intraband transitions are not possible because there are no empty final states in the valence bands and no occupied initial states in the conduction bands.

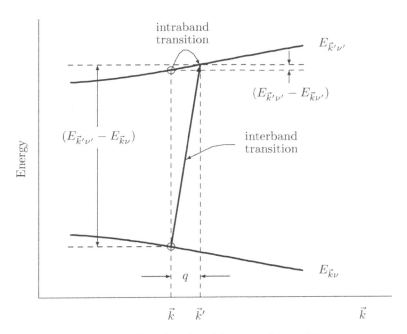

Figure 7.1. Interband and intraband transitions.

For metallic perovskites such as alkali tungsten bronzes or ReO_3 or for n-type semiconducting materials the conduction band is partially occupied. This allows the possibility of intraband transitions at low photon energies; namely, $\hbar\omega \approx |E_{\vec{k}+\vec{q},\nu} - E_{\vec{k},\nu}|$. Such intraband transitions lead to what is called "free-carrier" absorption.

In addition, in metals, collective excitations of the conduction electrons in the form of plasmons can also absorb energy. For $Na_x WO_3$ or ReO_3, the plasmon energy is about $2\,eV$ while for n-type $SrTiO_3$ with 10^{19} electrons/cm^3 it is only a few hundredths of an eV.

In view of the preceding discussion it is clear that for $\hbar\omega \gtrsim 1\,eV$ the optical properties of the insulating perovskites are determined principally by interband transitions. The same is true for the metallic systems except that plasmon absorption must also be considered. In either case, the magnitude of the photon wavevector may be neglected when $\hbar\omega \lesssim 100\,eV$ and the interband transitions are essentially "vertical". Consequently, the transition matrix elements that are required are

$$\langle \vec{k}, \nu' | \vec{a}_0 \cdot \vec{p} | \vec{k}, \nu \rangle \qquad \nu \neq \nu'. \qquad (7.21)$$

The expression for the dielectric function due to interband transition takes the form

$$\varepsilon_2(\omega) \equiv \varepsilon_2(0,\omega) = \frac{1}{\pi}\left(\frac{e}{m\omega}\right)^2\left(\frac{2\pi}{2a}\right)^3\frac{1}{N}\sum_{\vec{k}\nu\nu'}|\langle\vec{k}\nu'|\,\vec{a}_0\cdot\vec{p}\,|\vec{k}\nu\rangle|^2$$

$$\times\,\delta(E_{\vec{k}\nu'}-E_{\vec{k}\nu}-\hbar\omega)\,f(E_{\vec{k}\nu})\,[1-f(E_{\vec{k}\nu'})]. \qquad (7.22)$$

In the following subsection we shall show that the function, $\varepsilon_2(\omega)$, for the perovskites has pronounced structure which reflects the nature of the van Hove singularities in the electronic density of states.

(a) Joint density of states

Before launching into a detailed calculation of the matrix elements needed for $\varepsilon_2(\omega)$ we shall explore a simple, but frequently employed model for the purpose of demonstrating how the structure of the electronic density of states manifests itself in the optical properties. We begin by expressing the matrix elements of the momentum in terms of those of $\vec{r}$ by making use of the relation

$$\langle f|\vec{p}|i\rangle = \frac{im}{\hbar}\,(E_f - E_i)\,\langle f|\vec{r}|i\rangle. \qquad (7.23)$$

Then we write (7.22) in the form

$$\varepsilon_2(\omega) = \frac{1}{(\hbar\omega)^2}\left(\frac{2\pi}{2a}\right)^3\frac{1}{N}\sum_{\vec{k}\nu\nu'}|M_{\nu\nu'}(\vec{k})|^2(E_{\vec{k}\nu'}-E_{\vec{k}\nu})^2$$

$$\times\,\delta(E_{\vec{k}\nu'}-E_{\vec{k}\nu}-\hbar\omega)\,f(E_{\vec{k}\nu})[1-f(E_{\vec{k}\nu'})] \qquad (7.24)$$

with

$$|M_{\nu\nu'}(\vec{k})|^2 = \frac{e^2}{\pi}\,|\langle\vec{k},\nu'|\vec{a}_0\cdot\vec{r}|\vec{k},\nu\rangle|^2. \qquad (7.25)$$

In (7.24) we have used $E_f - E_i = E_{\vec{k}\nu'} - E_{\vec{k}\nu}$. Because of the delta function in (7.24) we may replace $(E_{\vec{k}\nu'} - E_{\vec{k}\nu})^2$ by $(\hbar\omega)^2$ and obtain

$$\varepsilon_2(\omega) = \left(\frac{2\pi}{2a}\right)^3\frac{1}{N}\sum_{\vec{k}\nu\nu'}|M_{\nu\nu'}(\vec{k})|^2\,\delta(E_{\vec{k}\nu'}-E_{\vec{k}\nu}-\hbar\omega)\,f(E_{\vec{k}\nu})\,[1-f(E_{\vec{k}\nu'})]. \qquad (7.26)$$

Now consider an insulating perovskite for which the valence bands are filled and the conduction bands are empty. In this case the only allowed transitions are between an initial state in one of the valence bands to one of the unoccupied conduction bands. With the convention that ν refers to a valence band and ν' to a conduction band, $f(E_{\vec{k}\nu}) = [1 - f(E_{\vec{k}\nu'})] = 1$ and

$$\varepsilon_2(\omega) = \left(\frac{2\pi}{2a}\right)^3\frac{1}{N}\sum_{\vec{k}\nu\nu'}|M_{\nu\nu'}(\vec{k})|^2\,\delta(E_{\vec{k}\nu'}-E_{\vec{k}\nu}-\hbar\omega). \qquad (7.27)$$

In order to proceed further we need to know the transition matrix elements. The $\vec{k}$ dependence of $M_{\nu\nu'}(\vec{k})$ is discussed later in this chapter. For our purpose in this section we shall simply replace the matrix element by its average value over the Brillouin zone, $\langle M_{\nu\nu'} \rangle$. With this approximation we have

$$\varepsilon_2(\omega) \simeq \left\{ \sum_{\nu\nu'} \langle |M_{\nu\nu'}|^2 \rangle \right\} \left(\frac{2\pi}{2a} \right)^3 \frac{1}{N} \sum_{\vec{k}} \delta(E_{\vec{k}\nu'} - E_{\vec{k}\nu} - \hbar\omega). \tag{7.28}$$

The quantity $\frac{1}{N} \sum_k \delta[\hbar\omega - (E_{\vec{k}\nu'} - E_{\vec{k}\nu})]$ is similar to the density of states function defined in the preceding chapter by (6.8). However, in this case we have two energies, $E_{\vec{k}\nu'}$ and $E_{\vec{k}\nu}$ rather than a single energy. The function is called the *joint density of states*, abbreviated by "JDOS" and denoted by $[J(\omega)]_{\nu\nu'}$. It specifies the number of pairs of valence- and conduction-band states with an energy difference in the range between $\hbar\omega$ and $\hbar\omega + d(\hbar\omega)$.

We now show that the JDOS has the same van Hove singularities that the DOS has and therefore that these structures are expected to be reflected in the optical properties of the cubic perovskites.

The JDOS is easily calculated for interband transitions from non-bonding bands to conduction bands. The non-bonding state energy is constant (independent of $\vec{k}$) for the model developed in Chapter 4. The initial and final state energies to be considered are

$$\begin{aligned}
E_{\vec{k}\nu} &= E_{\parallel} & \text{for } \nu = \sigma^0 & \quad \text{(sigma non–bonding bands),} \\
&= E_{\perp} & \text{for } \nu = \pi^0 & \quad \text{(pi non–bonding bands),} \\
E_{\vec{k}\nu'} &= E_{\vec{k}\sigma^*} & \text{for } \nu' = \sigma^* & \quad \text{(sigma conduction bands),} \\
&= E_{\vec{k}\pi^*} & \text{for } \nu' = \pi^* & \quad \text{(pi conduction bands).}
\end{aligned} \tag{7.29}$$

The JDOS for such transitions is

$$[J(\omega)]_{\alpha^0\beta^*} = \frac{1}{N} \sum_{\vec{k}} \delta[E_{\vec{k}\nu'} - E_{\vec{k}\nu} - \hbar\omega] = \frac{1}{N} \sum_{\vec{k}} \delta[\hbar\omega - (E_{\vec{k}\nu'} - E_{\vec{k}\nu})]$$

$$= \rho_{\beta^*}(\hbar\omega + E_{\alpha^0}) = \rho_{\beta}(\hbar\omega + E_{\alpha^0}) \tag{7.30}$$

where $\alpha^0 = \pi^0$ or σ^0 and $\beta^* = \pi^*$ or σ^*. In particular, we have the JDOS for four interband transitions:

$$\begin{aligned}
[J(\omega)]_{\sigma^0\sigma^*} &= \rho_\sigma(\hbar\omega + E_{\parallel}) & \text{(threshold for } \sigma^0 \to \sigma^*\text{: } \hbar\omega = E_e - E_{\parallel}), \\
[J(\omega)]_{\sigma^0\pi^*} &= \rho_\pi(\hbar\omega + E_{\parallel}) & \text{(threshold for } \sigma^0 \to \pi^*\text{: } \hbar\omega = E_t - E_{\parallel}), \\
[J(\omega)]_{\pi^0\sigma^*} &= \rho_\sigma(\hbar\omega + E_{\perp}) & \text{(threshold for } \pi^0 \to \sigma^*\text{: } \hbar\omega = E_e - E_{\perp}), \\
[J(\omega)]_{\pi^0\pi^*} &= \rho_\pi(\hbar\omega + E_{\perp}) & \text{(threshold for } \pi^0 \to \pi^*\text{: } \hbar\omega = E_t - E_{\perp}).
\end{aligned}$$

$$\tag{7.31}$$

The functions ρ_π and ρ_σ are the DOS functions given in Chapter 6 by (6.28) and (6.62), respectively.

Equation (7.31) shows that in the case of transitions from the non-bonding bands, the JDOS degenerates into a DOS function and therefore possesses the same critical structure as the DOS. This result is not limited to transitions from the narrow, non-bonding bands, but is also true for transitions from the σ or π valence bands as well. For example, the JDOS for interband transitions from the $\pi(\alpha\beta)$ valence band to a $\pi^*(\alpha\beta)$ conduction band $(\alpha\beta = xy, xz, \text{ or } yz)$ is

$$[J^{\alpha\beta}(\omega)]_{\pi\pi^*} = \frac{1}{N}\sum_{\vec{k}} \delta[\hbar\omega - (E_{\vec{k}\pi^*}^{\alpha\beta} - E_{\vec{k}\pi}^{\alpha\beta})]$$

$$= -\frac{1}{\pi}\Im m \frac{1}{N}\sum_{\vec{k}} \frac{1}{\hbar\omega - (E_{\vec{k}\pi^*}^{\alpha\beta} - E_{\vec{k}\pi}^{\alpha\beta}) + i0^+}. \qquad (7.32)$$

The calculation is greatly simplified if we add a null term to (7.32); namely,

$$-\frac{1}{\pi}\Im m \frac{1}{N}\sum_{\vec{k}} \frac{1}{\hbar\omega + (E_{\vec{k}\pi^*}^{\alpha\beta} - E_{\vec{k}\pi}^{\alpha\beta}) + i0^+}. \qquad (7.33)$$

The term of (7.33) is zero for $\hbar\omega > 0$, because the denominator can not vanish. Now, combining (7.32) and (7.33) and using

$$(E_{\vec{k}\pi^*}^{\alpha\beta} - E_{\vec{k}\pi}^{\alpha\beta})^2 = (E_t - E_\perp)^2 + 16(pd\pi)^2(S_\alpha^2 + S_\beta^2), \qquad (7.34)$$

we find

$$[J^{\alpha\beta}(\omega)]_{\pi\pi^*} = -\frac{1}{\pi}\Im m \frac{1}{N}\sum_{\vec{k}} \frac{2\hbar\omega}{(\hbar\omega)^2 - (E_t - E_\perp)^2 - 16(pd\pi)^2(S_\alpha^2 + S_\beta^2) + i0^+}$$

$$= -\frac{1}{\pi}\Im m \frac{(\frac{1}{2}\hbar\omega)}{2(pd\pi)^2} \frac{1}{N}\sum_{\vec{k}} \frac{1}{W + C_{2\alpha} + C_{2\beta} + i0^+} \qquad (7.35)$$

where

$$W \equiv \frac{(\frac{1}{2}\hbar\omega)^2 - [\frac{1}{2}(E_t - E_\perp)]^2}{2(pd\pi)^2} - 2. \qquad (7.36)$$

The sum in (7.35) converges to an integral for large N and is easily evaluated by using (6.20) and (6.26). The result is

$$[J^{\alpha\beta}(\omega)]_{\pi\pi^*} = \frac{(\frac{1}{2}\hbar\omega)}{2(pd\pi)^2} \frac{1}{\pi^2} K\left(\sqrt{1 - \left(\frac{W}{2}\right)^2}\right) \Theta\left(1 - \left(\frac{W}{2}\right)^2\right). \qquad (7.37)$$

It will be recalled from Chapter 6 that the complete elliptic integral $K(x)$ has a jump discontinuity of $\pi/2$ at $x = 0$ and a logarithmic infinity at $x = 1$. For the K function in (7.37) this means that the jump discontinuity occurs at $W/2 = \pm 1$ and the logarithmic singularity at $W/2 = 0$. For $W/2 = \pm 1$ we have

$$\hbar\omega = 2\sqrt{[(E_t - E_\perp)/2]^2 + 8(pd\pi)^2} \qquad \text{(for the plus sign),}$$
$$\hbar\omega = (E_t - E_\perp) \qquad \text{(for the minus sign).} \qquad (7.38)$$

The first energy corresponds to the energy separation of the π and π^* band at R and the second energy of (7.38) to the energy difference between the π and π^* band at Γ. That is, at the top and the bottom of the bands. The singularity at $W/2 = 0$ occurs for $\hbar\omega = 2\sqrt{[(E_t - E_\perp)/2]^2 + 4(pd\pi)^2}$. This corresponds to vertical transitions from $\pi(\alpha\beta)$ to the $\pi^*(\alpha\beta)$ band states whose wavevector is such that $S_\alpha^2 + S_\beta^2 = 1$. Wavevectors satisfying this condition arise from the surface of a rod of square cross-section oriented along k_γ with corners at the X symmetry points in the k_α–k_β plane. Since the $\pi(\alpha\beta)$ and $\pi^*(\alpha\beta)$ bands are mirror image of each other with respect to the mid-gap, the shape of the JDOS is similar to the DOS functions, but twice as wide. A similar result can be obtained for the JDOS corresponding to the $\sigma \rightarrow \sigma^*$ transition. Figure 7.2(a) shows the JDOS for all the possible 45 interband transitions and the total of all transitions is indicated by the dashed line. Figure 7.2(b) shows a comparison of the (total JDOS)/ω^2 with the experimental result of $\varepsilon_2(\omega)$ for SrTiO$_3$ [1].

JDOS functions can be derived for the many different types of interband transitions and each contribution to ε_2 reflects a convolution of the critical structure of the bands involved. However, it should be remembered that the approximation which leads to the JDOS result does not take into account the $\vec{k}$ dependence of the transition matrix elements. Imbedded in the transition matrix elements are selection rules and shapes that can possibly modulate and smooth out sharp structure that occurs in the JDOS. In the next section we develop a more detailed model that takes the $\vec{k}$ dependence into account.

(b) LCAO transition matrix elements

The optical response of a cubic crystal is isotropic and consequently we may choose a particular direction of the polarization of the electric field without loss of generality. We assume that the polarization of $\vec{A}$ and $\vec{\mathcal{E}}$ is along the x-axis:

$$\vec{A} = A\,\vec{e}_x.$$

The interband matrix elements of concern are then

$$\left\langle \vec{k}, \nu' \left| -i\hbar\frac{\partial}{\partial x} \right| \vec{k}, \nu \right\rangle = \sum_{jj'\alpha\alpha'} [a_{j'\alpha'}(\vec{k}, \nu')]^* \, a_{j\alpha}(\vec{k}, \nu) \sum_{\vec{R}_m, \vec{R}_{m'}} e^{i\vec{k}\cdot(\vec{R}_{mj} - \vec{R}_{m'j'})}$$
$$\times \int d\vec{r}\, \xi_{\alpha'}^*(\vec{r}) \left(-i\hbar\frac{\partial}{\partial x} \right) \xi_\alpha [\vec{r} - (\vec{R}_m - \vec{R}_{m'}) - (\vec{\tau}_j - \vec{\tau}_{j'})]. \qquad (7.39)$$

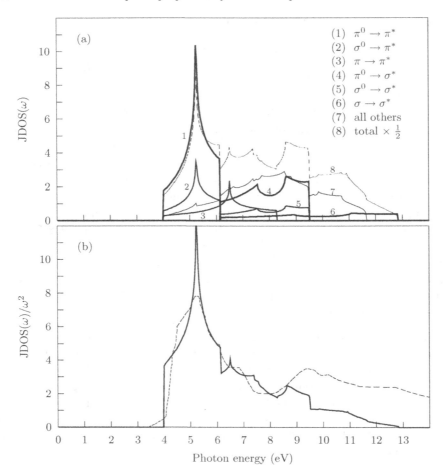

Figure 7.2. (a) Contributions to JDOS. (b) Comparison of $JDOS/\omega^2$ with the experimental result for $\varepsilon_2(\omega)$ (dashed line) for $SrTiO_3$ [1].

The integral on the right-hand side of (7.39) is the matrix element of the x-component of the momentum operator between Löwdin orbitals centered at the origin and at $(\vec{R}_{mj} - \vec{R}_{m'j'})$. The integral depends on the vector difference $\vec{R}_s = (\vec{R}_m - \vec{R}_{m'})$ but not on the individual unit-cell vectors, $\vec{R}_m$ and $\vec{R}_{m'}$. The symbol $\sum_{\vec{R}_s}$ denotes a sum over $\vec{R}_m - \vec{R}_{m'}$. It is convenient to define the matrix elements of the momentum operator between Löwdin orbitals by

$$P^x_{\alpha'j',\alpha j}(\vec{R}_s) \equiv \int d\vec{r}\, \xi^*_{\alpha'}(\vec{r})\left(-i\hbar\frac{\partial}{\partial x}\right)\xi_\alpha[\vec{r} - \vec{R}_s - (\vec{\tau}_j - \vec{\tau}_{j'})]. \qquad (7.40)$$

Then, the matrix element of (7.39) takes the form

$$\left\langle \vec{k}\nu' \right| -i\hbar\frac{\partial}{\partial x} \left| \vec{k}\nu \right\rangle = \sum_{jj'\alpha\alpha'} \left[a_{j'\alpha'}^{(\vec{k}\nu')}\right]^* a_{j\alpha}^{(\vec{k},\nu)} \sum_{\vec{R}_s} e^{i\vec{k}\cdot[\vec{R}_s - (\vec{\tau}_j - \vec{\tau}_{j'})]} P^x_{j'\alpha',j\alpha}(\vec{R}_s). \quad (7.41)$$

According to (7.39), the interband transition matrix element is a sum of contributions of electron transitions between localized Löwdin orbitals. The optical excitation of an electron between the Löwdin orbitals is, according to (7.40), dependent upon the overlap of the final state orbital with the derivative of the initial state orbital. Because of the localized character of the Löwdin orbitals one expects that this overlap will decrease rapidly with increasing distance between the two orbitals. Therefore the optical transitions are determined by electron excitation between Löwdin orbitals centered on sites near one another.

A qualitative LCAO theory similar to that employed for energy bands in Chapter 4 will be developed in the remainder of this chapter. This model considers only the optically induced transition between Löwdin orbitals centered on the same site and those between cation and anion orbitals on adjacent atomic sites.

There are two types of electron excitations to consider; site-diagonal and nearest-neighbor transitions. The site-diagonal transitions are between Löwdin orbitals on the same atomic site and involve the integrals

$$\int d\vec{r}\, \xi_{\alpha'}^*(\vec{r}) \left(-i\hbar\frac{\partial}{\partial x}\right) \xi_\alpha(\vec{r}) \quad (7.42)$$

while the nearest-neighbor transitions, involving p to d (or d to p) transitions, are of the form

$$\int d\vec{r}\, \xi_{\alpha'}^*(\vec{r}) \left(-i\hbar\frac{\partial}{\partial x}\right) \xi_\alpha(\vec{r} \pm a\vec{e}_j) \quad (7.43)$$

where $\vec{e}_j$ is a unit vector along any of the coordinate axes.

The simple LCAO band model we have been employing in previous chapters includes only the d orbitals on the B-ion sites and the p orbitals on the anion sites. Consequently, the only types of site-diagonal transitions that can occur in our model are "d to d" or "p to p" transitions. It can easily be shown by symmetry arguments that the matrix elements of (7.42) for such transitions vanish. If other orbitals, representing core states or higher-lying atomic states, were included in the basis set for the energy bands then site-diagonal transitions would occur. For example, the matrix elements for nd to $(n+1)p$ or $(n-1)p$ to nd orbital transitions on a B-cation site do not vanish. Such transitions are indeed important for the optical properties of the perovskites; however, the photon energy at which they occur is usually higher than for the p to d transitions between adjacent ions. The effect of neglecting these higher-energy site-diagonal transitions is to limit the validity of our description of $\varepsilon_2(\omega)$ to a photon energy range below the threshold for such transitions. The method we shall use in the remainder of this chapter to discuss interband transitions can easily be generalized to include site-diagonal transitions, but we shall not include them in our discussion.

We now focus our attention on the p to d (or d to p) transitions between nearest-neighbor ions. To determine the character of these matrix elements we first consider their symmetry properties. The Löwdin orbitals (as in Table 3.1) are of the form:

$$\xi_{p(\alpha)}(\vec{r}) = \sqrt{\frac{3}{4\pi}} \left(\frac{r_\alpha}{r}\right) r R_p(r)$$

$$\xi_{d(\alpha\beta)}(\vec{r}) = \sqrt{\frac{5}{4\pi}} \sqrt{3} \left(\frac{r_\alpha r_\beta}{r^2}\right) r^2 R_d(r) \qquad (t_{2g} \text{ orbitals}) \qquad (7.44)$$

$$\xi_{d(z^2)}(\vec{r}) = \sqrt{\frac{5}{4\pi}} \frac{1}{2} \left(\frac{3z^2 - r^2}{r^2}\right) r^2 R_d(r)$$

$$\xi_{d(x^2)}(\vec{r}) = \sqrt{\frac{5}{4\pi}} \frac{\sqrt{3}}{2} \left(\frac{x^2 - y^2}{r^2}\right) r^2 R_d(r)$$

$\left.\right\}$ $(e_g \text{ orbitals})$

where r_α is the αth Cartesian component of $\vec{r}$, $r = |\vec{r}|$ and $R_p(r)$ and $R_d(r)$ are spherically symmetric radial functions normalized so that $\int \xi^* \xi \, d\vec{r} = 1$.

We are concerned with integrals of the type

$$\int d\vec{r} \, \xi_d^*(\vec{r}) \left(-i\hbar \frac{\partial}{\partial x}\right) \xi_p(\vec{r} \pm a\vec{e}_j), \qquad (7.45)$$

$$\int d\vec{r} \, \xi_p^*(\vec{r}) \left(-i\hbar \frac{\partial}{\partial x}\right) \xi_d(\vec{r} \pm a\vec{e}_j), \qquad (7.46)$$

where ξ_d and ξ_p are any of the d or p orbitals, respectively.

Integration by parts shows that the matrix elements are Hermitian so that

$$\int d\vec{r} \, \xi_p^*(\vec{r}) \left(-i\hbar \frac{\partial}{\partial x}\right) \xi_d(\vec{r} \pm a\vec{e}_j) = \left[\int d\vec{r} \, \xi_d^*(\vec{r} \pm a\vec{e}_j) \left(-i\hbar \frac{\partial}{\partial x}\right) \xi_p(\vec{r})\right]^*. \quad (7.47)$$

Also by a shift of the origin:

$$\int d\vec{r} \, \xi_d^*(\vec{r}) \left(-i\hbar \frac{\partial}{\partial x}\right) \xi_p(\vec{r} \pm a\vec{e}_j) = \int d\vec{r} \, \xi_d^*(\vec{r} \mp a\vec{e}_j) \left(-i\hbar \frac{\partial}{\partial x}\right) \xi_p(\vec{r}), \qquad (7.48)$$

and again by symmetry conditions

$$\int d\vec{r} \, \xi_d^*(\vec{r} + a\vec{e}_j) \left(-i\hbar \frac{\partial}{\partial x}\right) \xi_p(\vec{r}) = \int d\vec{r} \, \xi_d^*(\vec{r} - a\vec{e}_j) \left(-i\hbar \frac{\partial}{\partial x}\right) \xi_p(\vec{r}). \qquad (7.49)$$

Therefore we need only investigate the symmetry properties of the integral on the right-hand side of (7.48). To proceed further in the analysis it is convenient to define a fictitious set of localized orbitals which have the angular symmetries of atomic orbitals. We define a set of "bar" orbitals with s and d atomic-orbital symmetries by

$$\bar{R}_p(r) \equiv r \frac{\partial}{\partial r} R_p(r) \equiv \frac{\sqrt{5}}{2} \bar{R}_d(r) \tag{7.50}$$

in terms of which

$$\bar{s}(\vec{r}) \equiv \sqrt{\frac{3}{4\pi}} \left(R_p(r) + \frac{1}{3} \bar{R}_p(r) \right) \tag{7.51}$$

$$\bar{d}_{\alpha\beta}(\vec{r}) \equiv \sqrt{\frac{3}{4\pi}} \, 2 \, \frac{r_\alpha r_\beta}{r^2} \, \bar{R}_p(r) \tag{7.52}$$

$$\bar{d}_{z^2}(\vec{r}) \equiv \sqrt{\frac{3}{4\pi}} \frac{1}{\sqrt{3}} \frac{(3z^2 - r^2)}{r^2} \bar{R}_p(r) \tag{7.53}$$

$$\bar{d}_{x^2}(\vec{r}) \equiv \sqrt{\frac{3}{4\pi}} \frac{(x^2 - y^2)}{r^2} \bar{R}_p(r) \,. \tag{7.54}$$

Now consider the derivative of the Löwdin orbital;

$$\frac{\partial}{\partial x} \xi_{p(\alpha)}(\vec{r}) = \sqrt{\frac{3}{4\pi}} \left(R_p(r) \, \delta_{\alpha x} + \frac{x r_\alpha}{r^2} \, \bar{R}_p(r) \right). \tag{7.55}$$

Then using the bar orbitals of (7.51)–(7.54) we have

$$\frac{\partial}{\partial x} \xi_{p(x)}(\vec{r}) = \bar{s}(\vec{r}) - \frac{1}{2\sqrt{3}} \bar{d}_{z^2}(\vec{r}) + \frac{1}{2} \bar{d}_{x^2}(\vec{r}) \tag{7.56}$$

$$\frac{\partial}{\partial x} \xi_{p(y)}(\vec{r}) = \frac{1}{2} \bar{d}_{xy}(\vec{r}) \tag{7.57}$$

$$\frac{\partial}{\partial x} \xi_{p(z)}(\vec{r}) = \frac{1}{2} \bar{d}_{xz}(\vec{r}) \,. \tag{7.58}$$

Equations (7.56)–(7.58) show that the derivatives of the $\xi_{p(\alpha)}$ orbitals have the angular symmetries of s and d atomic orbitals. Consequently, we can use the Slater–Koster results in Table 3.2 to determine which matrix elements vanish by symmetry and also to express the various non-vanishing elements in terms of a minimum set of parameters. To do this we need to define four parameters which are analogous to the Slater–Koster overlap parameters. We define

$$\frac{1}{a}(\bar{s}d\sigma) \equiv \int \xi^*_{d(z^2)}(\vec{r} \pm a\vec{e}_z) \, \bar{s}(\vec{r}) \, d\vec{r} \,, \tag{7.59}$$

$$\frac{1}{a}(\bar{d}d\sigma) \equiv \int \xi^*_{d(z^2)}(\vec{r} \pm a\vec{e}_z) \, \bar{d}_{z^2}(\vec{r}) \, d\vec{r} \,, \tag{7.60}$$

$$\frac{1}{a}(\bar{d}d\pi) \equiv \int \xi^*_{d(xy)}(\vec{r} \pm a\vec{e}_x) \, \bar{d}_{xy}(\vec{r}) \, d\vec{r} \,, \tag{7.61}$$

$$\frac{1}{a}(\bar{d}d\delta) \equiv \int \xi^*_{d(xy)}(\vec{r} \pm a\vec{e}_z) \, \bar{d}_{xy}(\vec{r}) \, d\vec{r} \,. \tag{7.62}$$

All of the integrals of the type in (7.48) can be expressed in terms of the four parameters $(\bar{s}d\sigma), (\bar{d}d\sigma), (\bar{d}d\pi)$, and $(\bar{d}d\delta)$, with the help of Table 3.2. As examples,

$$\int \xi_{d(xz)}^*(\vec{r} \pm a\vec{e}_x) \left(-i\hbar\frac{\partial}{\partial x} \right) \xi_{p(z)}(\vec{r}) \, d\vec{r} = -i\hbar \int \xi_{d(xz)}^*(\vec{r} \pm a\vec{e}_x) \frac{1}{2} \bar{d}_{xz}(\vec{r}) \, d\vec{r}$$

$$= \frac{-i\hbar}{2a}(\bar{d}d\pi)$$

$$\int \xi_{d(z^2)}^*(\vec{r} \pm a\vec{e}_z) \left(-i\hbar\frac{\partial}{\partial x} \right) \xi_{p(x)}(\vec{r}) \, d\vec{r} = -i\hbar \int \xi_{d(z^2)}^*(\vec{r} \pm a\vec{e}_z)$$

$$\times \left[\bar{s}(\vec{r}) - \frac{1}{2\sqrt{3}}\bar{d}_{z^2}(\vec{r}) + \frac{1}{2}\bar{d}_{x^2}(\vec{r}) \right] \, d\vec{r}$$

$$= -\frac{i\hbar}{a} \left[(\bar{s}d\sigma) - \frac{1}{2\sqrt{3}}(\bar{d}d\sigma) \right]$$

$$\int \xi_{d(x^2)}^*(\vec{r} \pm a\vec{e}_y) \left(-i\hbar\frac{\partial}{\partial x} \right) \xi_{p(y)}(\vec{r})d\vec{r} = -i\hbar \int \xi_{d(x^2)}^*(\vec{r} \pm a\vec{e}_y) \bar{d}_{xy}(\vec{r}) \, d\vec{r} = 0 \,.$$

Proceeding in this manner it is easily found that there are only four non-zero types of integrals:

$$\int \xi_{d(xy)}^*(\vec{r} \pm a\vec{e}_{j'}) \left(-i\hbar\frac{\partial}{\partial x} \right) \xi_{p(y)}(\vec{r}) \, d\vec{r}, \tag{7.63}$$

$$\int \xi_{d(xz)}^*(\vec{r} \pm a\vec{e}_{j'}) \left(-i\hbar\frac{\partial}{\partial x} \right) \xi_{p(z)}(\vec{r}) \, d\vec{r}, \tag{7.64}$$

$$\int \xi_{d(z^2)}^*(\vec{r} \pm a\vec{e}_{j'}) \left(-i\hbar\frac{\partial}{\partial x} \right) \xi_{p(x)}(\vec{r}) \, d\vec{r}, \tag{7.65}$$

$$\int \xi_{d(x^2)}^*(\vec{r} \pm a\vec{e}_{j'}) \left(-i\hbar\frac{\partial}{\partial x} \right) \xi_{p(x)}(\vec{r}) \, d\vec{r}. \tag{7.66}$$

The integrals of (7.63)–(7.66) involve a d orbital of $(\alpha'\beta')$ type located at $\vec{r} = \pm a\vec{e}_{j'}$ and a p orbital of the γ type at the origin, where $(\alpha'\beta') = xy, xz, x^2$, or z^2 and $\gamma = x, y$, or z. Therefore, dropping also the superscript x of P^x, the notation for the momentum matrix elements can be simplified as follows:

$$\int \xi_{d(\alpha'\beta')}^*(\vec{r} - a\vec{e}_{j'}) \left(-i\hbar\frac{\partial}{\partial x} \right) \xi_{p(\gamma)}(\vec{r}) \, d\vec{r} \equiv P_{(\alpha'\beta')j',\gamma} \,. \tag{7.67}$$

The phase factors appearing in (7.41) are simply $e^{\pm ik_{j'}a}$ since we are including only the nearest-neighbor interactions. Finally, the transition matrix elements of (7.41) may be written as

$$\left\langle \vec{k}\nu' \left| \left(-i\hbar\frac{\partial}{\partial x} \right) \right| \vec{k}\nu \right\rangle = \sum_{(\alpha'\beta')j',\gamma} \left[a_{(\alpha'\beta')}^{(\vec{k}\nu')} \right]^* a_{(j'\gamma)}^{(\vec{k}\nu)} P_{(\alpha'\beta')j',\gamma} \, 2\cos(k_{j'}a) \,. \tag{7.68}$$

Table 7.1. *Momentum operator matrix elements $P_{(\alpha'\beta')j',\gamma}$ (defined in (7.67)) for nearest-neighbor Löwdin orbitals, in units of $(-i\hbar/2a)$.*

$(\alpha'\beta')$	γ	$j' = x$	$j' = y$	$j' = z$
(xy)	y	$(\bar{d}d\pi)$	$(\bar{d}d\pi)$	$(\bar{d}d\delta)$
(xz)	z	$(\bar{d}d\pi)$	$(\bar{d}d\delta)$	$(\bar{d}d\pi)$
(z^2)	x	$-(\bar{s}d\sigma) - \frac{1}{4\sqrt{3}}(\bar{d}d\sigma)$	$-(\bar{s}d\sigma) - \frac{1}{4\sqrt{3}}(\bar{d}d\sigma) - \frac{\sqrt{3}}{2}(\bar{d}d\delta)$	$2(\bar{s}d\sigma) - \frac{1}{\sqrt{3}}(\bar{d}d\sigma)$
(x^2)	x	$\sqrt{3}(\bar{s}d\sigma) + (\bar{d}d\sigma)$	$-\sqrt{3}(\bar{s}d\sigma) + (\bar{d}d\sigma) + \frac{1}{2}(\bar{d}d\delta)$	$(\bar{d}d\delta)$

In order to proceed with the analysis it is necessary to specify the bands involved in the interband transition. There are a large number of possible transitions corresponding to just the p to d and d to p transitions. For example, for an insulating perovskite the nine valence bands (three π, two σ, three π^0, and one σ^0) are occupied and the five conduction bands (three π^* and two σ^*) are empty. There are 45 possible interband transitions; transitions from any of the nine valence bands to any of the five conduction bands. The contributions to $\varepsilon_2(\omega)$ due to some interband transitions are shown in Fig. 7.3. For a given polarization of the electric field many of the 45 transitions are forbidden and many are symmetry equivalent so that the number of transition matrix elements that must be considered is greatly reduced. Nevertheless, the number of distinct transitions is still quite substantial.

In the following sections we consider the character of different interband transitions.

7.3 Interband transitions from non-bonding bands

There are four types of interband transitions from the non-bonding bands to the conduction bands: $\pi^0 \to \pi^*$, $\pi^0 \to \sigma^*$, $\sigma^0 \to \pi^*$, and $\sigma^0 \to \sigma^*$.

(a) $\pi^0 \to \pi^*$ interband transition

There are three π^0 and three π^* bands; $\pi^0(\alpha\beta)$ and $\pi^*(\alpha\beta)$ where $\alpha\beta = xy$, xz, or yz. Thus there are nine possible $\pi^0 \to \pi^*$ transitions; arising from $\pi^0(\alpha\beta)$ and $\pi^*(\alpha\gamma)$. For an x-polarized electric field, Table 7.1 shows that non-zero matrix elements occur only for transitions between an d_{xy} orbital and a p_y orbital and between an d_{xz} orbital and a p_z orbital. Consequently, there are only four allowed

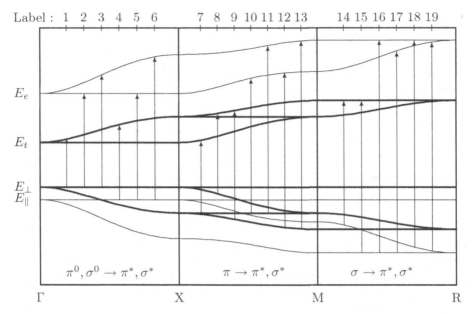

Figure 7.3. Major contributions to optical transitions.

$\pi^0 \rightarrow \pi^*$ transitions:

$$\pi^0(xy) \rightarrow \pi^*(xy) \tag{7.69a}$$

$$\pi^0(xz) \rightarrow \pi^*(xz) \tag{7.69b}$$

$$\pi^0(yz) \rightarrow \pi^*(xy) \tag{7.69c}$$

$$\pi^0(yz) \rightarrow \pi^*(xz). \tag{7.69d}$$

The restriction of the transitions to those in (7.69) shows that selection rules for optical transitions are contained in the transition matrix elements.

Transitions (a) and (b) of (7.69) are "unmixed" transitions in that they involve initial and final band states derived from the same 3×3 block of the Hamiltonian (see Chapter 4). Transitions (7.69c) and (7.69d) are "mixed" transitions that involve initial and final states from different 3×3 blocks of the Hamiltonian. The unmixed transitions, (7.69a) and (7.69b), are symmetry-equivalent and make identical contributions to $\varepsilon_2(\omega)$. Similarly, the mixed transitions (7.69c) and (7.69d) are also symmetry-equivalent. Consequently, it is necessary only to consider (a) and (c).

The π^0 band wavefunctions have zero amplitudes for the d orbitals. Therefore the only non-vanishing matrix element between Löwdin orbitals that contributes to (a) is $P^x_{(xy)x,y}$. According to (7.68) the transition matrix element for (a) is

$$\left\langle \vec{k}\pi^*(xy) \middle| \left(-i\hbar\frac{\partial}{\partial x} \right) \middle| \vec{k}\pi^0(xy) \right\rangle = \left[a^{[\vec{k}\pi^*(xy)]}_{xy} \right]^* a^{[\vec{k}\pi^0(xy)]}_y \, P_{(xy)x,y} \, 2\,C_x \, . \tag{7.70}$$

For the mixed interband transitions, (c), we find

$$\left\langle \vec{k}\pi^*(xy) \middle| \left(-i\hbar\frac{\partial}{\partial x} \right) \middle| \vec{k}\pi^0(yz) \right\rangle = \left[a^{[\vec{k}\pi^*(xy)]}_{xy} \right]^* a^{[\vec{k}\pi^0(yz)]}_y \, P_{(xy)z,y} \, 2\,C_z \, . \tag{7.71}$$

In calculating $\varepsilon_2(\omega)$ the matrix elements of (7.70) and (7.71) should be multiplied by 2 in order to account for the equivalent transitions (7.69b) and (7.69d).

(b) $\pi^0 \rightarrow \sigma^*$ interband transitions

There are six possible $\pi^0 \rightarrow \sigma^*$ interband transitions; $\pi^0(\alpha\beta) \rightarrow \sigma^*(\pm)$, where $\sigma^*(+)$ and $\sigma^*(-)$ are the two distinct branches of the σ^* conduction bands.

The wavefunctions for the σ^* band states involve the x^2 and z^2 type of d orbitals and hence, according to Table 7.1, have matrix elements only with the x type of p orbital. This limits the allowed transitions to those between $\pi^0(xy)$ or $\pi^0(xz)$ and $\sigma^*(\pm)$. The four allowed transitions are:

$$\pi^0(xy) \rightarrow \sigma^*(\pm) \tag{7.72a}$$
$$\pi^0(xz) \rightarrow \sigma^*(\pm). \tag{7.72b}$$

The two types of transitions represented by (7.72a) and (7.72b) are equivalent and hence we need only consider (7.72a). The interband transition matrix elements are:

$$\left\langle \vec{k}\sigma^*(\pm) \middle| \left(-i\hbar\frac{\partial}{\partial x} \right) \middle| \vec{k}\pi^0(xy) \right\rangle$$
$$= \left\{ \left[a^{[\vec{k}\sigma^*(\pm)]}_{z^2} \right]^* P_{(z^2)x,x} + \left[a^{[\vec{k}\sigma^*(\pm)]}_{x^2} \right]^* P_{(x^2)x,x} \right\} a^{[\vec{k}\pi^0(xy)]}_x \, 2\,C_x \, . \tag{7.73}$$

In calculating $\varepsilon_2(\omega)$ these matrix elements should be multiplied by 2 to account for the symmetry-equivalent transitions of (7.72b).

(c) Tabulation of interband transition matrix elements

Proceeding in the manner described above, the matrix elements for all of the interband transitions can easily be found. The results for all of the distinct transitions are given in Table 7.2. The first two columns specify the initial and final bands.

Table 7.2. *Interband transition matrix elements. Labels refer to the assignments in Fig. 7.3. $\vec{k}$ dependence of amplitudes are dropped for short-hand.*

ν	ν'	Label	Weight	$\langle \vec{k}\nu' \vert -i\hbar\frac{\partial}{\partial x} \vert \vec{k}\nu \rangle$
$\pi^0(xy)$	$\pi^*(xy)$	1	2	$\left(a_{xy}^{\pi*(xy)}\right)^* P_{(xy)z,y} \left(a_y^{\pi 0(xy)}\right) 2C_x$
$\pi^0(yz)$	$\pi^*(xy)$	1	2	$\left(a_{xy}^{\pi*(xy)}\right)^* P_{(xy)z,y} \left(a_y^{\pi 0(yz)}\right) 2C_z$
$\pi^0(xy)$	$\sigma^*(\pm)$	3, 2	2	$\left\{\left(a_{z^2}^{\sigma*(\pm)}\right)^* P_{(z^2)x,x} + \left(a_{x^2}^{\sigma*(\pm)}\right)^* P_{(x^2)x,x}\right\} \left(a_x^{\pi 0(xy)}\right) 2C_x$
σ^0	$\pi^*(xy)$	4	2	$\left(a_{xy}^{\pi*(xy)}\right)^* P_{(xy)y,y} \left(a_y^{\sigma 0}\right) 2C_y$
σ^0	$\sigma^*(\pm)$	6, 5	1	$\left\{\left(a_{z^2}^{\sigma*(\pm)}\right)^* P_{(z^2)x,x} + \left(a_{x^2}^{\sigma*(\pm)}\right)^* P_{(x^2)x,x}\right\} \left(a_x^{\sigma 0}\right) 2C_x$
$\pi(xy)$	$\pi^*(xy)$	9	2	$\left\{\left(a_{xy}^{\pi*(xy)}\right)^* P_{(xy)x,y} \left(a_y^{\pi(xy)}\right) + \left(a_{xy}^{\pi(xy)}\right) P_{(xy)x,y}^* \left(a_y^{\pi*(xy)}\right)^*\right\} 2C_x$
$\pi(xy)$	$\pi^*(yz)$	7	2	$\left(a_{xy}^{\pi*(xy)}\right)^* P_{(xy)z,y} \left(a_y^{\pi*(yz)}\right)^* 2C_z$
$\pi(yz)$	$\pi^*(xy)$	8	2	$\left(a_{xy}^{\pi*(xy)}\right)^* P_{(xy)z,y} \left(a_y^{\pi*(yz)}\right) 2C_z$
$\pi(xy)$	$\sigma^*(\pm)$	13, 12	1	$\left\{\left(a_{z^2}^{\sigma*(\pm)}\right)^* P_{(z^2)y,x} + \left(a_{x^2}^{\sigma*(\pm)}\right)^* P_{(x^2)y,x}\right\} \left(a_x^{\pi(xy)}\right) 2C_y + \left(a_{xy}^{\pi(xy)}\right) P_{(xy)y,y}^* \left(a_y^{\sigma*(\pm)}\right)^* 2C_y$
$\pi(xz)$	$\sigma^*(\pm)$	11, 10	1	$\left\{\left(a_{z^2}^{\sigma*(\pm)}\right)^* P_{(z^2)z,x} + \left(a_{x^2}^{\sigma*(\pm)}\right)^* P_{(x^2)z,x}\right\} \left(a_x^{\pi(xz)}\right) 2C_z + \left(a_{xz}^{\pi(xz)}\right) P_{(xz)z,z}^* \left(a_z^{\sigma*(\pm)}\right)^* 2C_z$
$\sigma(\pm)$	$\pi^*(xy)$	15, 14	1	$\left(a_{z^2}^{\sigma(\pm)} + a_{x^2}^{\sigma(\pm)}\right) P_{(z^2)y,x}^* \left(a_x^{\pi*(xy)}\right)^* 2C_y + \left(a_{xy}^{\pi*(xy)}\right)^* P_{(xy)y,y} \left(a_y^{\sigma(\pm)}\right) 2C_y$
$\sigma(\pm)$	$\sigma^*(\pm)$	16-19	1	$\left(a_{z^2}^{\sigma*(\pm)}\right)^* P_{(z^2)x,x} \left(a_x^{\sigma(\pm)}\right) 2C_x + \left\{\left(a_{z^2}^{\sigma(\pm)}\right) P_{(z^2)x,x}^* + \left(a_{x^2}^{\sigma(\pm)}\right) P_{(x^2)x,x}^*\right\} \left(a_x^{\sigma*(\pm)}\right)^* 2C_x$

156

The third column labels the inequivalent transitions and the forth column, labeled "weight", gives the number of symmetry-equivalent transitions. In calculating $\varepsilon_2(\omega)$ the matrix element should be multiplied by the weight factor. In Table 7.2 it should be noted that the row corresponding to $\pi^0(xy) \rightarrow \sigma^*(\pm)$ specifies two inequivalent transitions and the row corresponding to $\sigma(\pm) \rightarrow \sigma^*(\pm)$ specifies four inequivalent transitions.

7.4 Frequency dependence of $\varepsilon_2(\omega)$ for insulating and semiconducting perovskites

In the preceding section we obtained the forms of the matrix elements for interband transitions in terms of the parameters $(\bar{s}d\sigma)$, $(\bar{d}d\sigma)$, $(\bar{d}d\delta)$, and $(\bar{d}d\pi)$. The matrix elements may be used to calculate $\varepsilon_2(\omega)$ from (7.22). However, it should be kept in mind that the model we are employing includes only matrix elements between $2p$ and nd orbitals located on adjacent cation and anion sites.[1]

The goal of this section is to describe the dominant optical properties of the insulating and semiconducting cubic perovskites. For photon energies less than about $10\,\mathrm{eV}$ these properties are dominated by interband transitions. That is, transitions in which an electron in one of the valence-band states absorbs a photon and is promoted to one of the conduction-band states. There are numerous other processes that contribute to the optical properties. These include exciton, defect, impurity, and free-carrier absorption. In addition, processes in which an electron and a collective excitation such as a phonon are simultaneously excited. However, for the range of photon energies being considered these other processes only add fine detail or tend to broaden and smear out sharp structure associated with the interband transitions. We do not discuss these other processes here.

The band-gap energy, E_g, for typical insulating perovskites (e.g., $SrTiO_3$, $BaTiO_3$ or $KTaO_3$) is 3–3.5 eV. The lowest-energy interband transitions are the $\pi^0 \rightarrow \pi^*$ and $\sigma^0 \rightarrow \pi^*$ type transitions. These occur for a photon energy of $\hbar\omega$ in the range $E_g \leq \hbar\omega \leq E_g + W_\pi$, where W_π is the π^*-band width; 3–4 eV. Interband transitions from the top of the π and σ bands to the π^* band also occur in this range. In addition, transitions from π^0 and σ^0 to the bottom of the σ^* band may

---

[1] The reader is cautioned not to take the language of this description literally. The model does not imply that the actual optical transitions consist of transferring an electron from one ion to its neighbor. The optical transitions involve electrons making transitions from one delocalized energy band state to another delocalized energy band state. The method of evaluating the momentum integrals in terms of nearest-neighbor Löwdin orbitals is a mathematical convenience which leads to a real-space description of the contributions to the transition matrix elements. The same can be said for the band model. Even though only nearest-neighbor matrix elements of the Hamiltonian are included, the resulting states are delocalized energy band states. By contrast, in a strong correlation model, hopping of a localized electrons between neighboring ions involves a substantial Coulomb repulsion energy and a consideration of the orbital and spin quantum numbers of the initial and final atomic states.

occur in the upper portion of the previously mentioned photon energy range. Some of these transitions are illustrated in Fig. 7.2.

At higher photon energies, $E_g + W_\pi \leq \hbar\omega \leq E_g + W_2$, transitions from the bottom of the π or σ bands to the π^* and σ^* bands become possible, where W_2 is 10–12 eV. In this range, however, other types of transitions not included in our model can also occur. For example, interband transitions from the π^0 and σ^0 bands into bands associated with the $(n+1)p$ and $(n+1)s$ states of the B ions and the s states of the A ions can occur. Also, at even higher $\hbar\omega$, transitions from the core states of the oxygen and B ions become possible. For SrTiO$_3$ these transitions would be into the energy bands derived from the Ti(4p), Ti(4s), and Sr(4s) orbitals.

At lower energies, $\hbar\omega < E_g$, there are a number of other processes that can contribute to optical absorption, including excitonic state absorption, defect and impurity state absorption, and low-frequency plasmon absorption. These processes will add additional fine structure to the absorption spectrum, particularly in the energy gap region.

In the following sections we shall derive approximate results for the frequency dependence of $\varepsilon_2(\omega)$ using our model based on the $2p$–nd transitions. The effect of neglecting other types of transitions is to limit the validity of our description of $\varepsilon_2(\omega)$ to values of $\hbar\omega$ less than the threshold for these other processes. The $2p$–nd transitions will dominate the optical properties for $\hbar\omega \leq 10$ eV. For perovskite metals, such as Na$_x$WO$_3$ or ReO$_3$, interband transitions can initiate from the partially filled π^* bands.

(a) Band-edge behavior of $\varepsilon_2(\omega)$

The onset of strong optical absorption in perovskite insulators or semiconductors for $\hbar\omega \gtrsim E_g$ is due to interband transitions from the non-bonding bands into the empty π^* conduction bands. Usually the π^0 band is higher than the σ^0 band so that for $\hbar\omega$ just above E_g the absorption is due to the $\pi^0 \rightarrow \pi^*$ interband transitions. In this section we shall show that $\varepsilon_2(\omega)$ has an abrupt rise from zero to a large value at $\hbar\omega = E_g$. We shall also show that the band edge value of $\varepsilon_2(\omega)$ is determined by the parameter $(\bar{d}d\pi)$.

With the help of (7.22) and Table 7.1 we find for the (unmixed) $\pi^0(xy) \rightarrow \pi^*(xy)$ transitions per spin state that

$$[\varepsilon_2(\omega)]_{\pi^0(xy) \rightarrow \pi^*(xy)} = \frac{2}{\pi} \left(\frac{e}{m\omega}\right)^2 \left(\frac{2\pi}{2a}\right)^3 \frac{1}{N}$$
$$\times \sum_{\vec{k}} \left| \left\langle \vec{k}, \pi^*(xy) \right| \left(-i\hbar\frac{\partial}{\partial x}\right) \left| \vec{k}, \pi^0(xy) \right\rangle \right|^2$$

$$\times \delta \left[E_{\vec{k},\pi^*(xy)} - E_{\vec{k},\pi^0(xy)} - \hbar\omega \right]$$

$$= \frac{2}{\pi} \left(\frac{e}{m\omega} \right)^2 \int_{BZ} d\vec{k} \left| a_{xy}^{[\vec{k},\pi^*(xy)]} a_y^{[\vec{k},\pi^0(xy)]} \right|^2$$

$$\times \left(\frac{\hbar}{2a} \right)^2 (\bar{d}d\pi)^2 (2\,C_x)^2 \, \delta \left[E_{\vec{k},\pi^*(xy)} - (E_\perp + \hbar\omega) \right], \qquad (7.74)$$

where the integral is over the first Brillouin zone. In obtaining the final result in (7.74) we have set

$$f\left(E_{\vec{k},\pi^0(xy)} \right) = 1 - f\left(E_{\vec{k},\pi^*(xy)} \right) = 1, \qquad \frac{1}{N} \sum_{\vec{k}} = \left(\frac{2a}{2\pi} \right)^3 \int_{BZ} d\vec{k},$$

and have used the flat band approximation $E_{\vec{k},\pi^0} = E_\perp$ described in Chapter 4. The initial factor of 2 on the right-hand side of (7.74) accounts for the two symmetry-equivalent transitions and the factor of $(\hbar/2a)^2$ results from the units employed in Table 7.1.

The required wavefunction amplitudes were determined in Chapter 4. They are

$$a_{xy}^{[\vec{k},\pi^*(xy)]} = \frac{\left[E_\perp - E_{\vec{k},\pi^*(xy)} \right]}{\sqrt{\left[E_\perp - E_{\vec{k},\pi^*(xy)} \right]^2 + 4(pd\pi)^2(S_x^2 + S_y^2)}}, \qquad (7.75)$$

$$a_y^{[\vec{k},\pi^0(xy)]} = -\frac{S_y}{\sqrt{S_x^2 + S_y^2}}; \qquad (S_\alpha = \sin k_\alpha a). \qquad (7.76)$$

Because of the δ function, the amplitude $a_{xy}^{\vec{k},\pi^*(xy)}$ may be evaluated at $E_{\vec{k},\pi^*(xy)} = E_\perp + \hbar\omega$ in the integrand of (7.74). Then we obtain the result

$$[\varepsilon_2(\omega)]_{\pi^0(xy)\to\pi^*(xy)} = \frac{2}{\pi} \left(\frac{e}{m\omega} \right)^2 \frac{2\hbar^2}{a^2} (\bar{d}d\pi)^2$$

$$\times \int_{BZ} d\vec{k} \, g_{xy,xy}(\vec{k},\omega) \, \delta \left[E_{\vec{k},\pi^*(xy)} - (E_\perp + \hbar\omega) \right] \qquad (7.77)$$

where

$$g_{xy,xy}(\vec{k},\omega) \equiv \frac{C_x^2 S_y^2}{(S_x^2 + S_y^2)} \left[\frac{(\hbar\omega)^2}{(\hbar\omega)^2 + 4(pd\pi)^2(S_x^2 + S_y^2)} \right]. \qquad (7.78)$$

The integral of (7.77) is similar to those encountered in calculating the DOS (density of states) functions in Chapter 6. For example, the DOS function for the π^* band is

$$\rho_{\pi^*}(E) = \left(\frac{2a}{2\pi} \right)^3 \int_{BZ} d\vec{k} \, \delta \left[E_{\vec{k},\pi^*} - E \right]. \qquad (7.79)$$

For $E > E_\perp$, $\rho_{\pi^*}(E)$ is the same as the function $\rho_\pi(E)$ given by (6.28). It is evident from (7.77) that $g_{xy,xy}(\vec{k}, \omega)$ contains the k dependence of the interband transition matrix elements and as may be seen from (7.78), it is not constant over the Brillouin zone. On the other hand, $g_{xy,xy}(\vec{k}, \omega)$ is a very smoothly varying function of $\vec{k}$ and consequently any sharp structure in $\rho_\pi(E)$ will be replicated in $\varepsilon_2(\omega)$. As was shown in Chapter 6, $\rho_\pi(E)$ has very pronounced structure due to the two-dimensional behavior of the π bands. In particular, $\rho_\pi(E)$ possesses jump discontinuities at the edges of the π bands and a logarithmic singularity at the center of the bands. These van Hove singularities will also appear in $\varepsilon_2(\omega)$.

Let us consider in detail the behavior of $\varepsilon_2(\omega)$ for $\hbar\omega$ very near to the band-gap energy, $E_g = E_t - E_\perp$. Under this condition, the only possible $\pi^0(xy) \to \pi^*(xy)$ transitions are those which arise from a small cylinder oriented along the k_z-axis from the Γ to the X point in the Brillouin zone. Within this small cylinder, $k_x a$ and $k_y a$ are small so that

$$g_{xy,xy}(\vec{k}, \omega) \simeq \frac{S_y^2}{S_x^2 + S_y^2}. \tag{7.80}$$

The integral of (7.77) is unchanged if $k_x a$ and $k_y a$ are interchanged so that we may make the substitution

$$\frac{S_y^2}{S_x^2 + S_y^2} \to \frac{1}{2}\left(\frac{S_x^2}{S_x^2 + S_y^2} + \frac{S_y^2}{S_x^2 + S_y^2}\right) = \frac{1}{2}. \tag{7.81}$$

This gives

$$[\varepsilon_2(\omega)]_{\pi^0(xy) \to \pi^*(xy)} \to \left(\frac{2\pi}{2a}\right)^3 \frac{\hbar^4 e^2}{\pi m^2 a^2} \frac{(\bar{d}d\pi)^2}{(\hbar\omega)^2}$$

$$\times \left\{\left(\frac{2a}{2\pi}\right)^3 \int_{BZ} d\vec{k}\, \delta\left[E_{\vec{k}\pi^*} - (E_\perp + \hbar\omega)\right]\right\}. \tag{7.82}$$

In (7.82) the arrow signifies an equality in the limit as $\hbar\omega$ tends to E_g. Using (7.79) we obtain

$$[\varepsilon_2(\omega)]_{\pi^0(xy) \to \pi^*(xy)} \to \left(\frac{2\pi}{2a}\right)^3 \frac{\hbar^4 e^2}{\pi m^2 a^2} \frac{(\bar{d}d\pi)^2}{E_g^2} \rho_{\pi^*}(E_\perp + \hbar\omega),$$

$$= \frac{1}{2}\chi_\pi (\bar{d}d\pi)^2\, \Theta(\hbar\omega - E_g), \tag{7.83}$$

with

$$\chi_\pi \equiv \frac{\pi\hbar^4(e^2/a)}{2m^2 a^4 E_g (pd\pi)^2} \tag{7.84}$$

where we have used (6.28) to evaluate ρ_{π^*}:

$$\rho_{\pi^*}(E_\perp + \hbar\omega) = \rho_\pi(E_\perp + \hbar\omega) \rightarrow \frac{1}{\pi^2} \frac{\frac{1}{2}E_g}{(pd\pi)^2} K(0) \, \Theta(\hbar\omega - E_g)$$

$$= \frac{E_g}{4\pi(pd\pi)^2} \Theta(\hbar\omega - E_g).$$

Equation (7.83) shows that $\varepsilon_2(\omega)$ due to the unmixed $\pi^0 \rightarrow \pi^*$ interband transition possesses a jump at $\hbar\omega = E_g$, the magnitude of which is determined by the optical parameter $(\bar{d}d\pi)$ and by the energy band parameters E_g and $(pd\pi)$. The jump in $\varepsilon_2(\omega)$ decreases with increasing band gap. This is the opposite of the behavior of the JDOS, which increases with increasing band gap. On the other hand, $\varepsilon_2(\omega)$ decreases as the square of $(pd\pi)$ which is the same dependence that the JDOS has. Thus, we see that matrix-element effects are significant [2].

To obtain the total jump in $\varepsilon_2(\omega)$ at the band edge we must also include the contributions from the mixed $\pi^0 \rightarrow \pi^*$ transitions. A calculation similar to that described above, using (7.71) and Tables 7.1 and 7.2, gives for the mixed transitions

$$[\varepsilon_2(\omega)]_{\pi^0(yz)\rightarrow\pi^*(xy)} = \frac{2}{\pi} \left(\frac{e}{m\omega}\right)^2 \frac{\hbar^2}{a^2} (\bar{d}d\delta)^2$$

$$\times \int_{BZ} d\vec{k} \; g_{yz,xy}(\vec{k},\omega) \; \delta\left[E_{\vec{k}\pi^*(xy)} - (E_\perp + \hbar\omega)\right] \quad (7.85)$$

with

$$g_{yz,xy}(\vec{k},\omega) = \frac{S_y^2 C_z^2}{(S_y^2 + S_z^2)} \left[\frac{(\hbar\omega)^2}{(\hbar\omega)^2 + 4(pd\pi)^2(S_x^2 + S_y^2)}\right]. \quad (7.86)$$

For $\hbar\omega$ very near to the band-gap energy, E_g, the transitions arise from the same cylindrical region described above. In that region S_x and S_y are small

$$g_{yz,xy}(\vec{k},\omega) \rightarrow \frac{S_y^2 C_z^2}{(S_y^2 + S_z^2)}.$$

In the integral of (7.85) only $g_{yz,xy}(\vec{k},\omega)$ contains a k_z dependence and

$$\int_{-\pi/2a}^{\pi/2a} dk_z \, g_{yz,xy}(\vec{k},\omega) = S_y^2 \int_{-\pi/2a}^{\pi/2a} \frac{dk_z \, \cos^2(k_z a)}{S_y^2 + \sin^2(k_z a)}$$

$$= \frac{\pi}{a} \left[\sqrt{S_y^4 + S_y^2} - S_y^2\right]. \quad (7.87)$$

As S_y tends to zero, the result of (7.87) vanishes and hence the mixed transitions make no contribution to $\varepsilon_2(\omega)$ as $\hbar\omega$ tends to E_g. Therefore, the contributions of all of the $\pi^0 \rightarrow \pi^*$ interband transitions at the band gap, $\hbar\omega = E_g$, produce a jump in $\varepsilon_2(\omega)$ given by (7.83).

Later in this chapter (see (7.132)) we show that the contribution to $\varepsilon_2(\omega)$ due to the $\pi \to \pi^*$ transitions at $\hbar\omega = E_g$ is exactly one-half of the contribution from $\pi^0 \to \pi^*$. Moreover, if $E_\perp$ and $E_\parallel$ are close to each other then the contributions by the $\sigma^0 \to \pi^*$ (see (7.113)) and $\sigma(-) \to \pi^*$ (see (7.139)) transitions at $\hbar\omega = E_g' \simeq E_g$ should also be added. Therefore the total jump at $\hbar\omega = E_g$ will be 2.5 – 3 times that given by (7.83). It is of interest to estimate the jump in $\varepsilon_2(\omega)$. Using $(e^2/a) \simeq \frac{1}{2}(e^2/2a_H) = (13.6/2)\,\mathrm{eV}$, $E_g = 3.85\,\mathrm{eV}$, $(pd\pi) = 1.15\,\mathrm{eV}$, $E_{g\sigma} = 5.7\,\mathrm{eV}$, $(pd\sigma) = -2.6\,\mathrm{eV}$, $m = 9.11 \times 10^{-28}$ g, and $\hbar = 1.055 \times 10^{-27}$ erg s we obtain a total jump (~ 2.78 times that of (7.83), and an extra factor of two for the two spin states) at the band edge,

$$[\varepsilon_2(\omega)]_{\pi,\pi^0 \to \pi^*}\big|_{\hbar\omega = E_g} \simeq 21.1\,(\bar{d}d\pi)^2 . \tag{7.88}$$

To obtain a numerical value we need to estimate $(\bar{d}d\pi)$. This dimensionless parameter is the overlap integral between a Löwdin d orbital and a neighboring fictitious d orbital derived from the derivative of a Löwdin p orbital. Therefore, it should be a number between zero and one. Experimentally, Cardona [1] has observed a very sharp rise in the optical absorption and reflectivity of $BaTiO_3$ and $SrTiO_3$. His result for $\varepsilon_2(\omega)$ is shown as dashed line in Fig. 7.4(b). The behavior of $\varepsilon_2(\omega)$ for these two perovskites is quite similar, in agreement with our simple model. There is an abrupt rise in $\varepsilon_2(\omega)$ for $\hbar\omega$ between 3 and 4 eV. It is not actually a jump discontinuity because the π^0 bands are not actually dispersionless (flat) as we have assumed here. The magnitude of $\varepsilon_2(\omega)$ at 4 eV is about 6 for either $BaTiO_3$ or $SrTiO_3$. According to (7.88) this implies that $(\bar{d}d\pi)$ is about 0.5. This value is large for a nearest-neighbor overlap. However, the Löwdin orbitals are more extended than atomic orbitals and the empirical parameter $(\bar{d}d\pi)$ has to compensate for the interactions of more distant neighbors that have been neglected in our model.

Experimentally, the index of refraction (real part) is about 2.5 at $\hbar\omega = E_g$ for $BaTiO_3$ or $SrTiO_3$ [1]. According to (7.15) the absorption coefficient α is given by

$$\alpha(\omega) = \frac{\omega}{c} \frac{\varepsilon_2(\omega)}{n(\omega)} .$$

If we set $\hbar\omega = E_g$, $n(E_g/\hbar) \simeq 2.5$ and use the value $\varepsilon_2(E_g/\hbar) \simeq 6$ we find that $\alpha \simeq 4.67 \times 10^5\,\mathrm{cm}^{-1}$ which is also in good agreement with Fig. 7 of [1].

(b) Frequency dependence of $\varepsilon_2(\omega)$ from $\pi^0 \to \pi^*$ transitions

In the preceding section we obtained the contribution of the $\pi^0 \to \pi^*$ transition to $\varepsilon_2(\omega)$ for $\hbar\omega \simeq E_g$. To obtain ε_2 as a function of ω we must evaluate the integrals in (7.77) and (7.85) including the functions $g_{xy,xy}(\vec{k},\omega)$ and $g_{yz,xy}(\vec{k},\omega)$ which contain the $\vec{k}$ dependence of the transition matrix elements. Consider first the unmixed

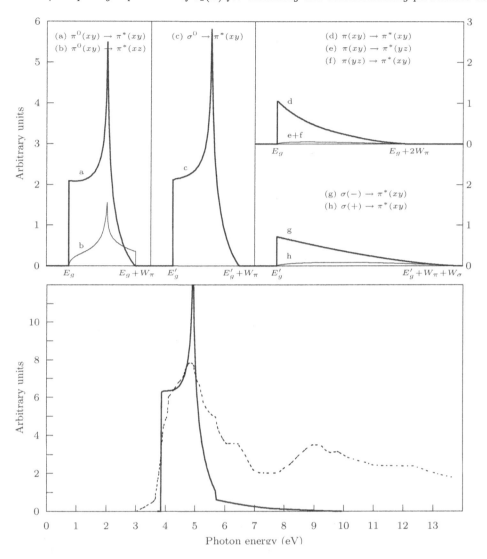

Figure 7.4. (a) Major contributions to $\varepsilon_2(\omega)$ and (b) comparison with the experimental result (dashed line) for SrTiO$_3$ [1].

$\pi^0 \to \pi^*$ transitions. Referring to (7.77) one sees that the integrand is evaluated only on the surface in $\vec{k}$-space where $\hbar\omega = E_{\vec{k}\pi^*(xy)} - E_\perp$. On this surface,

$$S_x^2 + S_y^2 = \frac{\hbar\omega(\hbar\omega - E_g)}{4(pd\pi)^2} \equiv \Omega^2(\omega) \tag{7.89}$$

so that

$$[\varepsilon_2(\omega)]_{\pi^0(xy)\to\pi^*(xy)} = \frac{2}{\pi}\left(\frac{e}{m\omega}\right)^2 \frac{\hbar^2}{a^2} (\tilde{d}d\pi)^2 \left[\frac{(\hbar\omega)^2}{(\hbar\omega)^2 + 4(pd\pi)^2\Omega^2(\omega)}\right] \frac{1}{\Omega^2(\omega)}$$

$$\times \int_{BZ} d\vec{k}\, C_x^2\, S_y^2\, \delta\left[E_{\vec{k}\pi^*(xy)} - (E_\perp + \hbar\omega)\right]. \tag{7.90}$$

The factor, $C_x^2 S_y^2$ in the integrand may be written as

$$C_x^2 S_y^2 = 2S_y^2 - (S_x^2 + S_y^2)S_y^2 - \frac{1}{4} + \frac{1}{4}C_{2y}^2$$

$$= \left(2 - \Omega^2(\omega)\right)S_y^2 - \frac{1}{4} + \frac{1}{4}C_{2y}^2 \tag{7.91}$$

where $S_\alpha = \sin k_\alpha a$, $C_\alpha = \cos k_\alpha a$, and $C_{2\alpha} = \cos 2k_\alpha a$. From symmetry considerations, it follows that S_y^2 may be replaced by $\frac{1}{2}(S_x^2 + S_y^2)$ in the integrand of (7.90),
thus we find the integral of (7.90) is

$$\left\{\frac{1}{2}\left(2 - \Omega^2(\omega)\right)\Omega^2(\omega) - \frac{1}{4}\right\} \int_{BZ} d\vec{k}\, \delta\left[E_{\vec{k}\pi^*(xy)} - (E_\perp + \hbar\omega)\right]$$

$$+\frac{1}{4}\int_{BZ} d\vec{k}\, C_{2y}^2\, \delta\left[E_{\vec{k}\pi^*(xy)} - (E_\perp + \hbar\omega)\right]. \tag{7.92}$$

Using the expression

$$\lambda(\omega) \equiv \frac{\hbar\omega(\hbar\omega - E_g)}{2(pd\pi)^2} - 2 = 2\left[\Omega^2(\omega) - 1\right], \tag{7.93}$$

which is obtained from (6.19) at $(\hbar\omega + E_\perp)$, the first term of (7.92) may be immediately evaluated from the definition of the density of states, $\rho_{\pi^*}(E)$, to give

$$\left\{\frac{1}{4}\left[1 - \frac{\lambda(\omega)^2}{2}\right]\right\}\left(\frac{2\pi}{2a}\right)^3 \rho_{\pi^*}(E_\perp + \hbar\omega). \tag{7.94}$$

Calculation of the remaining term of (7.92) must be evaluated directly. The calculation is simplified by employing a mathematical manipulation.

We note that for $\hbar\omega > 0$ the quantity $\delta[E_{\vec{k}\pi(xy)} - (E_\perp + \hbar\omega)]$ vanishes since
the valence-band states all have energy $E_{\vec{k}\pi(xy)} \leq E_\perp$. Therefore, we may add this
null term to the integrand of the second term of (7.92). We then write:

$$\int_{BZ} d\vec{k}\, C_{2y}^2\, \delta\left[E_{\vec{k}\pi^*(xy)} - (E_\perp + \hbar\omega)\right]$$

$$= \int_{BZ} d\vec{k}\, C_{2y}^2\left\{\delta\left[E_{\vec{k}\pi^*(xy)} - (E_\perp + \hbar\omega)\right] + \delta\left[E_{\vec{k}\pi(xy)} - (E_\perp + \hbar\omega)\right]\right\}$$

$$= -\Im\left\{\frac{2(\hbar\omega - E_g/2)}{(2a)^3(pd\pi)^2}\int_{-\pi}^{\pi} dr \int_{-\pi}^{\pi} dt\, \frac{\cos^2 r}{\lambda(\omega) + \cos t + \cos r + i0^+}\right\}$$

$$= \frac{8\pi(\hbar\omega - E_g/2)}{(2a)^3(pd\pi)^2} \int_0^\pi dr \int_0^\pi dt \, \cos^2 r \, \delta\left[\lambda(\omega) + \cos t + \cos r\right]. \tag{7.95}$$

In obtaining the result of (7.95) we have employed the expressions for $E_{\vec{k}\pi^*(xy)}$ and $E_{\vec{k}\pi(xy)}$, and defined $r = 2k_y a$, $t = 2k_x a$. The integration over t in (7.95) can be performed with the result that

$$\int_{BZ} d\vec{k} \, C_{2y}^2 \, \delta\left[E_{\vec{k}\pi^*(xy)} - (E_\perp + \hbar\omega)\right]$$

$$= \frac{8\pi(\hbar\omega - E_g/2)}{(2a)^3(pd\pi)^2} \int_{-1}^{1} \frac{\mu^2 \, d\mu \, \Theta[1 - (\lambda + \mu)^2]}{\sqrt{(1 - \mu^2)[1 - (\lambda + \mu)^2]}}. \tag{7.96}$$

Because of the Θ function the integral of (7.96) is non-vanishing only in the region $-2 < \lambda(\omega) < 2$. This range is equivalent to $E_g < \hbar\omega < E_g + W_\pi$, where $W_\pi = \sqrt{(\frac{1}{2}E_g)^2 + 8(pd\pi)^2} - \frac{1}{2}E_g$ is the π^*-band width.

The integral may be evaluated in terms of complete elliptic functions and for $|\lambda(\omega)| < 2$ we find

$$\int_{BZ} d\vec{k} \, C_{2y}^2 \, \delta\left[E_{\vec{k}\pi^*(xy)} - (E_\perp + \hbar\omega)\right]$$

$$= \left(\frac{2\pi}{2a}\right)^3 \frac{(\hbar\omega - E_g/2)}{(pd\pi)^2 \, \pi^2} \left\{\left[1 + \frac{\lambda^2(\omega)}{2}\right]K(k) - 2E(k)\right\}, \tag{7.97}$$

where $k^2 = 1 - \lambda^2(\omega)/4$. The functions $K(k)$ and $E(k)$ are the complete elliptic integrals of the first and second kind, respectively,

$$K(k) = \int_0^{\pi/2} \frac{d\alpha}{\sqrt{1 - k^2 \sin^2 \alpha}}, \tag{7.98}$$

$$E(k) = \int_0^{\pi/2} d\alpha \sqrt{1 - k^2 \sin^2 \alpha}. \tag{7.99}$$

Combining the results of (7.94) and (7.97) in (7.90), we obtain:

$$[\varepsilon_2(\omega)]_{\pi^0(xy)\to\pi^*(xy)} = \frac{1}{2} \chi_\pi \frac{E_g(\bar{d}d\pi)^2}{(\hbar\omega)^2 + 4(pd\pi)^2\Omega^2(\omega)}$$

$$\times \left[1 - \frac{E(k)}{K(k)}\right] \frac{4\pi(pd\pi)^2}{\Omega^2(\omega)} \, \rho_{\pi^*}(E_\perp + \hbar\omega). \tag{7.100}$$

As $\hbar\omega \to E_g$, $\Omega^2(\omega) \to 0$ and

$$\left[1 - \frac{E(k)}{K(k)}\right] \frac{1}{\Omega^2(\omega)} \to 1$$

so that $[\varepsilon_2(\omega)]_{\pi^0(xy)\to\pi^*(xy)}$ correctly tends to the result of (7.83). Approaching the center of the absorption band where $\hbar\omega \to E_g/2 + \sqrt{(E_g/2)^2 + 4(pd\pi)^2} \equiv \hbar\omega_c$, we have $\Omega^2(\omega) = 1$, $k^2 = 1$, $E(k)/K(k) = 0$ and

$$[\varepsilon_2(\omega)]_{\pi^0(xy)\to\pi^*(xy)} = \frac{1}{2}\chi_\pi \frac{(\bar{d}d\pi)^2 E_g}{(\hbar\omega_c)(2\hbar\omega_c - E_g)} \rho_{\pi^*}(E_\perp + \hbar\omega_c). \tag{7.101}$$

The function $\rho_{\pi^*}(E_\perp + \hbar\omega)$ produces a logarithmic infinity in $[\varepsilon_2(\omega)]_{\pi^0(xy)\to\pi^*(xy)}$ at $\hbar\omega(\hbar\omega - E_g) = 4(pd\pi)^2$.

At the top of the absorption band, $\hbar\omega \to E_g/2 + \sqrt{(E_g/2)^2 + 8(pd\pi)^2} = E_g + W_\pi$, for which $\Omega^2(\omega) = 2$, $k^2 = 0$, $E(k)/K(k) = 1$, so that $[\varepsilon_2(\omega)]_{\pi^0(xy)\to\pi^*(xy)} = 0$. Therefore the jump discontinuity in the DOS function at the top of the π^* band does *not* manifest itself in $[\varepsilon_2(\omega)]_{\pi^0(xy)\to\pi^*(xy)}$. The reason for this is that the momentum matrix element, which varies as C_x^2, vanishes for transitions to the top of the π^* band.

Next, we consider the contribution of the mixed $\pi^0(xz) \to \pi^*(xy)$ transition to $\varepsilon_2(\omega)$. Using (7.85)–(7.87), and the definitions (7.89) and (7.93), we have

$$[\varepsilon_2(\omega)]_{\pi^0(yz)\to\pi^*(xy)} = \frac{2}{\pi}\frac{\hbar^2}{a^2}(\bar{d}d\delta)^2\left(\frac{e}{m\omega}\right)^2 \frac{(\hbar\omega)^2}{(\hbar\omega)^2 + 4(pd\pi)^2\Omega^2(\omega)}$$

$$\times \frac{\pi}{a}\int_{-\pi/2a}^{\pi/2a} dk_x \int_{-\pi/2a}^{\pi/2a} dk_y \left[\sqrt{S_y^4 + S_x^2} - S_y^2\right]$$

$$\times \delta\left[E_{\bar{k}\pi^*(xy)} - (E_\perp + \hbar\omega)\right] \tag{7.102}$$

where the Brillouin zone integral in the second and third lines can be expressed as

$$\frac{\pi}{a}\frac{(\hbar\omega - E_g/2)}{(pd\pi)^2}\left(\frac{2}{2a}\right)^2 \int_0^\pi dr \int_0^\pi dt\, \delta[\lambda(\omega) + \cos r + \cos t]$$

$$\times \left\{\frac{1}{2}\sqrt{(3 - \cos r)(1 - \cos r)} - \frac{1}{2}\Omega^2(\omega)\right\}$$

with $r = 2k_y a$ and $t = 2k_x a$. This expression can be put into form

$$\frac{\pi^3}{2a^3}\left\{\frac{(\hbar\omega - E_g/2)}{\pi^2(pd\pi)^2}\int_{-1}^1 d\mu \frac{\sqrt{(3 - \mu)}\,\Theta[1 - (\lambda + \mu)^2]}{\sqrt{(1 + \mu)[1 - (\lambda + \mu)^2]}} - \Omega^2(\omega)\,\rho_{\pi^*}(E_\perp + \hbar\omega)\right\}. \tag{7.103}$$

The integral in the first term of (7.103) can be solved in terms of incomplete elliptic integrals of the first and third kind as

$$\mathfrak{Re}\int_{-1}^1 d\mu \frac{\sqrt{(3 - \mu)}\,\Theta[1 - (\lambda + \mu)^2]}{\sqrt{(1 + \mu)[1 - (\lambda + \mu)^2]}} = \frac{1}{\sqrt{2}}\left[(\lambda + 4)F(\varphi, k_0) - \lambda\, \Pi(\varphi, \alpha^2, k_0)\right]$$

with $\alpha^2 = 1 - \lambda/2$, $k_0^2 = \alpha^2(1 + \lambda/4)$, and $\sin\varphi = \sqrt{(1 + \lambda/2)/(1 + \lambda/4)}$ within the interval $-2 < \lambda < 0$, and $\sin\varphi = 1$ for $0 < \lambda < 2$. Hence, combining this with (7.103) we get

$$[\varepsilon_2(\omega)]_{\pi^0(yz)\rightarrow\pi^*(xy)} = \chi_\pi(\bar{d}d\delta)^2 \frac{2\pi(pd\pi)^2 E_g}{(\hbar\omega)^2 + 4(pd\pi)^2 \Omega^2(\omega)}$$

$$\times \left\{ \frac{(\hbar\omega - E_g/2)}{\sqrt{2\pi^2(pd\pi)^2}} \left[(\lambda + 4)F(\varphi, k_0) - \lambda\, \Pi(\varphi, \alpha^2, k_0) \right] \right.$$

$$\left. - \Omega^2(\omega)\, \rho_{\pi^*}(E_\perp + \hbar\omega) \right\}. \tag{7.104}$$

As $\hbar\omega \rightarrow E_g$, $\Omega^2(\omega) \rightarrow 0$, $\lambda \rightarrow -2$, $\varphi \rightarrow 0$, $F(0, k_0) \rightarrow 0$, and $\Pi(0, \alpha^2, k_0) \rightarrow 0$. Therefore, the transition from $\pi^0(yz)$ to $\pi^*(xy)$ does not contribute any jump at $\hbar\omega = E_g$.

As $\hbar\omega \rightarrow E_g + W_\pi$, i.e., transition to the top of the π^* band, $\Omega^2(\omega) \rightarrow 2$, $\lambda \rightarrow 2$, $\varphi = \pi/2$, $F(\pi/2, 0) = \Pi(\pi/2, 0, 0) = \pi/2$, and a jump of

$$(\sqrt{2} - 1)\, \chi_\pi\, (\bar{d}d\delta)^2 \left(\frac{E_g}{E_g + W_\pi} \right) \tag{7.105}$$

occurs at $\hbar\omega = E_g + W_\pi$ (see Table 7.3).

7.5 Frequency dependence of $\varepsilon_2(\omega)$
from $\sigma^0 \rightarrow \pi^*$ transitions

The σ^0 band lies at $E = E_\parallel$ according to our simple energy band model. Usually $E_\parallel$ is about the same as $E_\perp$ or $1\,\mathrm{eV}$ or so lower in energy. Thus the $\sigma^0 \rightarrow \pi^*$ interband transitions are important in contributing to $\varepsilon_2(\omega)$ near the band-gap energy, $\hbar\omega = E_t - E_\perp$, or within an eV of it. In this section we derive expressions for the contributions of the $\sigma^0 \rightarrow \pi^*$ transitions to $\varepsilon_2(\omega)$.

We begin by calculating the contribution at the threshold energy $\hbar\omega = E'_g \equiv E_t - E_\parallel$. From Table 7.2 we have

$$[\varepsilon_2(\omega)]_{\sigma^0 \rightarrow \pi^*} = 2\,[\varepsilon_2(\omega)]_{\sigma^0 \rightarrow \pi^*(xy)}$$

$$= \frac{2}{\pi} \left(\frac{e}{m\omega} \right)^2 \left(\frac{2\pi}{2a} \right)^3 \frac{1}{N} \sum_{\vec{k}} \left| a_{xy}^{[\vec{k},\pi^*(xy)]}\, a_y^{[\vec{k},\sigma^0]} \right|^2 \left(\frac{\hbar}{2a} \right)^2 (\bar{d}d\pi)^2\, (2C_y)^2$$

$$\times \delta \left[E_{\vec{k}\pi^*(xy)} - (E_\parallel + \hbar\omega) \right]. \tag{7.106}$$

Making use of (4.42) and (4.58) for the wavefunction amplitudes and using Table

7.1, we can rewrite (7.106) as

$$[\varepsilon_2(\omega)]_{\sigma^0 \to \pi^*} = \frac{2}{\pi} \left(\frac{e}{m\omega}\right)^2 \int_{BZ} d\vec{k} \frac{\left(E_{\vec{k}\pi^*(xy)} - E_\perp\right)^2}{\left(E_{\vec{k}\pi^*(xy)} - E_\perp\right)^2 + 4(pd\pi)^2(S_x^2 + S_y^2)}$$

$$\times \left(\frac{S_x^2 S_z^2}{S_x^2 S_y^2 + S_y^2 S_z^2 + S_z^2 S_x^2}\right) \left(\frac{\hbar^2}{a^2}(\bar{d}d\pi)^2 C_y^2\right)$$

$$\times \delta\left[E_{\vec{k}\pi^*(xy)} - (E_\parallel + \hbar\omega)\right]. \tag{7.107}$$

Because of the δ function we may make the following replacements:

$$E_{\vec{k}\pi^*(xy)} \to E_\parallel + \hbar\omega = E_\perp + \hbar\bar{\omega}, \tag{7.108}$$

$$(S_x^2 + S_y^2) \to \Omega^2(\bar{\omega}) = \frac{\hbar\bar{\omega}(\hbar\bar{\omega} - E_g)}{4(pd\pi)^2}, \tag{7.109}$$

$$\hbar\bar{\omega} \equiv \hbar\omega + E_\parallel - E_\perp, \tag{7.110}$$

and by symmetry,

$$C_y^2 \to \frac{1}{2}(C_y^2 + C_x^2) = 1 - \frac{1}{2}(S_y^2 + S_x^2) = 1 - \frac{1}{2}\Omega^2(\bar{\omega}). \tag{7.111}$$

Only the factor $F \equiv S_x^2 S_z^2 / [S_x^2 S_y^2 + S_z^2(S_x^2 + S_y^2)] = \frac{1}{2}S_z^2\Omega^2 / [S_x^2 S_y^2 + S_z^2\Omega^2]$ has a dependence on k_z and its integral $I_F = \int dk_z a\, F = \pi/2\{1 - \sqrt{A/(1+A)}\}$ with $A = S_x^2 S_y^2 / \Omega^2$.

The analytical evaluation of (7.107) can be accomplished but is rather tedious. Therefore, let us first consider an approximation for I_F that simplifies the calculation.

For the integration in (7.107) over k_x–k_y plane, the values of k_x and k_y are constrained to lie on an curve on k_x–k_y plane defined by the δ function. This energy surface is determined by Ω^2 and therefore we look for an approximation for I_F in terms of Ω^2. We note that $I_F \to \pi/2$ for near Γ and X, and near M it tends to $(\pi/2)(\sqrt{3} - 1)/\sqrt{3}$. An approximation that matches these results is $I_F \cong (\pi/2)[1 + (\Omega^2 - \Omega^4)/(2\sqrt{3})]$. With this approximation the integral of (7.107) is easily evaluated as

$$[\varepsilon_2(\omega)]_{\sigma^0 \to \pi^*} \cong \frac{\pi^2\hbar^4(e^2/a)(\bar{d}d\pi)^2}{m^2 a^4(\hbar\omega)^2} \left[\frac{(\hbar\bar{\omega})^2}{(\hbar\bar{\omega})^2 + 4(pd\pi)^2\Omega^2(\bar{\omega})}\right]$$

$$\times \left[1 - \frac{1}{2}\Omega^2(\bar{\omega})\right]\left[1 + \frac{\Omega^2 - \Omega^4}{2\sqrt{3}}\right]\rho_{\pi^*}(\hbar\bar{\omega} + E_\perp). \tag{7.112}$$

Just above threshold, as $\hbar\bar{\omega} \to E_g$, $\hbar\omega \to E_g'$, and $\Omega^2(\bar{\omega}) = 0$. This leads to a jump discontinuity contribution to ε_2 of

$$[\varepsilon_2(\omega)]_{\sigma^0 \to \pi^*} \to \frac{\pi \hbar^4 (e^2/a)(\bar{d}d\pi)^2}{4m^2 a^4 E_g (pd\pi)^2} \left(\frac{E_g}{E_g + E_\perp - E_\parallel} \right)^2$$

$$= \frac{1}{2} \chi_\pi (\bar{d}d\pi)^2 \left(\frac{E_g}{E_g'} \right)^2. \qquad (7.113)$$

At $\hbar\omega = E_m + \sqrt{(E_g/2)^2 + 8(pd\pi)^2} - E_\parallel = E_g' + W_\pi$, corresponding to the transition from σ^0 to the top of the π^* band, the contribution vanishes (see Table 7.3) because the factor C_y^2 in (7.107) vanishes at that energy.

Returning to (7.107), the analytical solution can be found as follows:

$$[\varepsilon_2(\omega)]_{\sigma^0 \to \pi^*} = \frac{2}{\pi} \frac{\hbar^2}{a^2} (\bar{d}d\pi)^2 \left(\frac{e}{m\omega} \right)^2 \left[\frac{(\hbar\bar{\omega})^2}{(\hbar\bar{\omega})^2 + 4(pd\pi)^2 \Omega^2(\bar{\omega})} \right]$$

$$\times \int_{BZ} d\vec{k} \frac{C_y^2 S_x^2 S_z^2}{S_x^2 S_y^2 + S_z^2 \Omega^2} \delta \left[E_{\vec{k}\pi^*(xy)} - (E_\parallel + \hbar\omega) \right]. \qquad (7.114)$$

Integration over the variable k_z gives

$$\int_{-\pi/2a}^{\pi/2a} dk_z \frac{C_y^2 S_x^2 S_z^2}{S_x^2 S_y^2 + S_z^2 \Omega^2} = \frac{\pi}{a} \frac{C_y^2 S_x^2}{\Omega^2} \left[1 - \frac{S_x^2 S_y^2}{\sqrt{\Omega^2 S_x^2 S_y^2 + S_x^4 S_y^4}} \right]. \qquad (7.115)$$

Therefore the contribution of the first term to $\varepsilon_2(\omega)$ can be obtained using the same procedure in Subsection 7.4(b) as

$$[\varepsilon_2(\omega)]_{\sigma^0 \to \pi^*(xy)} = \frac{\pi^2 \hbar^4 (e^2/a)(\bar{d}d\pi)^2}{m^2 a^4 (\hbar\bar{\omega})^2} \left[\frac{(\hbar\bar{\omega})^2}{(\hbar\bar{\omega})^2 + 4(pd\pi)^2 \Omega^2(\bar{\omega})} \right]$$

$$\times \left[1 - \frac{E(k)}{K(k)} \right] \frac{1}{\Omega^2(\bar{\omega})} \rho_{\pi^*}(E_\perp + \hbar\bar{\omega}) \qquad (7.116)$$

with $k^2 = 1 - \lambda^2/4$ and

$$\lambda(\bar{\omega}) = \frac{(\hbar\bar{\omega})(\hbar\bar{\omega} - E_g)}{2(pd\pi)^2} - 2 = 2 \left[\Omega^2(\bar{\omega}) - 1 \right].$$

The second term of (7.115) can be put into form

$$\frac{\pi}{a\Omega^2} \int_{-\pi/2a}^{\pi/2a} dk_x \int_{-\pi/2a}^{\pi/2a} dk_y \frac{C_y^2 S_y^2 S_x^4}{\sqrt{\Omega^2 S_x^2 S_y^2 + S_x^4 S_y^4}} \delta \left[E_{\vec{k}\pi^*(xy)} - (E_\perp + \hbar\bar{\omega}) \right]$$

$$= \frac{2\pi}{(2a)^3} \frac{1}{\Omega^2(\bar{\omega})} \frac{(\hbar\bar{\omega} - E_g/2)}{(pd\pi)^2} I(\lambda)$$

where

$$I(\lambda) = \int_{-1}^{1} dt \, \Theta\left[1 + (\lambda + t)^2\right] \frac{t^2 + (\lambda - 2)t + (1 - \lambda)}{\sqrt{(1 + t)(1 - \lambda - t)(r - t)(\lambda + r + t)}} \tag{7.117}$$

with

$$r = -\frac{\lambda}{2} + \frac{1}{2}\sqrt{(\lambda + 2)(\lambda + 10)} \,.$$

The integral in (7.117) can be solved in terms of the incomplete elliptic integrals of the first and second kind to give

$$I(\lambda) = 2\left[F'(\varphi, k_0) - E'(\varphi, k_0)\right] + (\lambda + 2) \cos \varphi$$

$$+ (1 - r)\left[E'(\varphi, k_0) - k_0 \, F'(\varphi, k_0)\right] \tag{7.118}$$

with $k_0 = (\lambda + r - 1)/(1 + r)$, $\sin \varphi = 1$ within the interval $0 < \lambda < 2$, and

$$\sin \varphi = \frac{(\lambda + 2)(1 + r)}{2(\lambda + r - 1) - \lambda(r - 1)} \qquad \text{for} \quad -2 < \lambda < 0 \,.$$

Therefore, the total contribution of the $\sigma^0 \to \pi^*(xy)$ transition to $\varepsilon_2(\omega)$ can be written as

$$[\varepsilon_2(\omega)]_{\sigma^0 \to \pi^*(xy)} = \chi_\pi \frac{(\bar{d}d\pi)^2}{(\hbar\omega)^2} \left[\frac{(\hbar\bar{\omega})^2}{(\hbar\bar{\omega})^2 + 4(pd\pi)^2\Omega^2(\bar{\omega})}\right] \frac{2\pi(pd\pi)^2 E_g}{\Omega^2(\bar{\omega})}$$

$$\times \left\{\left[1 - \frac{E(k)}{K(k)}\right] \rho_{\pi^*}(E_\perp + \hbar\bar{\omega}) + \frac{(\hbar\bar{\omega} - E_g/2)}{2\pi^2(pd\pi)^2} I(\lambda)\right\}. \tag{7.119}$$

As $\hbar\bar{\omega} \to E_g$, $\Omega^2(\bar{\omega}) \to 0$, $\lambda(\bar{\omega}) \to -2$, $k \to 0$, $k^2/\Omega^2(\bar{\omega}) \to 2$, $[1 - E/K]/\Omega^2(\bar{\omega}) \to 1$, $r \to 1$, $\varphi \to 0$, $k_0 \to 1$, and $I(-2) = 0$. Hence, as before (in (7.113)) we get a jump of

$$[\varepsilon_2(\omega)]_{\sigma^0 \to \pi^*(xy)} \to \frac{1}{2} \chi_\pi \, (\bar{d}d\pi)^2 \left(\frac{E_g}{E_g'}\right)^2 \qquad \text{per spin state.} \tag{7.120}$$

As $\hbar\bar{\omega} \to E_g + W_\pi$, $\Omega^2(\bar{\omega}) \to 2$, $\lambda(\bar{\omega}) \to 2$, $k \to 0$, $k^2/\Omega^2(\bar{\omega}) \to 2$, $[1 - E/K]/\Omega^2(\bar{\omega}) \to 1$, $r \to 1$, $\varphi \to \pi/2$, $k_0 \to 1$, and $I(2) = 0$. Hence, we get no jump from the transitions to the top of the π^* band. The results for the two limiting cases are the same as those obtained previously using the approximation for I_F (see (7.113)).

7.6 Frequency dependence of $\varepsilon_2(\omega)$ from $\pi \to \pi^*$ transitions

The top of the π valence band coincides with the π^0 band at $E = E_\perp$. Consequently, transitions from the top of the π band to the bottom of the π^* band will contribute to $\varepsilon_2(\omega)$ for $\hbar\omega$ near to the band-gap energy. In this section we consider the contributions of the $\pi \to \pi^*$ interband transitions to $\varepsilon_2(\omega)$.

According to Table 7.2, there are three types of such interband transitions: $\pi(xy) \to \pi^*(xy)$, $\pi(xy) \to \pi^*(yz)$, and $\pi(yz) \to \pi^*(xy)$. For the unmixed transitions, $\pi(xy) \to \pi^*(xy)$, we find from (7.22) and Tables 7.1 and 7.2 that

$$[\varepsilon_2(\omega)]_{\pi(xy)\to\pi^*(xy)} = \frac{2}{\pi} \left(\frac{e}{m\omega}\right)^2 \frac{\hbar^2}{a^2} (\bar{d}d\pi)^2 \int_{BZ} d\vec{k}\, C_x^2\, \delta\left[E_{\vec{k}\pi^*(xy)} - E_{\vec{k}\pi(xy)} - \hbar\omega\right]$$

$$\times \left|\left(a_{xy}^{[\vec{k},\pi^*(xy)]}\right)^* a_y^{[\vec{k},\pi(xy)]} - a_{xy}^{[\vec{k},\pi(xy)]}\left(a_y^{[\vec{k},\pi^*(xy)]}\right)^*\right|^2. \quad (7.121)$$

The minus sign within the absolute signs comes from the fact that $P^* = -P$, since the Löwdin orbitals, ξ, are real. The amplitude, $a_{xy}^{[\vec{k},\pi^*(xy)]}$ is given by (7.75) and $a_{xy}^{[\vec{k},\pi(xy)]}$ is obtained by substituting $E_{\vec{k}\pi(xy)}$ for $E_{\vec{k}\pi^*(xy)}$ in that equation. The remaining amplitudes are determined from

$$a_y^{[\vec{k},\nu]} = \frac{2i(pd\pi)S_x}{\sqrt{(E_\perp - E_{\vec{k}\nu})^2 + 4(pd\pi)^2(S_x^2 + S_y^2)}} \quad (7.122)$$

for $\nu = \pi(xy)$ or $\pi^*(xy)$. Using (7.75) and (7.122) together with the expressions for $E_{\vec{k}\pi^*(xy)}$ and $E_{\vec{k}\pi(xy)}$ gives

$$[\varepsilon_2(\omega)]_{\pi(xy)\to\pi^*(xy)} = \frac{2}{\pi} \left(\frac{e}{m\omega}\right)^2 \left[\frac{\hbar^2}{a^2}(\bar{d}d\pi)^2\right] 4(pd\pi)^2 E_g^2$$

$$\times \int_{BZ} d\vec{k}\, \frac{S_x^2\, C_x^2}{4(pd\pi)^2(S_x^2 + S_y^2)\left[E_g^2 + 16(pd\pi)^2(S_x^2 + S_y^2)\right]}$$

$$\times \delta\left[E_{\vec{k}\pi^*(xy)} - E_{\vec{k}\pi(xy)} - \hbar\omega\right]. \quad (7.123)$$

The δ function is satisfied when

$$\sqrt{\left(\frac{1}{2}E_g\right)^2 + 4(pd\pi)^2(S_x^2 + S_y^2)} = \frac{\hbar\omega}{2} \quad (7.124)$$

or

$$\frac{(\hbar\omega)^2 - E_g^2}{16(pd\pi)^2} = (S_x^2 + S_y^2) \equiv \Omega^2(\omega). \quad (7.125)$$

These relations allow the integrand to be simplified considerably and the result is

$$[\varepsilon_2(\omega)]_{\pi(xy)\to\pi^*(xy)} = \frac{2}{\pi} \left(\frac{e}{m\omega}\right)^2 \frac{\hbar^2}{a^2} (\bar{d}d\pi)^2 \frac{E_g^2}{(\hbar\omega)[(\hbar\omega)^2 - E_g^2]}$$

$$\times \int_{BZ} d\vec{k}\, (2\, C_x\, S_x)^2\, \delta\left[\lambda(\omega) + C_{2x} + C_{2y}\right], \quad (7.126)$$

where

$$\lambda(\omega) \equiv \frac{(\hbar\omega)^2 - E_g^2}{8(pd\pi)^2} - 2 = 2\left[\Omega^2(\omega) - 1\right]. \tag{7.127}$$

In obtaining (7.126) we have employed the relation

$$\frac{\hbar\omega}{4(pd\pi)^2}\,\delta\left[\lambda(\omega) + C_{2x} + C_{2y}\right] = \delta\left[E_{\vec{k}\pi^*(xy)} - E_{\vec{k}\pi(xy)} - \hbar\omega\right], \tag{7.128}$$

which is valid for $\hbar\omega > 0$. Another useful relation is

$$\rho_{\pi^*}\left(\frac{\hbar\omega}{2} + \frac{1}{2}(E_t + E_\perp)\right) = \left(\frac{2a}{2\pi}\right)^3 \frac{\hbar\omega}{2(pd\pi)^2} \int_{BZ} d\vec{k}\, \delta\left[\lambda(\omega) + C_{2x} + C_{2y}\right]$$

$$= \frac{1}{\pi^2}\frac{\hbar\omega}{2(pd\pi)^2}\, K(k)\,\Theta(k) \tag{7.129}$$

with $k^2 \equiv 1 - \lambda^2(\omega)/4$. Use of (7.129) gives

$$[\varepsilon_2(\omega)]_{\pi(xy)\to\pi^*(xy)} = \frac{\pi^2\hbar^4(e^2/a)}{4m^2a^4\Omega^2(\omega)}\,(\bar{d}d\pi)^2\,\frac{E_g^2}{(\hbar\omega)^4}\left\{\rho_{\pi^*}\left(\frac{\hbar\omega}{2} + \frac{1}{2}(E_t + E_\perp)\right)\right.$$

$$\left. - \frac{\hbar\omega}{2\pi^2(pd\pi)^2}\int_{-1}^1 \frac{\mu^2\,\Theta[1 - (\lambda + \mu)^2]\,d\mu}{\sqrt{(1 - \mu^2)[1 - (\lambda + \mu)^2]}}\right\}. \tag{7.130}$$

The second term of (7.130) can be immediately evaluated by using the result of (7.97). The final result for $\varepsilon_2(\omega)$ is

$$[\varepsilon_2(\omega)]_{\pi(xy)\to\pi^*(xy)} = \frac{1}{4}\chi_\pi(\bar{d}d\pi)^2\frac{4\pi(pd\pi)^2\,E_g^3}{(\hbar\omega)^4\Omega^2(\omega)}$$

$$\times \left[\frac{E(k)}{K(k)} - \frac{\lambda^2(\omega)}{4}\right]\rho_{\pi^*}\left(\frac{\hbar\omega}{2} + \frac{1}{2}(E_t + E_\perp)\right). \tag{7.131}$$

In the limit as $\hbar\omega \to E_g$, the quantity

$$\left[\frac{E(k)}{K(k)} - \frac{\lambda^2(\omega)}{4}\right] \to \Omega^2(\omega)$$

and

$$\rho_{\pi^*}\left(\frac{\hbar\omega}{2} + \frac{1}{2}(E_t + E_\perp)\right) \to \frac{E_g}{4\pi(pd\pi)^2}$$

so that,

$$[\varepsilon_2(\omega)]_{\pi(xy)\to\pi^*(xy)}\big|_{\hbar\omega\to E_g} = \frac{1}{4}\chi_\pi\,(\bar{d}d\pi)^2\,\Theta(\hbar\omega - E_g). \tag{7.132}$$

It is exactly one-half of the $\pi^0(xy) \to \pi^*(xy)$ band-edge contribution (per spin state) given by (7.83). As $\hbar\omega \to 2\sqrt{(\frac{1}{2}E_g)^2 + 8(pd\pi)^2} = 2W_1$, namely the top of the absorption band, one has

$$\lambda(\omega) \to 2 - \alpha,$$

$$k^2 \to \alpha,$$

$$\left[\frac{E(k)}{K(k)} - \frac{\lambda^2(\omega)}{4}\right] \to \frac{\alpha}{2},$$

where

$$\alpha = \frac{W_1(2W_1 - \hbar\omega)}{(pd\pi)^2}.$$

One finds that apart from a numerical factor

$$[\varepsilon_2(\omega)]_{\pi(xy) \to \pi^*(xy)} \propto \chi_\pi (\bar{d}d\pi)^2 \left(\frac{E_g}{2W_1}\right)^2 \frac{E_g(2W_1 - \hbar\omega)}{2(pd\pi)^2}. \tag{7.133}$$

Thus, the contribution to $\varepsilon_2(\omega)$ vanishes linearly in this limit. This behavior of $\varepsilon_2(\omega)$ for transitions to the top of the band is not what is expected on the basis of the constant-matrix-elements approximation.

All of the mixed transitions have coefficients in terms of $(\bar{d}d\delta)^2$, which is much smaller relative to $(\bar{d}d\pi)^2 \ll (pd\pi)^2 < (pd\sigma)^2$. Therefore they do not lead to considerable contributions, especially at the band edge.

7.7 σ → π* interband transitions

Calculation of the contributions of $\sigma \to \pi^*$ transitions to $\varepsilon_2(\omega)$ are difficult to treat analytically because they involve complicated transition matrix elements. From Table 7.2 one may see that the $\sigma(\pm) \to \pi^*(xy)$ transition involves three terms in the matrix element. In addition, the σ band amplitudes are themselves quite complex as may be seen from Subsection 4.5(b) in Chapter 4.

Fortunately, to obtain the threshold behaviors at Γ and R we need only the dominant term of the transition matrix element, the term involving $\left|a_{xy}^{\pi^*(xy)} a_y^{\sigma(\pm)}\right|^2$. The other two terms in Table 7.2 are negligible at the thresholds. The dominant term of the transition matrix element for $\sigma(\pm) \to \pi^*(xy)$ is

$$\left|a_{xy}^{\pi^*(xy)} P_{(xy)y,y} \, 2 \, C_y \, a_y^{\sigma(\pm)}\right|^2 = \frac{\hbar^2}{a^2}(\bar{d}d\pi)^2 C_y^2 \left|a_{xy}^{\pi^*(xy)} a_y^{\sigma(\pm)}\right|^2. \tag{7.134}$$

Both $a_{xy}^{\pi^*(xy)}$ and $a_y^{\sigma(\pm)}$ are finite as $\vec{k} \to 0$. For the two transitions we have approx-

Table 7.3. *Band-edge behavior of interband contributions to $\varepsilon_2(\omega)$ including a factor of 2 for the two spin states.* $[\varepsilon_2(\omega)]_{\nu \to \nu'} = \chi_\pi \, T_{\nu\nu'} \, f(\hbar\omega)$.

ν	ν'	Edge (E_0)	$T_{\nu\nu'}$	$f(\hbar\omega)$
$\pi^0(xy)$	$\pi^*(xy)$	E_g	$(\bar{d}d\pi)^2$	$\Theta(\hbar\omega - E_0)$
		$E_g + W_\pi$	$\left(\frac{E_g}{E_0}\right)(\bar{d}d\pi)^2$	$\left(\frac{W_1(E_0 - \hbar\omega)}{(pd\pi)^2}\right)$
$\pi^0(yz)$	$\pi^*(xy)$	E_g	$(\bar{d}d\delta)^2$	$\left(\frac{(\hbar\omega - E_0)E_0}{2(pd\pi)^2}\right)^{1/2}$
		$E_g + W_\pi$	$2(\sqrt{2} - 1)\left(\frac{E_g}{E_0}\right)(\bar{d}d\delta)^2$	$\Theta(E_0 - \hbar\omega)$
σ^0	$\pi^*(xy)$	E_g'	$\left(\frac{E_g}{E_0}\right)^2(\bar{d}d\pi)^2$	$\Theta(\hbar\omega - E_0)$
		$E_g' + W_\pi$	$\frac{E_g(E_g + W_\pi)}{E_0^2}(\bar{d}d\pi)^2$	$\left(\frac{W_1(E_0 - \hbar\omega)}{2(pd\pi)^2}\right)$
$\pi(xy)$	$\pi^*(xy)$	E_g	$\frac{1}{2}(\bar{d}d\pi)^2$	$\Theta(\hbar\omega - E_0)$
		$E_g + 2W_\pi$	$\left(\frac{E_g}{E_0}\right)^2(\bar{d}d\pi)^2$	$\left(\frac{E_g(E_0 - \hbar\omega)}{2(pd\pi)^2}\right)$
$\pi(xy)$	$\pi^*(yz)$	E_g	$\left(\frac{pd\pi}{E_g}\right)^4(\bar{d}d\delta)^2$	$\left(\frac{(\hbar\omega - E_0)E_0}{2(pd\pi)^2}\right)^{5/2}$
		$E_g + 2W_\pi$	$\frac{(pd\pi)^2}{W_1^2}\left(\frac{E_g}{E_0}\right)(\bar{d}d\delta)^2$	$\left(\frac{E_0(E_0 - \hbar\omega)}{2(pd\pi)^2}\right)^{3/2}$
$\pi(yz)$	$\pi^*(xy)$	E_g	$(\bar{d}d\delta)^2$	$\left(\frac{(\hbar\omega - E_0)E_0}{2(pd\pi)^2}\right)^{1/2}$
		$E_g + 2W_\pi$	$\frac{(E_g + W_\pi)(pd\pi)^2}{W_\pi W_1^2}\left(\frac{E_g}{E_0}\right)(\bar{d}d\delta)^2$	$\left(\frac{E_0(E_0 - \hbar\omega)}{2(pd\pi)^2}\right)^{3/2}$
$\sigma(-)$	$\pi^*(xy)$	E_g'	$\frac{4\xi}{(4\xi + 3)}\left(\frac{E_g}{E_0}\right)^2(\bar{d}d\pi)^2$	$\Theta(\hbar\omega - E_0)$
		$E_g' + W_\pi + W_\sigma$	$C_1\,\eta\,J_f(\eta)\,(\bar{d}d\pi)^2$	$\left(\frac{W_2(E_0 - \hbar\omega)}{(pd\sigma)^2}\right)^{3/2}$
$\sigma(+)$	$\pi^*(xy)$	E_g'	$2\xi\,I_f(\xi)\left(\frac{E_g}{E_0}\right)^2(\bar{d}d\pi)^2$	$\left(\frac{(\hbar\omega - E_0)E_{g\sigma}}{2(pd\pi)^2}\right)^{1/2}$
		$E_g' + W_\pi + W_\sigma$	$\frac{2C_1\,\eta}{(4\eta + 3)^2}(\bar{d}d\pi)^2$	$\left(\frac{W_2(E_0 - \hbar\omega)}{(pd\sigma)^2}\right)$

(a) $\chi_\pi = \pi\hbar^4(e^2/a)/[2m^2 a^4 E_g(pd\pi)^2]$,   (b) $E_g = E_t - E_\perp$,   (c) $W_1 = \sqrt{(\frac{1}{2}E_g)^2 + 8(pd\pi)^2}$,
(d) $W_\pi = W_1 - \frac{1}{2}E_g$,   (e) $E_g' = E_t - E_\parallel$,   (f) $E_{g\sigma} = E_e - E_\parallel$,   (g) $W_2 = \sqrt{(\frac{1}{2}E_{g\sigma})^2 + 6(pd\sigma)^2}$,
(h) $W_\sigma = W_2 - \frac{1}{2}E_{g\sigma}$,   (i) The ratios ξ, η, C_1 and the functions $I_f(\xi)$, $J_f(\eta)$ are defined in the text.
(j) $T_{\nu\nu'}$ are exact only for the cases with jumps. For the cases with power law behavior, the numerical factors that vary according to the shape of the small volume chosen for integration, are not included.

imately

$$[\varepsilon_2(\omega)]_{\sigma(\pm) \to \pi^*(xy)} \simeq \frac{1}{\pi}\left(\frac{e}{m\omega}\right)^2\left[\frac{\hbar^2}{a^2}(\bar{d}d\pi)^2\right]$$

$$\times \int_{BZ} d\vec{k}\,\left|a_{xy}^{\pi^*(xy)} a_y^{\sigma(\pm)}\right|^2 C_y^2\,\delta\left[E_{\vec{k}\pi^*(xy)} - E_{\vec{k}\sigma(\pm)} - \hbar\omega\right]. \quad (7.135)$$

The $\sigma(\pm)$ energy bands are given by

$$E_{\vec{k}\sigma(\pm)} = \frac{1}{2}(E_e + E_\parallel) - \sqrt{\left[\frac{1}{2}(E_e - E_\parallel)\right]^2 + 2(pd\sigma)^2 \left[S_x^2 + S_y^2 + S_z^2 \pm S^2\right]},$$

(7.136)

$$S^2 = \sqrt{(S_x^4 + S_y^4 + S_z^4) - (S_x^2 S_y^2 + S_y^2 S_z^2 + S_z^2 S_x^2)}.$$

(7.137)

Equation (7.136) shows that the $\sigma(-)$ branch is flat along the k_x, k_y or k_z-axis. Since $E_{\vec{k}\pi^*(xy)}$ is also flat along the k_z-axis there will be a jump in $\varepsilon_2(\omega)$ at $\hbar\omega = E_g' \equiv E_t - E_\parallel$ arising from transitions where the $\vec{k}$-vectors are located in a small cylinder about the k_z-axis. To calculate the jump contribution, we write, for $\vec{k}$ near the k_z-axis:

$$E_{\vec{k}\pi^*(xy)} \rightarrow E_t + \frac{4(pd\pi)^2}{E_g}[(k_x a)^2 + (k_y a)^2],$$

$$S^2 \rightarrow S_z^2 - \frac{1}{2}[(k_x a)^2 + (k_y a)^2],$$

$$E_{\vec{k}\sigma(-)} \rightarrow E_\parallel - \frac{3(pd\sigma)^2}{E_{g\sigma}}[(k_x a)^2 + (k_y a)^2],$$

$$E_{g\sigma} \equiv E_e - E_\parallel, \qquad E_g' \equiv E_t - E_\parallel,$$

$$\left|a_{xy}^{\pi^*(xy)}\right|^2 \rightarrow 1, \qquad \left|a_y^{\sigma(\pm)}\right|^2 \rightarrow 1, \qquad C_y^2 \rightarrow 1.$$

It then follows that

$$\int_{BZ} d\vec{k}\, C_y^2\, \delta\left[E_{\vec{k}\pi^*(xy)} - E_{\vec{k}\sigma(-)} - \hbar\omega\right]$$

$$\rightarrow \frac{\pi}{a^3} \int_0^{2\pi} d\phi \int_0^\infty \rho\, d\rho\, \delta\left\{(\hbar\omega - E_g') - \left[\frac{4(pd\pi)^2}{E_g} + \frac{3(pd\sigma)^2}{E_{g\sigma}}\right]\rho^2\right\}$$

$$= \frac{\pi^2}{a^3}\left[\frac{4(pd\pi)^2}{E_g} + \frac{3(pd\sigma)^2}{E_{g\sigma}}\right]^{-1} \Theta(\hbar\omega - E_g'),$$

(7.138)

and hence

$$[\varepsilon_2(\omega)]_{\sigma(-)\rightarrow\pi^*(xy)}\Big|_{\hbar\omega\rightarrow E_g'} = \chi_\pi(\bar{d}d\pi)^2\left(\frac{2\xi}{4\xi+3}\right)\left(\frac{E_g}{E_g'}\right)^2 \Theta(\hbar\omega - E_g')$$

with $\qquad \xi \equiv \dfrac{(pd\pi)^2 E_{g\sigma}}{(pd\sigma)^2 E_g}.$

(7.139)

The $\sigma(+)$ branch behaves quite differently; the energy $E_{\vec{k}\sigma(+)}$ depends on all three components of $\vec{k}$. As $|\vec{k}| \to 0$,

$$E_{\vec{k}\sigma(+)} \to E_{\|} - \frac{2(pd\sigma)^2}{E_{g\sigma}}[(k_x a)^2 + (k_y a)^2 + (k_z a)^2 + S^2]. \tag{7.140}$$

The contribution to $\varepsilon_2(\omega)$ is

$$[\varepsilon_2(\omega)]_{\sigma(+)\to\pi^*(xy)} \to \frac{1}{\pi}\left(\frac{e}{m\omega}\right)^2 \frac{\hbar^2}{a^5}(\bar{d}d\pi)^2 \int_0^\infty r^2\,dr \int_0^{2\pi} d\varphi \int_0^\pi \sin\theta\,d\theta$$

$$\times\; \delta\left\{(\hbar\omega - E_g') - r^2\left(\frac{2(pd\sigma)^2}{E_{g\sigma}}\right)f(\theta,\varphi)\right\} \tag{7.141}$$

where we have used spherical coordinates defined by

$$k_x a \to x = r\,\sin\theta\,\cos\varphi$$
$$k_y a \to y = r\,\sin\theta\,\sin\varphi$$
$$k_z a \to z = r\,\cos\theta \tag{7.142}$$

and

$$f(\theta,\varphi) = 2\xi\,\sin^2\theta + 1 + \sqrt{1 - 3\sin^2\theta\,\cos^2\theta - 3\sin^4\theta\,\cos^2\varphi\,\sin^2\varphi}\;.$$

One then obtains

$$[\varepsilon_2(\omega)]_{\sigma(+)\to\pi^*(xy)}\big|_{\hbar\omega\to E_g'} \to \chi_\pi\,(\bar{d}d\pi)^2\left(\frac{E_g}{E_g'}\right)^2 \xi\,I_f(\xi)\left[\frac{(\hbar\omega - E_g')E_{g\sigma}}{2(pd\pi)^2}\right]^{1/2} \tag{7.143}$$

where,

$$I_f(\xi) = \frac{1}{2\pi^2}\int_0^{2\pi} d\varphi \int_0^\pi \frac{\sin\theta\,d\theta}{f(\theta,\varphi)^{3/2}}. \tag{7.144}$$

The integral, $I_f(\xi)$, can not be evaluated analytically but is approximately given by

$$I_f(\xi) \simeq \frac{4.01/\pi}{(\xi + 2.4869)^{3/2}}, \tag{7.145}$$

or a quadratic fit will give

$$I_f(\xi) \simeq 0.05526\,\xi^2 - 0.1834\,\xi + 0.3247. \tag{7.146}$$

The analytical behavior of the contributions to $\varepsilon_2(\omega)$ for energies near the pi and sigma band gaps are summarized in Table 7.3. Inspection of the table shows that the contributions to $\varepsilon_2(\omega)$ for $\hbar\omega$ near a particular threshold (edge), $E_0 = \hbar\omega_0$, obey a power law of the form $|E_0 - \hbar\omega|^p$, where $p = 0$ (jump), an integer, or $n/2$ ($n = $ odd integer). We can understand how these power laws arise. A particular contribution to $\varepsilon_2(\omega)$ has the form

$$\varepsilon_2(\omega) \propto |P_{(\alpha'\beta')j',\gamma}|^2 \int d\vec{k} \, |[a^{\vec{k}\nu'}_{(\alpha'\beta')}]^* a^{\vec{k}\nu}_{j'\gamma}|^2 \cos^2(k_{j'}a) \, \delta[E_{\vec{k}(\alpha'\beta')} - E_{\vec{k}\gamma} - \hbar\omega]$$

(7.147)

where the 'P-factor' and the 'a-factors' are given in Table 7.2. The 'P-factor' is a constant and therefore may be ignored for our purpose here. The power laws of Table 7.3 apply to the thresholds at Γ or R. Each of the factors of the integrand of (7.147) are functions of $\sin^2(k_j a)$ and $\cos^2(k_j a)$, ($j = x, y,$ and z). For a threshold at Γ, $\cos^2(k_j a) \to 1$ and $\sin^2(k_j a) \to (k_j a)^2$. Therefore each of the factors will be a non-zero constant or expressible as a power series in the small parameters, $(k_j a)^2$. If the integrand involves all three components of the wavevector, $\vec{k}$, the integrand is conveniently expressed in spherical coordinates defined by (7.142). The delta function will then assume the form $\delta[\hbar\omega - E_0 - A(\theta, \varphi) r^2]$ and the remainder of the integrand can be written as $B(\theta, \varphi) r^s$ so that,

$$\varepsilon_2(\omega) \propto \int \sin\theta \, d\theta \int d\varphi \int r^2 \, dr \, \{B(\theta, \varphi) \, r^s \, \delta[\hbar\omega - E_0 - A(\theta, \varphi) r^2]\}$$

$$= \int \sin\theta \, d\theta \int d\varphi \left[\frac{B(\theta, \varphi)}{A(\theta, \varphi)}\right] \int dr \, r^{s+2} \, \delta\left[\frac{(\hbar\omega - E_0)}{A(\theta, \varphi)} - r^2\right]$$

$$= \left\{\int_0^\pi \sin\theta \, d\theta \int_0^{2\pi} d\varphi \left[\frac{B(\theta, \varphi)}{A(\theta, \varphi)^{(s+3)/2}}\right]\right\} (\hbar\omega - E_0)^{(s+1)/2}$$

$$= \text{constant} \times (\hbar\omega - E_0)^{(s+1)/2}, \quad (\text{as } \hbar\omega \to E_0).$$

(7.148)

For a threshold at R in the Brillouin zone a similar procedure can be used by writing a power series for the integrand factors in terms of the small parameters, x, y, z, where $k_x a = \pi/2 - x$, $k_y a = \pi/2 - y$, $k_z a = \pi/2 - z$. In this case the delta function assumes the form $\delta[E_0 - \hbar\omega - A(\theta, \varphi) r^2]$ and the result is

$$\varepsilon_2(\omega) \propto \text{constant} \times (E_0 - \hbar\omega)^{(s+1)/2}, \quad (\text{as } \hbar\omega \to E_0).$$

(7.149)

When $[E_{\vec{k}(\alpha'\beta')} - E_{\vec{k}\gamma}]$ depends on only two components of the wavevector, $\vec{k}$, cylindrical coordinates may be used to arrive at similar results. However, in this case the power-law exponent will be an integer. It may also be zero, meaning that there is a jump in the contribution to $\varepsilon_2(\omega)$ as $\hbar\omega \to E_0$. The transition from $\sigma(-)$ to $\pi^*(xy)$ for $E_0 = E'_g$ (at Γ) is an example for such a singular case (see (7.139)).

The above general rule can be applied easily to the behavior of the contributions by the $\sigma(\pm) \to \pi^*$ transitions to $\varepsilon_2(\omega)$ at R, namely, as $\hbar\omega \to E_0 = W_\pi + W_\sigma + E'_g$.

At this limit one has

$$E_{\vec{k}\pi^*(xy)} \rightarrow \frac{1}{2}(E_t + E_\perp) + W_1 - \frac{2(pd\pi)^2}{W_1}(x^2 + y^2),$$

$$E_{\vec{k}\sigma(\pm)} \rightarrow \frac{1}{2}(E_e + E_\parallel) - W_2 + \frac{(pd\sigma)^2}{W_2}[x^2 + y^2 + z^2 \mp S^2],$$

$$S^2 \rightarrow \sqrt{(x^4 + y^4 + z^4) - (x^2y^2 + y^2z^2 + z^2x^2)},$$

$$E_{g\sigma} \equiv E_e - E_\parallel, \qquad C_y^2 \rightarrow y^2 = r^2 \sin^2\theta \sin^2\varphi, \ (s = 2),$$

$$\left|a_{xy}^{\pi^*(xy)}\right|^2 \rightarrow \frac{E_g + W_\pi}{2W_1}, \qquad \left|a_y^{\sigma(\pm)}\right|^2 \rightarrow \frac{E_{g\sigma} + W_\sigma}{2W_2}.$$

The contribution to $\varepsilon_2(\omega)$ is

$$[\varepsilon_2(\omega)]_{\sigma(\pm)\rightarrow\pi^*(xy)} \rightarrow \frac{1}{\pi}\left(\frac{e}{m\omega}\right)^2 \frac{\hbar^2}{a^5}(\bar{d}d\pi)^2 \int_0^\infty r^2\, dr \int_0^{2\pi} d\varphi \int_0^\pi \sin\theta\, d\theta$$

$$\times \left(\frac{E_g + W_\pi}{2W_1}\right)\left(\frac{E_{g\sigma} + W_\sigma}{2W_2}\right) r^2 \sin^2\theta \sin^2\varphi$$

$$\times \delta\left\{(E_0 - \hbar\omega) - r^2\left(\frac{(pd\sigma)^2}{W_2}\right)f(\theta,\varphi)\right\} \tag{7.150}$$

where

$$f^{(\pm)}(\theta,\varphi) = 2\eta\sin^2\theta + 1 \mp \sqrt{1 - 3\sin^2\theta\cos^2\theta - 3\sin^4\theta\cos^2\varphi\sin^2\varphi},$$

with
$$\eta \equiv \frac{(pd\pi)^2 W_2}{(pd\sigma)^2 W_1}. \tag{7.151}$$

First, let us define

$$A(\theta,\varphi) = \left(\frac{(pd\sigma)^2}{W_2}\right)f^{(-)}(\theta,\varphi) \tag{7.152}$$

and

$$B(\theta,\varphi) = \left(\frac{E_g + W_\pi}{2W_1}\right)\left(\frac{E_{g\sigma} + W_\sigma}{2W_2}\right)\sin^2\theta\sin^2\varphi. \tag{7.153}$$

One then obtains

$$[\varepsilon_2(\omega)]_{\sigma(-)\rightarrow\pi^*(xy)}\Big|_{\hbar\omega\rightarrow E_0} \rightarrow \frac{C_1}{2}\chi_\pi(\bar{d}d\pi)^2\,\eta\,J_f(\eta)\left[\frac{W_2(E_0 - \hbar\omega)}{(pd\sigma)^2}\right]^{3/2} \tag{7.154}$$

where

$$C_1 = \frac{E_g(E_g + W_\pi)(E_{g\sigma} + W_\sigma)}{E_0^2 W_2}, \tag{7.155}$$

$$J_f(\eta) = \frac{1}{2\pi^2} \int_0^{2\pi} d\varphi \int_0^\pi \sin\theta \, d\theta \, \frac{\sin^2\theta \sin^2\varphi}{f^{(-)}(\theta,\varphi)^{5/2}} \,. \tag{7.156}$$

The integral, $J_f(\eta)$, can be evaluated numerically and an approximate expression is

$$J_f(\eta) \simeq \frac{0.669/\pi}{(\eta + 2.042)^{5/2}} \,, \tag{7.157}$$

or a quadratic fit will give

$$J_f(\eta) \simeq 0.0163\,\eta^2 - 0.0389\,\eta + 0.0356 \,. \tag{7.158}$$

For the case of $\sigma(+) \rightarrow \pi^*(xy)$ transition near R, the function $f^{(+)}(\theta,\varphi) = 0$ when $\theta = 0$ and this will cause the integral over the spherical angles to diverge. To treat this case we can use a small cylinder around the k_z-axis at R instead of a sphere around R. Near R this gives

$$S^2 \rightarrow 1 - S_z^2 - \frac{1}{2}\rho^2 \,,$$

$$E_{\vec{k}\sigma(+)} \rightarrow \frac{1}{2}(E_e + E_\parallel) - W_2 + \frac{3(pd\sigma)^2}{2W_2}\rho^2 \,.$$

The BZ integral will then be

$$\int_{BZ} d\vec{k}\, C_y^2\, \delta\left[E_{\vec{k}\pi^*(xy)} - E_{\vec{k}\sigma(+)} - \hbar\omega\right]$$

$$\rightarrow \frac{\pi}{a^3} \int_0^{2\pi} d\phi \int_0^\infty \rho\, d\rho\, \rho^2 \sin^2\phi$$

$$\times \delta\left\{(W_\pi + W_\sigma + E_g' - \hbar\omega) - \left[\frac{2(pd\pi)^2}{W_1} + \frac{3(pd\sigma)^2}{2W_2}\right]\rho^2\right\}$$

$$= \frac{\pi^2}{2a^3}\left[\frac{2(pd\pi)^2}{W_1} + \frac{3(pd\sigma)^2}{2W_2}\right]^{-2}(W_\pi + W_\sigma + E_g' - \hbar\omega) \,, \tag{7.159}$$

and hence the behavior is linear as $\hbar\omega \rightarrow E_0 = W_\pi + W_\sigma + E_g'$

$$[\varepsilon_2(\omega)]_{\sigma(+)\rightarrow\pi^*(xy)}\big|_{\hbar\omega\rightarrow E_0} = \chi_\pi\,(\bar{d}d\pi)^2\,\frac{C_1\eta}{(4\eta+3)^2}\left[\frac{W_2(E_0 - \hbar\omega)}{(pd\sigma)^2}\right] \tag{7.160}$$

where the parameters η and C_1 are defined in (7.152) and (7.156), respectively.

7.8 Summary

Light interacts with the electrons of a solid through its electromagnetic field causing a variety of electronic processes. absorption of photons by the solid is accomplished by direct and indirect transitions between electronic band states, excitation of optical phonons, magnons, and plasmons. For photon energies between 1 and 10 eV, absorption of light by insulating, cubic, d-band perovskites is determined principally by interband electronic transitions of electrons from occupied valence-band states to unoccupied conduction-band states.

In this chapter a *semiclassical* theory of the optical properties was described in which classical electromagnetic theory was joined with quantum theory by equating the classical energy loss to the quantum mechanical rate of transitions between quantum states. The resulting theory provided a description of the dielectric function, $\varepsilon_2(\omega)$, in terms of the electronic transition between energy band states of the solid. The frequency-dependent optical constants were then calculated from the dielectric function.

A qualitative theory of $\varepsilon_2(\omega)$ based on replacing the transition matrix elements by their average value led to a description of the optical properties in terms of joint density of states (JDOS) functions. It was shown that the JDOS possessed the same characteristic structures as the DOS, including jump discontinuities at the band edges and a logarithmic singularity in the center of the band. As a result these structures are apparent in the frequency-dependent dielectric function, $\varepsilon_2(\omega)$.

A method based on the LCAO approximation was developed to calculate the matrix element for interband transitions. This method involved the use of overlap integrals between fictitious localized orbitals and led to a description of the transition matrix elements in terms of three fundamental parameters: $(\bar{d}d\pi)$, $(\bar{d}d\delta)$, and $(\bar{s}d\sigma)$. Detailed calculations were carried out for the contributions to $\varepsilon_2(\omega)$ from the various interband transitions. It was shown that the DOS singularities show up in $\varepsilon_2(\omega)$ even when the frequency dependence of the matrix elements is included. It was also shown that matrix-element effects lead to significant changes in $\varepsilon_2(\omega)$ from what was calculated assuming a constant matrix element. Explicit results for $\varepsilon_2(\omega)$ and the absorption coefficient for $BaTiO_3$ and $SrTiO_3$ for $\hbar\omega = E_g$ (the band-gap energy) were found to be in reasonable agreement with the experimental observed values.

References

[1] M. Cardona, *Phys. Rev.* A **140**, 651 (1964).

[2] T. Wolfram and Ş. Ellialtıoğlu, *Appl. Phys.* **22**, 11 (1980).

Problems for Chapter 7

1. The theory presented in this chapter is called a semiclassical theory, meaning that it combines classical electromagnetic theory with quantum theory. Identify the key steps that connect the classical theory with the quantum theory.

2. Prove (7.23).

3. Derive the results of (7.55)–(7.58).

4. Explain why the optical properties of insulators or doped insulators are determined principally by interband transitions for $\hbar\omega \geq E_g$. For infrared light what new processes might be expected to play an important role?

5. Let the initial state belong to the energy band described by $E_i(\vec{k}) = \alpha(k_x a)^2$, and the final state belong to the band described by $E_f(\vec{k}) = E_0 + \beta(k_y a)^2$. Calculate the joint density of states, $J(\omega)$, defined by

$$J(\omega) = \left(\frac{2a}{\pi}\right)^3 \int_{-\pi/2a}^{\pi/2a} dk_x \int_{-\pi/2a}^{\pi/2a} dk_y \int_{-\pi/2a}^{\pi/2a} dk_z \, \delta\Big(\hbar\omega - [(E_f(\vec{k}) - E_i(\vec{k}))]\Big).$$

Evaluate the behavior of $J(\omega)$ for $\hbar\omega \to E_0^+$ and $\hbar\omega \to E_0 + \beta(\pi/2)^2$.

6. Derive the result for $P^x_{(xy)z,y}$ shown in Table 7.1.

7. If a complex function $f(z) = f_1(z) + if_2(z)$ is analytic in the upper half-plane, the real and imaginary parts of $f(z)$ are related by

$$f_\alpha(z) = \frac{1}{\pi} P \int_{-\infty}^{\infty} \frac{f_\beta(z') \, dz'}{z - z'} \qquad (\alpha, \beta = 1, 2),$$

given that $f(-z) = f^*(z)$ show that

$$f_1(\omega) = \frac{2}{\pi} P \int_0^{\infty} \frac{\omega' f_2(\omega') \, d\omega'}{\omega^2 - \omega'^2},$$

$$f_2(\omega) = -\frac{2\omega}{\pi} P \int_0^{\infty} \frac{f_1(\omega') \, d\omega'}{\omega^2 - \omega'^2}.$$

8. Show that the contribution to $\varepsilon_2(\omega)$ due to the transition $\pi(xy) \to \pi^*(yz)$ obeys the power law $(\hbar\omega - E_0)^{5/2}$ in the limit as $\hbar\omega \to E_0 = E_g$, and $(E_0 - \hbar\omega)^{3/2}$ in the limit as $\hbar\omega \to E_0 = 2W_1$ (see Table 7.3).

8

Photoemission from perovskites

In Chapter 7 we discussed interband transitions in which an electron occupying an energy band state was excited to a final state in a higher band by a photon of energy $\hbar\omega$. The relation between the electronic structure of the solid and optical properties such as reflectivity or absorption was established through the optical dielectric function.

If the photon energy is sufficiently large, the final state of an excited electron may be above the vacuum level of the solid. That is, the final state may be an unbound or continuum state in which the electron can escape from the solid. Such processes are called *electron photoemission*.

The kinetic energy of a photoemitted electron, E_{kin}, is

$$E_{\text{kin}} = \hbar\omega - |E_i| - \Phi, \qquad (8.1)$$

where E_i is the initial (bound) state energy and Φ is the work function of the solid. A considerable amount of information about the electronic states of the solid can be obtained by analyzing the kinetic energy distribution of the photoemitted electrons. Consequently, photoelectron spectroscopy has become a very important method for studying the electronic structure of solids and solid surfaces.

Recently, high-resolution electron-energy analyzers have become available which allow finer detail of the emitted electrons to be measured in using ultraviolet photoelectron spectroscopy (UPS). This advance coupled with the development of tunable, polarized synchrotron radiation sources have made it possible to track both the kinetic energy and the angle of photoemitted electrons. Angle-resolved photoemission spectroscopy or ARPES is now routinely used to determine the initial state energy and the wavevector with some precision. An energy resolution of about 2 meV and an angular resolution of 0.2° (about 1% of the Brillouin zone reciprocal lattice vector) are obtained experimentally.

The application of ARPES to the study of high-temperature superconductors is one of the most important methods of probing the electronic structures of the superconducting cuprates. Detailed features of the superconducting gap and the Fermi surface have been mapped out for many of the high-T_c materials [1] (see Chapter 11).

The range of electronic states that can be probed by photoemission depends upon the energy of the incident photon. UPS experiments make use of photon energies in the range 5–50 eV. The energy bands associated with the outer (or valence) electrons of the solid can be investigated by UPS.

To probe the deep core levels of the solid the photon energy must be in the x-ray region. X-ray photoelectron spectroscopy (XPS) studies employ x-ray photon sources such as the aluminum K_α line at 1486 eV. XPS experiments can probe both the valence electron energy bands and the deep core levels of the solid.

In this chapter we shall be mainly concerned with the interpretation of photoemission data for the valence electron energy bands.

8.1 Qualitative theory of photoemission

In photoemission the excitation of an electron proceeds through the same mechanism as in ordinary optical excitation. According to (7.10) the rate at which electrons make transitions due to photons of energy $\hbar\omega$ is

$$\frac{dW}{dt} = \frac{4\pi}{\hbar}|A|^2\left(\frac{e}{mc}\right)^2 \sum_{i,f} |\langle f|e^{i\vec{q}\cdot\vec{r}}\vec{a}_0 \cdot \vec{p}|i\rangle|^2 \delta(E_f - E_i - \hbar\omega)f(E_i)[1 - f(E_f)],$$

(8.2)

where A is the amplitude of the vector potential, i and f denote initial and final states, $\vec{q}$ is the wavevector of the photon and $f(E)$ is the Fermi distribution function.

In the usual optical experiment one is interested in how the rate of transitions varies with the photon energy $\hbar\omega$. On the other hand, in a typical photoemission experiment, the photon energy is held constant and the energies of the photoemitted electrons are analyzed. Thus in photoemission, one measures the number of electrons with kinetic energy in a range between E and $E + dE$ (per second). It is seen from (8.1) that the kinetic energy of a photoemitted electron is directly related to the initial state energy, E_i. Photoemission data are usually presented in the form of an electron energy distribution, $I(E)$, where E is the energy of the initial state from which the electron was emitted.

The rate of transitions from an initial state of energy E to a final state of energy $E + \hbar\omega$ is

$$I(E) = \sum_{if} \frac{dW_{if}(\hbar\omega)}{dt}\delta(E - E_i),$$

(8.3)

where the δ function, $\delta(E - E_i)$, projects out transitions from the initial states with energy E_i [2]. The emission rate per unit cell, $I(E)$, is

$$I(E) = \frac{1}{N^2} \sum_{\vec{k}\vec{k}'} \sum_{\nu\nu'} |M_{\nu\nu'}(\vec{k}, \vec{k}')|^2 \delta(\hbar\omega - E_{\vec{k}'\nu'} + E_{\vec{k}\nu})\delta(E - E_{\vec{k}\nu})f(E), \qquad (8.4)$$

where we have set $1 - f(E_{\vec{k}'\nu'}) = 1$, since the final states are unoccupied prior to emission and the matrix element is

$$|M_{\nu\nu'}(\vec{k}, \vec{k}')|^2 = \frac{4\pi}{\hbar}|A|^2 \left(\frac{e}{mc}\right)^2 |\langle k'\nu'|e^{i\vec{q}\cdot\vec{r}} \vec{a}_0 \cdot \vec{p}|\vec{k}\nu\rangle|^2. \qquad (8.5)$$

If we make the constant matrix element approximation (CMEA) so that $|M_{\nu\nu'}(\vec{k}, \vec{k}')|^2$ is independent of $\vec{k}$ and $\vec{k}'$, then,

$$I(E) \simeq f(E) \sum_{\nu\nu'} \langle|M_{\nu\nu'}|^2\rangle \frac{1}{N} \sum_{\vec{k}} \delta(E - E_{\vec{k}\nu}) \left\{ \frac{1}{N} \sum_{\vec{k}'} \delta[(\hbar\omega + E_{\vec{k}\nu}) - E_{\vec{k}'\nu'}] \right\}. \qquad (8.6)$$

The term in curly brackets may immediately be identified as the final state DOS function, $\rho_{\nu'}(\hbar\omega + E_{\vec{k}\nu})$. The dependence of this function on $E_{\vec{k}\nu}$ is very weak. The reason for this is that $\hbar\omega \simeq 1.5 \times 10^3$ eV (the Al K_α line, for example, is 1486 eV) while $|E_{\vec{k}\nu}| \lesssim 1.5 \times 10$ eV for a typical valence-band width. Thus, $\hbar\omega + E_{\vec{k}\nu}$ varies by only about 1% as the initial states range over the valence bands. Furthermore, since the final states have $E_{\vec{k}\nu} \simeq \hbar\omega$ they are continuum states which may be approximately described as plane waves that are slightly distorted by the periodic potential of the solid. Since plane-wave states will have a DOS function that varies as the square root of the energy it follows that

$$\rho_{\nu'}(\hbar\omega + E_{\vec{k}\nu}) \propto \sqrt{(\hbar\omega + E_{\vec{k}\nu})} \simeq \sqrt{\hbar\omega}. \qquad (8.7)$$

According to (8.7) $\rho_{\nu'}$ is approximately a constant that is independent of $E_{\vec{k}\nu}$. Therefore, writing $\rho_{\nu'}(\hbar\omega + E_{\vec{k}\nu}) = \rho_{\nu'}(\hbar\omega)$ we obtain from (8.6),

$$I(E) \simeq \sum_{\nu} \left(\sum_{\nu'} \rho_{\nu'}(\hbar\omega)\langle|M_{\nu\nu'}|^2\rangle \right) \frac{1}{N} \sum_{\vec{k}} \delta(E - E_{\vec{k}\nu})f(E)$$

$$= \sum_{\nu} M_\nu^2 \rho_\nu(E)f(E) \qquad (8.8)$$

where $M_\nu^2 \equiv \sum_{\nu'} \langle|M_{\nu\nu'}|^2\rangle \rho_{\nu'}(\hbar\omega)$ and $\rho_\nu = 1/N \sum \delta(E - E_{\vec{k}\nu})$ is the initial DOS.

Equation (8.8) shows that CMEA leads to the result that the energy distribution of photoelectrons in an XPS experiment is proportional to the density of filled states, $\rho_\nu(E)f(E)$. This conclusion appears to be approximately valid for a number of materials. A detailed comparison of $I(E)$ with theoretical DOS functions for the d bands of gold has been carried out by Shirley [3] and Freeouf *et al.* [4]. The agreement is remarkably good and suggests that the constant matrix element approximation is valid for gold.

The results of the CMEA are reasonably good for monatomic materials and for bands whose wavefunctions are composed of orbitals predominantly of one type; d orbitals for example in the case of gold. By contrast, application of (8.8) to the analysis of the XPS spectra of compounds such as the perovskites is not successful [5]. The principal difficulty is that the probabilities of exciting electrons from p and d orbitals are substantially different; at XPS energies the d-orbital probability appears to be 3–10 times larger than the p-orbital probability [6]. In order to take this effect into account it is necessary to formulate the CMEA in a different way.

An alternative approximation is to assume that the energy distribution is a sum of contributions arising from transitions from each of the basis orbitals which enter the wavefunctions. Then,

$$I(E) = \sum_{\nu} \sum_{j\alpha} n_{j\alpha\nu}(E), \tag{8.9}$$

$$n_{j\alpha\nu}(E) = \frac{C}{N^2} \sum_{\vec{k}\vec{k}'} \sum_{\nu'} |\langle \vec{k}, \nu; j\alpha| \, e^{-i\vec{q}\cdot\vec{r}}(\vec{a}_0 \cdot \vec{p})|\vec{k}'\nu'\rangle|^2$$
$$\times f(E)\, \delta(E - E_{\vec{k}\nu})\, \delta(\hbar\omega - E_{\vec{k}'\nu'} + E_{\vec{k}\nu}) \tag{8.10}$$

where C is a collection of constants and $\langle \vec{k}, \nu; j\alpha|$ denotes the amplitude of the $(j\alpha)$th orbital in the wavefunction $\psi_{\vec{k}\nu}$. For the LCAO wavefunctions the $(j\alpha)$th component is

$$\frac{1}{\sqrt{N}} \sum_{m} e^{i\vec{k}\cdot\vec{R}_{mj}} \, a_{j\alpha}(\vec{k}, \nu) \, \xi_\alpha(\vec{r} - \vec{R}_{mj}). \tag{8.11}$$

If (8.11) is employed in (8.10) and the final DOS taken to be constant, then one finds that

$$n_{j\alpha\nu}(E) \simeq \left(\sum_{\nu'} \rho_{\nu'}(\hbar\omega)\langle|M_{j\alpha\nu'}|^2\rangle \right) \frac{1}{N} \sum_{\vec{k}} |a_{j\alpha}(\vec{k}, \nu)|^2 \, \delta(E - E_{\vec{k}\nu}) f(E). \tag{8.12}$$

For this model $\langle|M_{j\alpha\nu'}|^2\rangle$ is an empirical parameter representing the average value of the square of the matrix element

$$M_{j\alpha\nu'}(\vec{k}, \vec{k}') = \frac{1}{\sqrt{N}} \sum_{m} e^{i\vec{k}\cdot\vec{R}_{mj}} \int d^3r \, \xi_\alpha(\vec{r} - \vec{R}_{mj}) \, e^{-i\vec{q}\cdot\vec{r}} \, (\vec{a}_0 \cdot \vec{p}) \, \psi_{\vec{k}'\nu'}(\vec{r}). \tag{8.13}$$

The quantity,

$$\sum_{\nu'} \rho_{\nu'}(\hbar\omega) \, \langle|M_{j\alpha\nu'}|^2\rangle \equiv \sigma_{j\alpha}(\hbar\omega), \tag{8.14}$$

is the effective cross-section for emission from the $(j\alpha)$th type orbital into the final continuum states. There are two cross-sections for the perovskites, $\sigma_p(\hbar\omega)$ and $\sigma_d(\hbar\omega)$ for the p and d orbitals, respectively. The partial energy distributions of photoelectrons of (8.12) can now be expressed in the form

$$n_{j\alpha\nu}(E) \simeq C \, \sigma_{j\alpha}(\hbar\omega) \, \rho_{j\alpha\nu}(E) \qquad (8.15)$$

where the quantity

$$\rho_{j\alpha\nu}(E) = \frac{1}{N} \sum_{\vec{k}} |a_{j\alpha}(\vec{k},\nu)|^2 \, \delta(E - \varepsilon_{\vec{k}\nu}), \qquad (8.16)$$

is the partial density of states function, abbreviated as PDOS. It specifies the portion of the DOS of the νth band that is associated with the Löwdin orbitals of the $(j\alpha)$th type. The PDOS functions satisfy the relations

$$\rho_\nu(E) = \sum_{j\alpha} \rho_{j\alpha\nu}(E), \qquad (8.17)$$

$$\rho(E) = \sum_{\nu} \rho_\nu(E), \qquad (8.18)$$

where ρ_ν is the DOS for the ν band and $\rho(E)$ is the total density of states. The sum of the PDOS functions for the d-symmetry orbitals ($d_{xy}, d_{yz}, d_{zx}, d_{x^2-y^2}$, and $d_{3z^2-r^2}$) is defined as $\rho_{d\nu}$ while the sum of the PDOS functions for the p-symmetry orbitals (x, y, z functions on each of the three oxygen sites of a unit cell) is designated by $\rho_{p\nu}$:

$$\rho_{d\nu}(E) = \sum_{d-\mathrm{symmetry}} \rho_{j\alpha\nu}(E),$$

$$\rho_{p\nu}(E) = \sum_{p-\mathrm{symmetry}} \rho_{j\alpha\nu}(E). \qquad (8.19)$$

With these definitions the energy distribution of photoelectrons can be written in the convenient form:

$$I(E) \simeq C \left\{ \sigma_d(\hbar\omega) \sum_\nu \rho_{d\nu}(E) + \sigma_p(\hbar\omega) \sum_\nu \rho_{p\nu}(E) \right\}. \qquad (8.20)$$

Equation (8.20) indicates that the electron distribution can be expressed as a weighted sum of the PDOS functions where the weighting factors are the effective photoionization cross-sections for the p and d orbitals.

The validity of (8.20) depends on, among other things, the assumptions implicit in writing (8.9); that there is no interference between emission from the different basis orbitals. However, when the squared matrix element is summed over the Brillouin zone (over all initial states) the cross-terms are subject to destructive interference while the diagonal terms always add constructively. This suggests that the diagonal contributions will dominate the cross-terms and therefore (8.9) is a reasonable approximation. In the latter portion of this chapter we shall show that

XPS experiments on perovskites appear to be in good agreement with the form of (8.20).

8.2 Partial density of states functions

In the preceding section it was found that the photoelectron energy distribution could be expressed approximately in terms of the PDOS functions. The PDOS functions can easily be obtained from the DOS functions given in Chapter 6.

The PDOS, $\rho_{p\pi^0}$ or $\rho_{p\sigma^0}$ associated with the oxygen orbitals of the non-bonding bands are the same as the corresponding DOS functions since there is no d-orbital component:

$$\rho_{p\pi^0}(E) = \rho_{\pi^0}(E) \qquad \text{(see (6.29))}, \qquad (8.21)$$

$$\rho_{p\sigma^0}(E) = \rho_{\sigma^0}(E) \qquad \text{(see (6.63))}. \qquad (8.22)$$

For the π (or π^*) bands we have from (8.16) that

$$\rho_{d\pi}(E) = \frac{1}{N} \sum_{\vec{k}} \frac{(E_\perp - E_{\vec{k}\nu}^{\alpha\beta})^2 \, \delta(E - E_{\vec{k}\nu})}{(E_\perp - E_{\vec{k}\nu}^{\alpha\beta})^2 + 4(pd\pi)^2(S_\alpha^2 + S_\beta^2)} \qquad (8.23)$$

$$\rho_{p\pi}(E) = \frac{1}{N} \sum_{\vec{k}} \frac{4(pd\pi)^2(S_\alpha^2 + S_\beta^2) \, \delta(E - E_{\vec{k}\nu})}{(E_\perp - E_{\vec{k}\nu}^{\alpha\beta})^2 + 4(pd\pi)^2(S_\alpha^2 + S_\beta^2)} . \qquad (8.24)$$

These PDOS can be obtained easily by using (4.34) and replacing $E_{\vec{k}\nu}$ by E in the coefficients of $\delta(E - E_{\vec{k}\nu})$ in (8.23) and (8.24). The result is that

$$\rho_{d\pi}(E) = \frac{\frac{1}{2}(E - E_\perp)}{[E - \frac{1}{2}(E_t + E_\perp)]} \rho_\pi(E) \qquad (8.25)$$

$$\rho_{p\pi}(E) = \frac{\frac{1}{2}(E - E_t)}{[E - \frac{1}{2}(E_t + E_\perp)]} \rho_\pi(E), \qquad (8.26)$$

where $\rho_\pi(E)$ is given by (6.28).

Similarly for the sigma bands we find

$$\rho_{d\sigma}(E) = \frac{1}{N} \sum_{\vec{k}} \frac{(E_\parallel - E_{\vec{k}\sigma})^2 X_\sigma^2}{C_\sigma^2(pd\sigma)^2} \delta(E - E_{\vec{k}\nu})$$

$$+ \frac{1}{N} \sum_{\vec{k}} \frac{[3(E_\parallel - E_{\vec{k}\sigma})(S_x^2 - S_y^2)]^2}{C_\sigma^2(pd\sigma)^2} \delta(E - E_{\vec{k}\nu})$$

$$= \frac{\frac{1}{2}(E - E_\parallel)}{[E - \frac{1}{2}(E_e + E_\parallel)]} \rho_\sigma(E) \qquad (8.27)$$

$$\rho_{p\sigma}(E) = \frac{1}{N}\sum_{\vec{k}}\frac{4S_z^2 X_\sigma^2}{C_\sigma^2}\,\delta(E - E_{\vec{k}\nu})$$

$$+\frac{1}{N}\sum_{\vec{k}}\frac{S_x^2[X_\sigma - 3(S_y^2 - S_x^2)]^2}{C_\sigma^2}\,\delta(E - E_{\vec{k}\nu})$$

$$+\frac{1}{N}\sum_{\vec{k}}\frac{S_y^2[X_\sigma + 3(S_y^2 - S_x^2)]^2}{C_\sigma^2}\,\delta(E - E_{\vec{k}\nu})$$

$$= \frac{\frac{1}{2}(E - E_e)}{[E - \frac{1}{2}(E_e + E_\|)]}\,\rho_\sigma(E)\,. \tag{8.28}$$

The quantities $E_{\vec{k}\sigma}$, X_σ, and $\rho_\sigma(E)$ are defined by (4.62)–(4.64), (6.54), and (6.62), respectively. C_σ is the normalization coefficient for the eigenstates given by (4.65).

The PDOS functions are more conveniently written in terms of the dimensionless DOS functions (see Sections 6.2 and 6.3):

$$\rho_{d\pi}(E) = \frac{(E - E_\perp)}{2(pd\pi)^2}\,\rho_\pi(\varepsilon_\pi)\,,$$

$$\rho_{p\pi}(E) = \frac{(E - E_t)}{2(pd\pi)^2}\,\rho_\pi(\varepsilon_\pi)\,,$$

$$\rho_{d\sigma}(E) = \frac{(E - E_\|)}{(pd\sigma)^2}\,\rho_\sigma(\varepsilon_\sigma)\,,$$

$$\rho_{p\sigma}(E) = \frac{(E - E_e)}{(pd\sigma)^2}\,\rho_\sigma(\varepsilon_\sigma)\,, \tag{8.29}$$

where

$$\rho_\pi(\varepsilon_\pi) = \rho_\pi(\varepsilon_\pi(E)) = \frac{1}{\pi^2}K\left(\sqrt{1 - [\varepsilon_\pi(E)/2]^2}\right)\Theta\left(1 - [\varepsilon_\pi(E)/2]^2\right)\,,$$

$$\rho_\sigma(\varepsilon_\sigma) = \rho_\sigma(\varepsilon_\sigma(E))$$

$$= \left[0.3183 + 0.1136\,x^2 - 0.0151\,(1 - x)\,\sqrt{x}\,\right]\Theta\left(1 - [\varepsilon_\sigma(E)/3]^2\right)$$

$$+ \left[0.432 - 0.1646\sqrt{1 - \varepsilon_\sigma^2} - 0.0151(1 - |\varepsilon_\sigma|)\,|\varepsilon_\sigma|\,\right]\Theta\left(1 - \varepsilon_\sigma(E)^2\right)\,,$$

with

$$x \equiv (3 - |\varepsilon_\sigma|)/2\,,$$

$$\varepsilon_\pi(E) = \frac{(E - E_t)(E - E_\perp)}{2(pd\pi)^2} - 2\,,$$

$$\varepsilon_\sigma(E) = \frac{(E - E_e)(E - E_\|)}{(pd\sigma)^2} - 3\,. \tag{8.30}$$

Finally, we can write the total (all-bands) PDOS functions as

$$\rho_d(E) = 3\rho_{d\pi}(E) + \rho_{d\sigma}(E),$$ (8.31)

$$\rho_p(E) = 3\rho_{p\pi}(E) + \rho_{p\sigma}(E) + 3\rho_{\pi^0}(E) + \rho_{\sigma^0}(E).$$ (8.32)

(Note that the function $\rho_{d\sigma}$ gives the DOS of both sigma bands.)

8.3 The XPS spectrum of SrTiO₃

The XPS spectrum of SrTiO₃ has been reported by Battye *et al.* [7] (and also by Kowalczyk *et al.* [8] as well as Sarma *et al.* [9]) and analyzed by Wolfram and Ellialtıoğlu [6]. The result of Battye *et al.* [7] is shown in Fig. 8.1 (dotted curve). Since there are no electrons occupying the conduction bands the emission arises entirely from the valence bands. The data represent $I(E)$ the rate of emission of electrons from initial state of energy E. Also shown in Fig. 8.1 (solid curve) is the theoretical $I(E)$ curve obtained from (8.20) with $\sigma_p(\hbar\omega)/\sigma_d(\hbar\omega) = 1/3$. The theoretical $I(E)$ possesses several features (labeled 1 through 6) which are related to the energy band structure. The peak centered at about $-1.7\,\text{eV}$ and labeled 1, is due to the π^0 and σ^0 non-bonding bands. The edges labeled 2 and 4 are the top and bottom of the π valence bands, respectively. Feature 3 is the logarithmic singularity

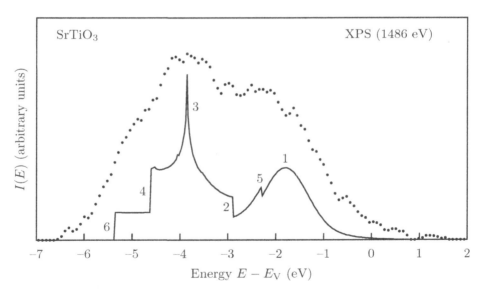

Figure 8.1. XPS photoelectron energy distribution $I(E)$ for SrTiO₃. The initial state energy is measured from the top of the valence band, E_V. ($\hbar\omega = 1486\,\text{eV}$ for Al K_α line.) The theoretical $I(E)$ is indicated by the solid curve. The XPS data is from [7].

in $\rho_\pi(E)$ at the center of the π band. The top and bottom of the σ valence bands correspond to features 5 and 6, respectively.

In order to make a direct comparison between $I(E)$ and the XPS data the theoretical curve should be broadened to account for the experimental resolution. The broadened curve, denoted by $\langle I(E)\rangle$, may be obtained from $I(E)$ in the following manner:

$$\langle I(E)\rangle = \frac{1}{R\sqrt{\pi}} \int_{-\infty}^{\infty} I(E') \exp\left\{-\left(\frac{E-E'}{R}\right)^2\right\} dE', \qquad (8.33)$$

where the parameter R determines the resolution. The resolution decreases as R increases. The FWHM (full-width at half-maximum)

$$\text{FWHM} = 2\,R\,\sqrt{\ln 2} = 1.665\,R. \qquad (8.34)$$

The instrumental resolution for most XPS data is about $0.55\,\text{eV}$ so that $R_{\text{inst.}} = 0.33\,\text{eV}$. Figure 8.2 shows $\langle I(E)\rangle$ (thin dashed line) compared with the experimental data for $SrTiO_3$ using $R = R_{\text{inst.}} = 0.33\,\text{eV}$. The solid curve passing through the data is $\langle I(E)\rangle$ with a resolution of $1.34\,\text{eV}$ ($R=0.8\,\text{eV}$). The agreement between the data and $\langle I(E)\rangle$ with $R=0.8\,\text{eV}$ is excellent and suggests that the effective experimental resolution is less (is not as good) than the instrumental resolution.

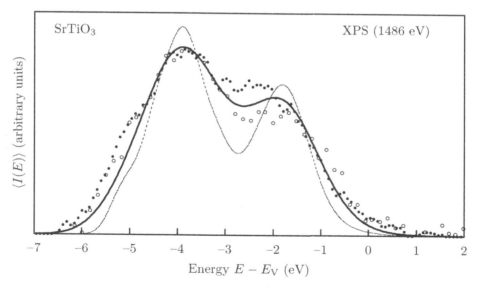

Figure 8.2. XPS photoelectron energy distributions (dots [7] and little circles [9]) compared with $\langle I(E)\rangle$ for resolution parameters of $R=0.33\,\text{eV}$ (FWHM $0.55\,\text{eV}$, thin dashed line) and $R=0.8\,\text{eV}$ (FWHM $1.34\,\text{eV}$, thick solid line). The cross-section ratio is $\sigma_p/\sigma_d \simeq 1/3$.

The theoretical fit, $\langle I(E) \rangle$ in Fig. 8.2 indicates that the cross-section ratio is $\sigma_p(1486\,\mathrm{eV})/\sigma_d(1486\,\mathrm{eV}) \simeq 1/3$. The cross-sections are dependent on the energy of the photons used in the photoemission experiment. For most XPS experiments with $10^3 \lesssim \hbar\omega \lesssim 3 \times 10^3$ eV the cross-sections probably do not vary much. However, UPS photoemission is performed with much lower photon energies (typically 5–50 eV) and σ_p/σ_d can differ substantially from that found in XPS. As an example of this effect consider the UPS spectrum of $SrTiO_3$ for $\hbar\omega = 21.2$ eV. The spectrum (open circles) reported by Henrich *et al.* [10] is shown in Fig. 8.3. The solid curve is $\langle I(E) \rangle$ calculated for the same parameters as used for the (solid curve) $\langle I(E) \rangle$ in Fig. 8.2 except that $\sigma_p/\sigma_d = 1$. The agreement between theory and experiment is essentially exact. Thus, it appears that $\sigma_p(21.2\,\mathrm{eV})/\sigma_d(21.2\,\mathrm{eV}) = 1$ and therefore the 21.2 eV UPS spectrum closely resembles the total valence-band DOS of $SrTiO_3$.

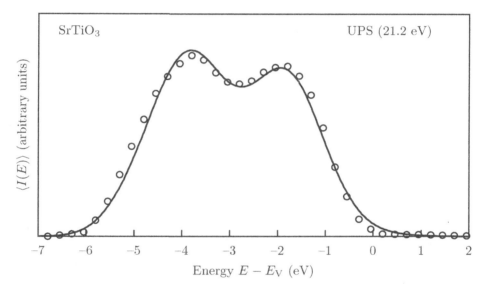

Figure 8.3. UPS spectrum of $SrTiO_3$ from [10] (open circles) compared with $\langle I(E) \rangle$ for $\sigma_p/\sigma_d = 1$ and $R = 0.8$ eV.

Some further comment on the UPS analysis is needed since 21.2 eV is not sufficiently large to use the arguments employed in Section 8.1. In particular, it can not be argued that the initial state energy is small compared to $\hbar\omega$, since $\hbar\omega = 21.2$ eV is comparable to the valence-band width.

In addition, modulation of the spectrum by varying matrix elements can also be expected. The reason that the partial DOS model still applies is that the UPS final states are energy bands derived from the Ti(4p), Ti(4s), and Sr(4s) orbitals. These bands are presumably very broad and produce an approximately constant final state DOS. The similarity of the UPS and XPS spectra tend to support this conclusion. Further evidence comes from the studies of Powell and Spicer [11], Derbenwick [12] and Henrich *et al.* [10] which suggest that the spectrum is not changing rapidly with $\hbar\omega$ for $12 \leq \hbar\omega \lesssim 21\,\text{eV}$.

8.4 Na_xWO_3

In Chapter 6, Section 6.4 we discussed the x dependence of the electronic proper-ties of Na_xWO_3. These properties also influence the photoelectron distributions. In addition to the x dependence the electronic parameters, the effective photoioniza-tion cross-section ratio, σ_d/σ_p is approximately 12, for high-energy XPS and of the order of 1 for UPS experiments [6].

For convenience the x-dependent parameters described in Section 6.4 (see (6.93)) are given below:

$$E_g(x) = 2.85 + 0.84\,x, \qquad \text{(in eV)} \qquad\qquad (8.35\text{a})$$

$$(pd\pi(x)) = 2.44 - 0.86\,x, \qquad \text{(in eV)} \qquad\qquad (8.35\text{b})$$

$$a(x) = \frac{1}{2}(3.785 + 0.0818\,x), \quad (\text{W}-\text{O distance in Å}) \qquad (8.35\text{c})$$

$$\varepsilon_F(x) = -2 + (2 - 1.086)\,x, \qquad \text{(in eV)} \qquad\qquad (8.35\text{d})$$

$$\rho_\pi(x) = \frac{6}{2\pi}(1 + 0.326\,x). \qquad\qquad\qquad (8.35\text{e})$$

Using (8.20), (8.33), and (8.35) the XPS photoelectron distributions as a function of x can be calculated. The results for $I(E)$ are shown in Fig. 8.4 for several values of x [13] and compared with the UPS experimental results of Hollinger *et al.* [14]. The scale factor, C, of (8.20) was chosen so that the theoretical peak intensity matched the experimental peak for $x = 0.4$. All other factors are known. The photo-emitted electron distributions arise from electrons occupying the π^* bands. The energies of these bands relative to the top of the valence band and the band widths change with x because of the x dependence of E_g and $(pd\pi)$. The shifting of the peak intensity toward higher energy as x increases is quite evident in Fig. 8.4.

The agreement between theory for XPS and the experimental UPS results is quite remarkable considering that strong transition matrix elements effects can modulate the UPS intensity curves. The areas under the theoretical curves compare well with the areas of the experimental curves even though no adjustment have been made to normalize the areas. Plasmon effects which contribute to the intensity tail in the band-gap region on the low-energy side are not included in the theoretical curves of Fig. 8.4. In addition, $\sigma_d/\sigma_p = 12$ for the theoretical curves while for UPS energies it should be approximately 1. However, this difference does not have a strong effect

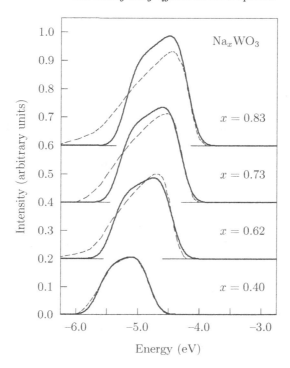

Figure 8.4. Comparison of the theoretical (solid curves) [13] and experimental (dashed curves) [14] photoemission energy distribution curves for several values of x.

on the $Na_x WO_3$ theoretical curves because the amount of p-orbital mixing into the band states of the lower part of the π^* bands is small. The contribution of the plasmons is discussed in the next section.

8.5 Many-body effects in XPS spectra

When an electron is photo-ejected from a metallic solid, there is simultaneously created a "hole" in the Fermi sea of electrons. The hole produces a number of effects which are observed in the photoemission spectra of a solid.

If an electron is instantaneously ejected from a solid the state of the remaining $(N-1)$-electron system will not in general be the ground state. If we define the binding energy E_{bind} of the emitted electron as the difference between the ground state energies of the N- and $(N-1)$-electron system, then it is clear that the energy, E, deduced from the kinetic energy of the emitted electron is not E_{bind}. In fact, $E \leq E_{bind}$ because the $(N-1)$-electron system is left in an excited state after photoemission process. The difference $E_r = |E_{bind} - E|$ is called the relaxation energy.

The quantity E_r is the decrease in the energy of the $(N-1)$-electron system after it "relaxes" to its ground state. Relaxation shifts of electrons emitted from core levels are often observed in photoemission experiments and have been the topic of theoretical discussions [15–18].

The particular manner in which a solid manifests hole relaxation in photoemission depends upon the photon energy, the electronic structure of the solid, and the state from which the electron is emitted. If $\hbar\omega$ is near to the photoionization threshold the kinetic energy of the photoelectron will be small. It will move slowly away from the ion core and consequently its dynamics will be strongly influenced by the attractive potential of the hole it leaves behind. If the emission were slow enough to justify adiabatic relaxation of the $(N-1)$-electron system then the kinetic energy of the emitted electron at the onset of photoemission would approach $E_{bind} - \Phi$. (However, this energy would not be that calculated by the usual one-electron energy band theory; that is, it is not the eigenvalue based on Koopman's theorem which was discussed in Section 2.3.) For XPS experiments the photoelectron kinetic energy is large and the "sudden approximation" is nearly valid. The hole potential appears to be suddenly switched-on. According to the sudden approximation of perturbation theory the $(N-1)$-electron system may be described as a superposition of the eigenstates of the new Hamiltonian; the original Hamiltonian for the N-electron system plus the hole potential. Therefore, there is a specific probability for each excited state of the $(N-1)$-electron system. If E_α is the excited state energy and E_0 is the ground state energy then there will be a distribution of peaks in the kinetic energies, $E_{kin,\alpha}$, of the emitted electrons at

$$E_{kin,\alpha} = \hbar\omega - \Phi - (E_\alpha - E_0). \tag{8.36}$$

For emission from core levels these series of peaks of (8.36) are called "shake-up" peaks. In addition, during relaxation from the αth excited state to the ground state there is a probability of a second electron being ejected. The latter spectrum is called a shake-off satellite.

In a metallic material the conduction electrons will move rapidly to neutralize a photohole. The adjustment of the conduction electrons to the hole has two important effects. First, electrons with energies near the Fermi energy will relax by making transitions from states just below the Fermi level to excited single-particle states above the Fermi level. A large number of low-energy electron–hole pairs can be generated and their effect is to produce a "tail" on the low-energy side of any characteristic peak in the XPS spectrum [19].

In addition to single-particle excitations, collective excitations in the form of plasma oscillations can also be stimulated by hole relaxation of the conduction electrons [20]. The excitation of plasmons produces satellite lines and structure in both core-level and valence-band spectra. These plasmon effects are particularly strong in metallic perovskites such as ReO_3 and alkali tungsten bronzes. For example, in Na_xWO_3 and H_xWO_3 intense core-level satellite lines and band-gap emission associated with plasmon creation have been observed [21, 22].

In order to analyze the XPS spectra of metallic perovskites it is necessary to include the effect of plasmon creation on the photoelectron distribution. When hole relaxation is accompanied by the creation of a plasmon an electron emitted from an initial state of energy E will appear to have originated from a state at $E - E_{pl}$, where E_{pl} is the plasmon energy. This effect can be included in the theoretical model of $I(E)$ by using an apparent distributions $I'(E)$ which includes the plasmon-shifted electrons. We define

$$I'(E) = (1 - \beta)\,I(E) + \frac{\beta}{\Gamma\sqrt{\pi}} \int_{-\infty}^{\infty} dE' \exp\left\{-\left(\frac{E - E'}{\Gamma}\right)^2\right\} I(E' + E_{pl}),$$

$$(8.37)$$

where β is the plasmon creation probability. The factor $(1 - \beta)$ is the fraction of emitted electrons not accompanied by plasmon creation. The second term accounts for the photoelectrons that are down-shifted in energy because of the plasmon effect. The plasmon band is represented by a Gaussian distribution centered at E_{pl}. The band width FWHM$= 2\,\Gamma\,\sqrt{\ln 2}$. In the limit as $\Gamma \to 0$, since then

$$I'(E) \to (1 - \beta)\,I(E) + \beta\,I(E + E_{pl}) \qquad (8.38)$$

which states that the apparent number of photoelectrons from initial states at E is the sum of contributions due to unshifted photoelectrons from states at E plus the number from states at $E + E_{pl}$ which were down-shifted in energy due to plasmon creation.

In comparing $I'(E)$ with experiment the distribution must be convolved with an experimental resolution function precisely as in (8.33) to produce the function $\langle I'(E)\rangle$.

A theoretical analysis of the XPS spectrum of Na_xWO_3 [21, 22] has been carried out by Wolfram and Ellialtıoğlu [6] using (8.37). Their results are shown in Fig. 8.5. The data is the dotted curve, the function $I'(E)$ is the dashed curve and $\langle I'(E)\rangle$ is the solid curve. The analysis provides a simple interpretation of the XPS data. The peak near the Fermi level, E_F, is due to electrons emitted from the partially filled π^* bands, which contain 0.8 electrons per unit cell. The small peak in the band-gap region (~ -1 to $3\,eV$) is due to conduction-band electrons shifted down in energy due to plasmon creation associated with hole relaxation. The plasmon energy $E_{pl} = 2.0\,eV$ [21, 22]. The peak in $I'(E)$ (the shoulder in the data) near $-4.8\,eV$ is due to emission of electrons from the non-bonding (π^0 and σ^0) bands. The large central peak is produced by the logarithmic peak in the π valence-band DOS. The lowest peak, near $-10\,eV$, arises from the jump discontinuity in the DOS

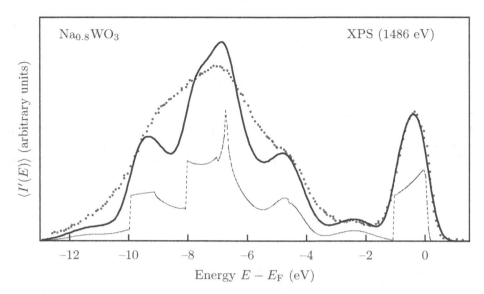

Figure 8.5. Comparison of the XPS valence-band spectrum of $Na_{0.8}WO_3$ with theory. Data is the dotted curve [22], function $I'(E)$ is the dashed curve and $\langle I'(E)\rangle$ is the solid curve.

at the bottom of the σ valence band. The tail from about -10 to $-12\,eV$ is due to σ band electrons shifted down in energy by the plasmon effect. The analysis indicates that $\sigma_p/\sigma_d(1486\,eV)$ is about $1/12$ and that β, the probability of plasmon creation, is 0.2. Similar results are obtained from the theoretical analysis of the XPS spectra of ReO_3 and H_xWO_3 [5, 21, 22].

The PDOS model, (8.20), appears to provide a useful method for analyzing the XPS spectra of the perovskites. Application of the model indicates σ_p/σ_d varies roughly between 0.3 and 0.1 for many of the perovskites. For the metallic perovskites the many-body plasmon excitation probability is about 0.2 for both core-level and valence-band emission.

References

[1] A. Damascelli, Z. Hussain, and Z.-X. Shen, *Rev. Mod. Phys.* **75**, 473 (2003), (A. Damascelli, Z.-X. Shen, and Z. Hussain, arXiv:cond-mat/0208504 v1 27 Aug 2002).

[2] N. V. Smith, *Phys. Rev. B* **3**, 1862 (1971).

[3] D. A. Shirley, *Phys. Rev. B* **5**, 4709 (1972).

[4] J. Freeouf, M. Erbudak, and D. E. Eastman, *Solid State Commun.* **13**, 771 (1973).

[5] G. K. Wertheim, L. F. Mattheiss, M. Campagna, and T. P. Pearsall, *Phys. Rev. Lett.* **32**, 997 (1974).

[6] T. Wolfram and Ş. Ellialtıoğlu, *Phys. Rev.* B **19**, 43 (1979).

[7] F. L. Battye, H. Höchst, and A. Goldmann, *Solid State Commun.* **19**, 269 (1976).

[8] S. P. Kowalczyk, F. R. McFeely, L. Ley, V. T. Gritsyna, and D. A. Shirley, *Solid State Commun.* **23**, 161 (1977).

[9] D. D. Sarma, N. Shanthi, S. R. Barman, N. Hamada, H. Sawada, and K. Terakura, *Phys. Rev. Lett.* **75**, 1126 (1995).

[10] V. E. Henrich, G. Dresselhaus, and H. J. Zeiger, *Bull. Am. Phys. (Soc. II)*, **22** (1977), *Solid State Research Report*, Lincoln Laboratory, MIT (1976), p.39.

[11] R. A. Powell and W. E. Spicer, *Phys. Rev.* B **13**, 2601 (1976).

[12] G. F. Derbenwick, *Ph.D. Thesis* (Stanford University, 1970) unpublished.

[13] T. Wolfram and L. Sutcu, *Phys. Rev.* B **31**, 7680 (1985).

[14] G. Hollinger, F. J. Himpsel, B. Reihl, P. Petrosa, and J. P. Doumerc, *Solid State Commun.* **44**, 1221 (1982).

[15] D. W. Davis and D. A. Shirley, *J. Electron Spectrosc.* **3**, 137 (1974).

[16] D. A. Shirley, *J. Electron Spectrosc.* **5**, 135 (1974).

[17] P. H. Citrin and D. R. Hamann, *Chem. Phys. Lett.* **22**, 301 (1973); *Phys. Rev.* B **10**, 4948 (1974).

[18] J. W. Gadzuk, *J. Vac. Sci. Technol.* **12**, 289 (1975).

[19] S. Doniach and M. Šunjić, *J. Phys.* C **3**, 285 (1970).

[20] M. Šunjić and D. Šokćević, *J. Electron Spectrosc.* **5**, 963 (1974).

[21] J-N. Chazalviel, M. Campagna, G. K. Wertheim, and H. R. Shanks, *Phys. Rev.* B **16**, 697 (1977).

[22] M. Campagna, G. K. Wertheim, H. R. Shanks, F. Zumsteg, and E. Banks, *Phys. Rev. Lett.* **34**, 738 (1975).

Problems for Chapter 8

1. Derive (8.25) and (8.26) from (8.23).

2. Make a graph of $\rho_{d\pi}(E)$, given in (8.25), using the parameters $(pd\pi) = 1\,\text{eV}$, $E_t = -5\,\text{eV}$, and $E_\perp = -8.2\,\text{eV}$. A table of $K(x)$ is given in Appendix B, and $\rho(E)$ is given in (8.30). Discuss the results in terms of covalency of the band states.

3. The angular frequency, ω_p, for plasma oscillations of a metal is given by the relation $\omega_p^2 \equiv 4\pi n_e e^2/m_e$, where n_e is the electron density. Show that in Gaussian (CGS) units $\omega_p = 5.65 \times 10^4\ n_e^{1/2}$ rad/s when n_e is the number of electrons per cm^3. Calculate the plasmon energy for $NaWO_3$ in eV.

4. The constant matrix element approximation for $\varepsilon_2(\omega)$ discussed in Chapter 7 leads to a description involving the JDOS. The constant matrix element approximation for the XPS energy distribution curve leads to a description involving the DOS. Explain the major factors that cause this difference.

5. Explain the following terms used in the description of photoemission:
   (a) hole relaxation;
   (b) shake-up peaks;
   (c) shake-off satellites.

9

Surface states on d-band perovskites

When a crystalline solid is terminated by a surface new types of energy bands can form that are localized at or near the surface. A geometrically perfect surface may have "intrinsic" surface states with energies lying within the band-gap region, above or below the bulk energy bands. These surface bands have wavefunctions that decrease exponentially with increasing distance into the crystal. The further the energy of the surface state is from the bulk band-edge energy, the more rapidly its wavefunction decreases with distance from the surface. Localized surface states associated with defects such as oxygen vacancies can also occur. Such surface bands and states can play an important role in chemisorption and catalysis in transition metal oxides.

In this chapter we will review the theoretical concepts that underlie the formation of surface bands and defect states based on our empirical LCAO model. The material is essential to the understanding of more fundamental and accurate calculation methods. A comprehensive review of the experiments on transition metal-oxide surfaces is available [1], but only those relevant to doped insulating perovskites and metallic $Na_x WO_3$ will be discussed here.

9.1 Perturbations at a surface

Figure 9.1 illustrates the two types of (001) surfaces for the perovskite structure. For the type I surface the B and oxygen ions are on the surface. For the type II surface the lattice starts with a layer of A and oxygen ions.

A number of different perturbations occur when a surface is formed even if the lattice is terminated in a geometrically perfect way. In many cases the atomic layers near the surface will "relax" by changing their interlayer and interatomic distances. For example, the distance between the first and second layers of $SrTiO_3$ differ by about 5% from the interior layer spacing [1–3]. The LCAO interaction parameters

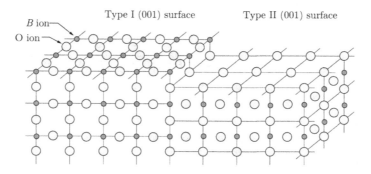

Figure 9.1. Type I and type II (001) surfaces of the cubic perovskite ABO_3 structure. The small circles represent B ions, and the large circles are the oxygen ions. The A ions are not shown.

such as $(pd\pi)$ and $(pd\sigma)$ depend exponentially on distance, therefore an expansion or contraction of 5% can lead to substantial changes in the parameters.

A surface can "reconstruct" by forming atomic patterns unlike those of the interior or by forming steps and terraces or a puckered surface. Patches of type I and type II surfaces can be expected on fractured surfaces. These perturbations alter not only the LCAO parameters, but also the electrostatic potentials. Occupied surface states occurring in the band gap can create a surface dipole layer, cause band bending, and may in some cases "pin" the Fermi energy.

The surface ions have missing neighbors (unsaturated bonds) that alter the potentials experienced by the ions at and near the surface. For example, at a perfectly terminated type I (001) surface of ABO_3, cubic perovskite, the Madelung potential at a surface B-ion site is a few eV less repulsive than at an interior site [4]. On the other hand, the Madelung potential for a surface oxygen ion is essentially unchanged from its bulk value. For a type II surface the change in the oxygen potential is large and that for the B ion much smaller (see Appendix D for details).

The point group symmetries of the surface ions are also different from that of the interior ions, causing additional splitting of the electronic levels. The point group of a B ion on a (001) surface is C_{4v} rather than O_h. As a result the t_{2g} group, threefold degenerate for O_h symmetry, is split into a doubly degenerate "e" level and a singlet, b_2 level. The point group symmetry of a surface oxygen is also different and splitting of the doubly degenerate oxygen pi level ($p_\perp$) occurs. Figure 9.2 shows schematically how the ion splittings are changed at the surface.

In general, the effects of the surface perturbations are largely confined to the first one or two layers of the solid, but some surface effects may extend for large distances. For example, for ferroelectrics such as $BaTiO_3$ the surface creates a large, long-range depolarization field.

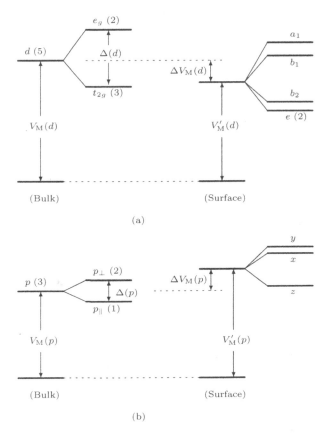

Figure 9.2. Electrostatic splitting of ion levels at a (001) surface. The left-hand side shows the bulk splitting and the right-hand side shows the splittings of ions on the surface for a type I surface: (a) d-orbital splittings and (b) p-orbital splittings. ΔV_{M} is the shift in the Madelung potential.

In addition to "natural" changes that occur when a surface is formed, the surface properties are also strongly influenced by the method of preparation and surface treatment. Surfaces are formed by cleaving, fracturing, or by chemical or epitaxial growth. In the performance of experiments a surface is often subjected to various surface treatments such as polishing, annealing, or bombardment with ions. Annealing in the presence of a reducing atmosphere (hydrogen for example) or an oxidizing agent results in n- or p-type doping. Bombardment of a surface with ions (argon ions, for example) is commonly used to clean the surface of impurities, but also results in surface oxygen vacancies, leading to surface defect states.

9.2 Surface energy band concepts

(a) The (001) surface

In this section we shall be concerned with an ideal, semi-infinite perovskite terminated by a (001) surface. As mentioned above there are two possible configurations
designated as *type I* and *type II*. The type I surface has the B ion and O ions
exposed on the first atomic layer forming a BO_2 surface layer. The second atomic
layer has A cations and oxygen ions present and forms an AO layer. Therefore,
the semi-infinite, type I, perovskite lattice consists of atomic layers parallel to the
surface in the sequence BO_2–AO–BO_2–AO– $\cdots$. The type II (001) surface begins
with the AO layer and forms the sequence AO–BO_2–AO–BO_2– $\cdots$.

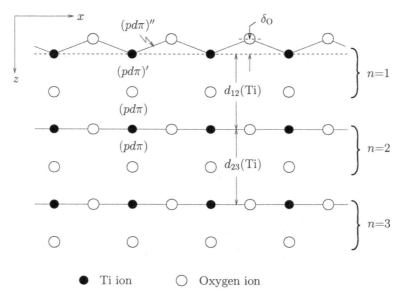

Figure 9.3. Schematic of a type I (001) surface showing the displacement of the surface
oxygen ions normal to the surface. The interlayer spacings are designated by $d_{12}(\mathrm{Ti})$ and
$d_{23}(\mathrm{Ti})$. Two atomic planes make up a unit-cell layer. They are indicated on the right-hand
side of the diagram. Also shown are the perturbed LCAO parameters $(pd\pi)'$ and $(pd\pi)''$.

A schematic type I (001) surface is shown in Fig. 9.3. For SrTiO$_3$ it is found
that the surface oxygen ions are slightly displaced perpendicular to the surface and
the first two layers deviate from the bulk interlayer spacing by a few percent.

Beginning at the surface, each pair of successive atomic layers forms a unit-
cell layer with the composition ABO_3. If we number the unit-cell layers starting
from $n = 1$ at the surface to $n = \infty$, the position of any atom, $\vec{R}_{j,m} + \vec{\tau}_{j,m}$, can be

specified by

$$\vec{R}_{j,m} = \vec{\rho}_{j,m} + \vec{z}_j(n) \tag{9.1}$$

where $\vec{\rho}_{j,m} = (x_{j,m} + \tau^x_{j,m})\,\vec{e}_x + (y_{j,m} + \tau^y_{j,m})\,\vec{e}_y$ is the projection of $\vec{R}_{j,m}$ on the xy-plane, and $\vec{z}_j(n) = [z_n + \tau^z_j(n)]\,\vec{e}_z$. Here z_n is the distance of a B ion from the surface, which for the infinite lattice is $2(n-1)a$, where $2a$ is the lattice spacing. For the semi-infinite lattice the interlayer spacing may not be uniform. The notation allows for the lattice spacing perpendicular to the surface to depend upon n, but assumes that the $x - y$ spacing is the same as in the bulk.

To begin our study of the surface energy bands we shall use the same nearest-neighbor model employed for the discussion of the bulk bands. Since the electronic and surface properties are principally determined by the π^* bands and the band-gap surface states we shall limit our discussion to the pi bands. A more complete discussion of the various types of surface bands that are possible can be found in references [4] and [5].

We need to specify the types of surface perturbations to be considered. It might be supposed that long-range Coulomb potentials such as the Madelung potentials could be altered over many atomic layers near the surface. However, as mentioned earlier, calculations [4] show that the Madelung potentials approach their bulk values after the first atomic layer.

The changes in the d-orbital site potentials and the electron–electron repulsion energy are the largest energies involved in the surface problem. For n-doped insulators the density of d electrons at the surface is much larger than for the interior ions when states form in the band-gap region. Therefore special attention must be paid to the Coulomb repulsion effects.

The next largest energies are the changes in LCAO two-center integrals such as $(pd\pi)$. We shall consider the perturbations in the first unit-cell layer, but assume that all other layers are described by the same parameters as for the infinite lattice.

For a geometrically perfect, (001) surface the pi and sigma bands do not mix and may be considered separately just as in the case of the bulk energy bands. When there are small displacements of the surface oxygen ions the pi and sigma orbitals are coupled by two-center integrals that are first order in the displacement, but the energy is affected only in second order. In this chapter we shall ignore the small mixing between the pi and sigma orbitals.

There are three different pi-type surface bands: those involving d_{yz} orbitals, those involving d_{xz} orbitals and those involving d_{xy} orbitals. We shall refer to these bands as the pi(yz), pi(xz), and pi(xy) surface bands. The parameters for the type

I, pi(yz) surface bands are:

$$E_t(n) = E_t + [\Delta E_t + UN_s]\delta_{n,1} \tag{9.2}$$

$$E_\perp(n) = E_\perp + \Delta E_\perp \delta_{n,1} \tag{9.3}$$

$$(pd\pi)_n = \begin{cases} (1 + \Delta'' \, \delta_{n,1})(pd\pi) & \text{between surface } d \text{ and } p \text{ orbitals} \\ (1 + \Delta' \, \delta_{n,1})(pd\pi) & \text{between surface } d \text{ and the} \\ & \text{first subsurface } p \text{ orbitals} \\ (pd\pi) & \text{otherwise} \end{cases} \tag{9.4}$$

ΔE_t and $\Delta E_\perp$ are the changes in diagonal matrix elements for the B and O sites, respectively, and Δ' and Δ'' are the fractional changes in the p–d interactions. In (9.2) U is the Coulomb repulsion among the electrons occupying the same surface d orbital and N_s is the number of electrons occupying the surface state per spin.

The surface energy bands and the potential, UN_s, must be calculated self-consistently. The total number of electrons occupying surface states is $N_s = N_s(xy) + N_s(yz) + N_s(xz)$, where $N_s(\alpha\beta)$ is the number of electrons occupying the pi($\alpha\beta$) surface band per spin state. The self-consistent solutions for the surface bands require that all three surface bands be considered simultaneously.

(b) Pi(yz) surface energy bands

Surface states involving the d_{yz}, $p_y(\vec{r} - a\vec{e}_z)$, and $p_z(\vec{r} - a\vec{e}_y)$ orbitals are symmetry equivalent to those states involving d_{xz}, $p_x(\vec{r} - a\vec{e}_z)$, and $p_z(\vec{r} - a\vec{e}_x)$, and therefore, we need only consider the pi(yz) states.

The LCAO equations that determine the eigenvalues and eigenvectors for the semi-infinite lattice are:

$$(\omega_t' - \omega)\, c_{yz}(1) + 2iS_y(1 + \Delta'')\, c_z(1) + (1 + \Delta')\, c_y(1) = 0, \tag{9.5}$$

$$(\omega_\perp' - \omega)\, c_z(1) - 2iS_y(1 + \Delta'')\, c_{yz}(1) = 0, \tag{9.6}$$

$$(\omega_\perp - \omega)\, c_y(1) + (1 + \Delta')\, c_{yz}(1) - c_{yz}(2) = 0, \tag{9.7}$$

and for $n > 1$

$$(\omega_t - \omega)\, c_{yz}(n) + 2iS_y\, c_z(n) + c_y(n) - c_y(n-1) = 0 \tag{9.8}$$

$$(\omega_\perp - \omega)\, c_z(n) - 2iS_y\, c_{yz}(n) = 0, \tag{9.9}$$

$$(\omega_\perp - \omega)\, c_y(n) + c_{yz}(n) - c_{yz}(n+1) = 0, \tag{9.10}$$

with $S_y = \sin k_y a$. The terms $c_{yz}(n)$, $c_z(n)$, and $c_y(n)$ are the amplitudes of the $d_{yz}(n)$, $p_z(n)$, and $p_y(n)$ orbitals, respectively. In (9.5)–(9.10) we have introduced a number of dimensionless quantities,

$$\omega = E/(pd\pi), \tag{9.11a}$$
$$\omega_t = E_t/(pd\pi), \tag{9.11b}$$
$$\omega_\perp = E_\perp/(pd\pi), \tag{9.11c}$$
$$\Delta\omega_t = \Delta E_t/(pd\pi), \tag{9.11d}$$
$$\Delta\omega_\perp = \Delta E_\perp/(pd\pi), \tag{9.11e}$$
$$u = U/(pd\pi), \tag{9.11f}$$
$$\omega_t' = \omega_t + \Delta\omega_t + uN_s, \tag{9.11g}$$
$$\omega_\perp' = \omega_\perp + \Delta\omega_\perp, \tag{9.11h}$$
$$\omega_g = E_g/(pd\pi). \tag{9.11i}$$

The amplitudes c_y and c_z can be expressed in terms of c_{yz} using (9.6), (9.7), (9.9) and (9.10). Substitution of these results into (9.5) and (9.8) yields the reduced secular equation,

$$\left[-2\cos\theta + \Delta p(\omega, k_y)\right] c_{yz}(1) + \left(1 + \Delta'\right) c_{yz}(2) = 0, \tag{9.12}$$
$$-2\cos\theta\, c_{yz}(n) + c_{yz}(n+1) + c_{yz}(n-1) = 0, \tag{9.13}$$

where

$$-2\cos\theta = (\omega_t - \omega)(\omega_\perp - \omega) - 4S_y^2 - 2 \tag{9.14}$$

and

$$\Delta p(\omega, k_y) = (\Delta\omega_t + uN_s)(\omega_\perp - \omega) - (1 + \Delta')^2 - 4S_y^2 \left\{ \frac{(\omega_\perp - \omega)(1 + \Delta'')^2}{(\omega_\perp' - \omega)} - 1 \right\} + 2, \tag{9.15a}$$
$$\Delta' = \left[(pd\pi)' - (pd\pi)\right]/(pd\pi), \tag{9.15b}$$
$$\Delta'' = \left[(pd\pi)'' - (pd\pi)\right]/(pd\pi). \tag{9.15c}$$

Equations (9.12) and (9.13) are simple second-order difference equations and the general solutions are:

$$c_{yz}(n) = \frac{1}{\sqrt{N}}\left(e^{in\theta} + \Lambda e^{-in\theta}\right), \tag{9.16}$$

$$\Lambda = -\left\{ \frac{1 - \Delta p(\omega, k_y)e^{i\theta} - \Delta' e^{2i\theta}}{1 - \Delta p(\omega, k_y)e^{-i\theta} - \Delta' e^{-2i\theta}} \right\} \tag{9.17}$$

where N is a normalization constant.

(c) Classification of the states

The solutions of (9.16) can be classified as *volume states* or *surface states* depending upon the behavior of the wavefunction amplitudes at large n. The volume states have wavefunctions whose amplitudes extend unattenuated throughout the entire semi-infinite lattice while the wavefunctions for the surface states have amplitudes that decrease exponentially with increasing distance into the solid.

Volume states

The conditions for a volume state are that its wavefunction remains bounded and non-vanishing as $n \rightarrow \infty$. These two conditions can be met only if the factor, θ, in equation (9.16) is a real number. For real θ the denominator in (9.17) is the negative complex conjugate of the numerator and hence Λ is a complex number with unit modulus. Therefore we may write $\Lambda = -e^{-i\delta}$, where δ is a real number and

$$c_{yz}(n) = \frac{1}{\sqrt{N}}(e^{in\theta} - e^{-i(n\theta+\delta)}). \tag{9.18}$$

Since θ is a real number it follows that $|\cos\theta| \leq 1$. Using this result with (9.14) yields the inequality:

$$|(\varepsilon_t - w)(\varepsilon_\perp - w) - 4S_y^2 - 2| \leq 2. \tag{9.19}$$

This equation may be solved to determine the possible values of the dimensionless energy, w, for which volume states can exist. One finds two regions of (w, k_y)-space which satisfy (9.19):

$$w_{\pi^*}(k_y, 0) \leq w_{vol}(k_y) \leq w_{\pi^*}\left(k_y, \frac{\pi}{2a}\right), \tag{9.20}$$

$$w_\pi(k_y, 0) \leq w_{vol}(k_y) \leq w_\pi\left(k_y, \frac{\pi}{2a}\right), \tag{9.21}$$

where $w_{vol}(k_y)$ is a volume state energy and $w_{\pi^*}(k_y, k_z)$ and $w_\pi(k_y, k_z)$ are the bulk (infinite lattice) dimensionless energy band dispersion relations,

$$w_{\left(\frac{\pi^*}{\pi}\right)}(k_y, k_z) = \frac{1}{2}(w_t + w_\perp) \pm \sqrt{\left[\frac{1}{2}(w_t - w_\perp)\right]^2 + 4(S_y^2 + S_z^2)}. \tag{9.22}$$

Thus the volume state energies are confined to the same $w - k_y$ regions as the bulk (infinite-lattice) energies. These regions are between the bottom and top of the $\pi^*(yz)$ and between the bottom and top of the $\pi(yz)$ bands as shown in Fig. 9.4. We shall refer to these regions as the "bulk continuum of states" or just the "bulk continuum". The volume states form the same continuum of energies as the infinite-lattice states. For every pair (w, k_y) for which there is a solution of the infinite lattice, there is also a volume state solution. The energies of the volume states are the same as those of the infinite lattice, but the wavefunctions are quite

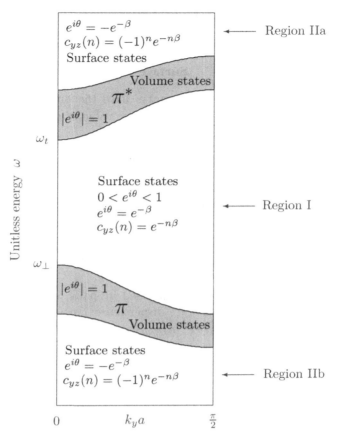

Figure 9.4. Surface energy band and volume energy band regions of (ω, k_y) space for the pi(yz) states.

different. For example, the square of the d-orbital amplitude is not uniform on the various layers since

$$|c_{yz}(n)|^2 = \frac{2}{N}\left[1 - \cos(2n\theta + \delta)\right]. \tag{9.23}$$

In fact, for a given volume state the square of the d-orbital amplitude will have maxima or minima on the nth layer whenever

$$n = \frac{j\pi}{2\theta} - \delta, \quad \text{where} \quad j = 0, \pm 1, \pm 2, \pm 3, \ldots . \tag{9.24}$$

When j is even the d-orbital probability on the nth layer vanishes, while for j odd the d-orbital amplitude is twice the average value. Since the positions of these maxima and minima vary with the particular volume state (i.e., vary with θ) the average d-orbital probability inside the lattice quickly approaches the infinite-lattice average.

Surface states

The wavefunctions for the surface states have the property that $c_{yz}(n) \to 0$ as $n \to \infty$, but is non-vanishing for at least one value of n. It is obvious from (9.16) that these conditions are met if

$$\Im \theta > 0,$$
$$\Lambda = 0. \tag{9.25}$$

The requirement that the imaginary part of $\theta > 0$ means that $e^{in\theta} \to 0$ with increasing n as $e^{-n(\Im\theta)}$. The second requirement, $\Lambda = 0$, imposes an eigenvalue condition, namely the *surface state condition* that

$$1 - \Delta p(\omega, k_y)\, e^{i\theta} - \Delta'\, e^{2i\theta} = 0. \tag{9.26}$$

The surface energy bands are specified by pairs, (ω, k_y), that satisfy (9.26). These pairs define the surface state dispersion curves, $\omega_s(k_y)$. The surface states are highly localized since $|c_{yz}(n+m)/c_{yz}(n)|^2 = e^{-2m(\Im\theta)}$.

According to (9.25), we may write

$$\theta = \alpha + i\beta,$$

where α and β are real numbers and $\beta > 0$. This gives

$$\cos\theta = \cos\alpha \cosh\beta - i \sin\alpha \sinh\beta. \tag{9.27}$$

However, for real energy, ω, (9.14) requires that $\cos\theta$ be real. This is compatible with (9.27) only if

$$\alpha = \ell\pi, \qquad (\ell = 0, \pm1, \pm2, \dots). \tag{9.28}$$

There are two distinct cases: ℓ is 0 or an even integer, and ℓ is an odd integer. We shall use $\ell = 0$ and $\ell = 1$. Other choices lead to equivalent results. We have

$$e^{i\theta} = e^{-\beta} \qquad \ell = 0 \tag{9.29}$$
$$e^{i\theta} = -e^{-\beta} \qquad \ell = 1. \tag{9.30}$$

In either case $e^{i\theta}$ is real and $0 < |e^{i\theta}| < 1$.

Consider the case for which $0 < e^{i\theta} < 1$. We have $c_{yz}(n) \propto e^{-n\beta}$ and decreases uniformly with increasing distance into the semi-infinite lattice.

Since $\cos\theta = \cosh\beta$, and $\cosh\beta > 1$ for $\beta > 0$, we have the inequality

$$(\omega_t - \omega_s)(\omega_\perp - \omega_s) - 4S_y^2 - 2 < -2. \tag{9.31}$$

If (9.31) is solved for w_s we find that these surface state energies must be in the band-gap region between the π^* and π volume bands (region I in Fig. 9.4)

$$w_\pi(k_y, 0) \leq w_s(k_y) \leq w_\pi^*(k_y, 0). \tag{9.32}$$

For the case that $-1 < e^{i\theta} < 0$, the d-orbital amplitude, $c_{yz}(n)$, alternates in sign from one unit layer to the next,

$$c_{yz}(n) \propto (-1)^n e^{-n\beta}. \tag{9.33}$$

The quantity $\cos \theta = -\cosh \beta < -1$ so that the surface states of this type have $w_s(k_y)$ that satisfies the equation

$$(w_t - w_s)(w_\perp - w_s) - 4S_y^2 - 2 > 2. \tag{9.34}$$

The allowed regions for $w_s(k_y)$ are

$$w_s(k_y) \geq w_{\pi^*}\left(k_y, \frac{\pi}{2a}\right) \tag{9.35}$$

and

$$w_s(k_y) \leq w_\pi\left(k_y, \frac{\pi}{2a}\right). \tag{9.36}$$

These amplitude-oscillating surface states occur only above the π^* band or below the π band. The regions are designated as regions IIa and IIb in Fig. 9.4.

(d) Pi(yz) density of surface states

The density of surface states (DOSS) can be found from the eigenvalue equation, (9.26), which may be written as

$$\frac{\Delta p - 2 \cos \theta}{1 + \Delta'} + \frac{1 + \Delta'}{\Delta p - 2 \cos \theta} + 2 \cos \theta = 0, \tag{9.37}$$

where $-2 \cos \theta$ and Δp are defined by (9.14) and (9.15), respectively. Equation (9.37) may be solved for the variable S_y^2 (which appears in both Δp and $-2 \cos \theta$) to obtain the quadratic equation

$$A(-4S_y^2)^2 + B(-4S_y^2) + C = 0, \tag{9.38}$$

with

$$A = \eta t + \eta^2, \tag{9.39a}$$
$$B = 2\gamma\eta + \xi\eta t + \gamma t, \tag{9.39b}$$
$$C = \gamma^2 + t^2 + \gamma\xi t, \tag{9.39c}$$
$$\gamma = (\Delta w_t + u N_s)(w_\perp - w) - t^2 + 2 + (1 - t)\xi, \tag{9.39d}$$

$$\eta = \frac{(\omega_\perp - \omega)r^2}{(\omega'_\perp - \omega)} - t, \tag{9.39e}$$

$$\xi = (\omega_t - \omega)(\omega_\perp - \omega) - 2, \tag{9.39f}$$

$$t = (1 + \Delta'), \tag{9.39g}$$

$$r = (1 + \Delta''). \tag{9.39h}$$

From (9.38) we obtain

$$\Omega(\omega, k_y) \equiv -\frac{1}{4A}\left\{-\frac{B}{2} \pm \sqrt{\left(\frac{B}{2}\right)^2 - AC}\right\} = S_y^2. \tag{9.40}$$

The DOSS is then

$$\rho_s(\Omega) = \frac{2a}{\pi}\left(\frac{d\Omega}{dk_y}\right)^{-1} = \frac{1}{\pi}\frac{1}{\sqrt{\Omega(1 - \Omega)}}, \tag{9.41}$$

and the DOSS, $\rho_s(\omega)$, as a function of ω, is given by

$$\rho_s(\omega) = \rho_s(\Omega)\frac{d\Omega}{d\omega} = \rho_s(\Omega)\frac{d}{d\omega}\left\{-\frac{1}{4A}\left\{-\frac{B}{2} \pm \sqrt{\left(\frac{B}{2}\right)^2 - AC}\right\}\right\}. \tag{9.42}$$

The surface band is one-dimensional, depending only on k_y, and therefore the DOSS has square-root singularities at $\Omega = 0$ and $\Omega = 1$.[1]

The number of electrons occupying the surface band can be written as

$$N_s(yz) = \int_0^{\Omega_F} \rho(\Omega)\,d\Omega = \frac{1}{2} - \frac{1}{\pi}\arcsin(1 - 2\Omega_F), \tag{9.43}$$

where Ω_F corresponds to the Fermi energy

$$\Omega_F = \Omega(\omega_F), \tag{9.44}$$

with

$$\omega_F = E_F/(pd\pi). \tag{9.45}$$

Consider the case of the "perfect" surface defined to be a surface for which all of the perturbation parameters, $\Delta\omega_t, \Delta\omega_\perp, \Delta'$, and Δ'' are zero. In this case $\Delta p(\omega, k_y) = 1$ and the eigenvalue condition, (9.26), gives $e^{i\theta} = 1$ which violates the surface state requirement that $e^{in\theta} \to 0$ as $n \to \infty$. Therefore we can conclude that there are no surface states on a "perfect" type I (001) surface. That does not mean that the states are the same as those of the infinite lattice. It means that all of the

---

[1] Equation (9.41) is valid so long as the quantity $D = (B/2)^2 - AC \geq 0$. If $D < 0$ it indicates that the surface energy band is truncated by intersecting the volume continuum. The truncation occurs at the value of k_y for which $D = 0$. $\Omega = 0$ and $\Omega = 1$ correspond to the bottom and top of the surface state band, respectively. In the case of a truncated surface band, one of the singularities occurs at the truncation energy.

states belong to the volume continuum. The energies are the same as those of an infinite lattice, but the wavefunctions are modified by the presence of the surface.

Of course, a "perfect" surface as we have defined it here is not the same as an "ideal" surface which is defined to be a surface that is geometrically perfect. The ideal surface will have non-zero perturbations even though it is atomically perfect. Furthermore, the energies of the surface states will depend upon the position of the Fermi energy and the number of electrons in the surface states. The Coulomb repulsion between electrons of opposite spin occupying the same surface d orbital will shift the energy calculated for the unoccupied surface state.

As a simple tutorial example consider the solution of (9.26) when the only non-zero parameter is $\kappa = (\Delta\omega_t + uN_s)$ and examine the solutions as $\kappa \to 0$. From (9.26) we obtain

$$e^{i\theta} = \frac{1}{\kappa(\omega_\perp - \omega) + 1}, \tag{9.46}$$

$$2\cos\theta = 2 + \kappa^2(\omega_\perp - \omega)^2 + \text{higher-order terms}. \tag{9.47}$$

Using (9.14) yields the surface state eigenvalue equation,

$$(\omega_t - \omega)(\omega_\perp - \omega) - 4S_y^2 + \kappa^2(\omega_\perp - \omega)^2 \cong 0. \tag{9.48}$$

For $\kappa = 0$, (9.48) describes the bottom of the π^* band as a function of k_y. Therefore, for small κ the surface band must lie near the π^* band edge, $\omega_{\pi^*}(k_y, 0)$. As a result, we may replace ω in the last term of (9.48) by $\omega_{\pi^*}(k_y, 0)$ so that to second order in κ the surface band is given by

$$\omega_{s\pi^*} \approx \frac{1}{2}(\omega_t + \omega_\perp) + \sqrt{\left[\frac{1}{2}(\omega_t - \omega_\perp)\right]^2 + 4S_y^2 - \kappa^2\left(\omega_\perp - \omega_{\pi^*}(k_y, 0)\right)^2}, \tag{9.49}$$

$$\approx \omega_{\pi^*}(k_y, 0) - \frac{\kappa^2\left(\omega_\perp - \omega_{\pi^*}(k_y, 0)\right)^2}{2\,\omega_{\pi^*}(k_y, 0) - (\omega_t + \omega_\perp)}. \tag{9.50}$$

The denominator in (9.50) is positive because the π^* band edge necessarily lies above the mid-gap energy, $\frac{1}{2}(\omega_t + \omega_\perp)$. Therefore it follows that there is *an entire* surface band below the π^* band edge, $\omega_{\pi^*}(k_y, 0)$, that lies in the band-gap region. This result is valid only for values of κ that are negative.[2]

For a doped, n-type perovskite, the Fermi level will be near the bottom of the π^* band edge at Γ, within a few meV of E_t (i.e., ω_F very close to ω_t) and the surface band will be partially occupied as indicated in Fig. 9.5(a). The parameter $\kappa = \Delta\omega_t + uN_s$ is a balance between the additional negative electrostatic potential and the positive electron–electron repulsion and must be calculated self-consistently. However, we can see that for negative values of κ there will always

---

[2] For positive κ the surface band is located above the top of the volume continuum. That is, above the top of the π^* band (Region IIa) and/or below the bottom of the π band (Region IIb).

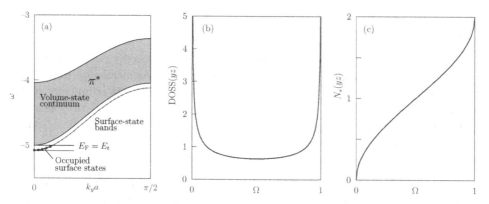

Figure 9.5. (a) Schematic showing the occupied surface states with the Fermi energy $E_F = E_t$, the bottom of the bulk conduction band. (b) DOSS and (c) N_s versus Ω for pi(yz) or pi(xz) surface states.

be a surface band below the bulk band in the band-gap region. If we keep $\Delta\omega_t$ fixed and increase the value of u, the value of κ becomes less negative. That will move the surface band closer to the bulk band-edge and reduce N_s. The effect of reducing N_s is to make κ more negative and to counteract the effect of increasing u. Therefore, the surface band will find a self-consistent solution between the energy of the unoccupied surface state ($N_s = 0$) and the lower edge of the π^* continuum. With the mean-field representation of the Coulomb repulsion we are using here, U would have to be infinite to force the surface band completely out of the band gap. Because of the canceling effects between uN_s and $\Delta\omega_t$ the surface bands tend to be near to the edge of the π^* continuum even when $\Delta\omega_t$ is a few eV negative.

It should be noted that for an n-doped insulator, the average number of electrons in a surface d orbital is much larger than for the interior d orbitals. For example, for a doping level of 10^{18} cm^{-3}, the average occupation (beyond that due to covalent bonding) of an interior $d(t_{2g})$ orbital is 6.4×10^{-5} electrons. On the other hand, a surface $d(t_{2g})$ orbital's occupation is of the order of unity when a surface band lies within the band gap. *Therefore, for the insulators, electron–electron correlation effects are more important for the surface energy bands and surface defect states than for the volume states.*

(e) Pi(xy) surface energy bands

The surface band involving the d_{xy} orbitals is easily derived since in the approximation of nearest-neighbor interactions, the unit-cell layers are uncoupled. Therefore we can immediately express the energy bands for the surface unit-cell layer as

$$E_s^{\pm}(k_x, k_y) = \left(\frac{E_t' + UN_s + E_{\perp}'}{2}\right) \pm \sqrt{\left(\frac{E_t' + UN_s - E_{\perp}'}{2}\right)^2 + 4[(pd\pi)'']^2(S_x^2 + S_y^2)} \ ,$$

$$(9.51)$$

where the primed and double primed symbols are the perturbed surface parameters. The form of the surface energy band dispersion is identical to the dispersion of the bulk π and π^* bands and therefore the DOSS, $\rho_s(E)$, can be obtained by making the following substitutions:

$$\rho_{\pi}(E) \rightarrow \rho_s(E), \qquad (9.52a)$$

$$E_t \rightarrow E_t' + UN_s, \qquad (9.52b)$$

$$(pd\pi) \rightarrow (pd\pi)'', \qquad (9.52c)$$

$$E_{\perp} \rightarrow E_{\perp}', \qquad (9.52d)$$

(here $A \rightarrow B$ means replace A by B),

into the expression for the pi DOS given by equation (6.28). This yields,

$$\rho_s(E) = \frac{1}{\pi^2} \frac{|E - \frac{1}{2}(E_t' + UN_s + E_{\perp}')|}{[(pd\pi)'']^2} K\left(\sqrt{1 - \left(\frac{\varepsilon(E)}{2}\right)^2}\right) \Theta\left[1 - \left(\frac{\varepsilon(E)}{2}\right)^2\right],$$

$$(9.53)$$

where

$$\varepsilon(E) = \frac{\left[E - \frac{1}{2}(E_t' + UN_s + E_{\perp}')\right]^2 - E_g'^2}{2[(pd\pi)'']^2} - 2. \qquad (9.54)$$

$E_t' = E_t + \Delta E_t$ is the perturbed site Madelung potential at the d ion and $E_g' = E_t' + UN_s - E_{\perp}'$ is the perturbed energy gap for the surface unit layer.

The DOSS for the pi(xy) surface band has jump discontinuities at its band edges and a logarithmic singularity at its band center just as in the case of the bulk states. The energies at which these discontinuities occur depend upon the perturbation parameters. The jump in the DOSS per spin state at $E = E_t'$ is,

$$\rho_s(E_t) = \frac{1}{2\pi} \frac{E_g'}{[(pd\pi)'']^2}. \qquad (9.55)$$

9.3 Self-consistent solutions for the band-gap surface states: SrTiO₃

The (001) surface of SrTiO₃ is typical of the surfaces of the insulating perovskites. Low-energy electron diffraction (LEED) and reflection high-energy electron diffraction (RHEED) experiments have been used to investigate the geometry of the surface [1, 2]. It is found that the surface oxygens move slightly upward, creating a puckered surface. LEED [1] experiments indicated 2% expansion of the distance $d_{12}(\text{Ti})$, but a contraction by 2% of $d_{23}(\text{Ti})$ (see the definitions in Fig. 9.3). The surface oxygen displacement, δ_O, is 4% of the Ti–O spacing. For the type II surface the first layer Sr–O distance, $d_{12}(\text{Sr})$, is contracted by 10%±2% and $d_{23}(\text{Sr})$ expanded by 4%±2%. The surface oxygen displacement was 8.2%±2%. Later RHEED [2] experiments confirmed the puckering due to displacement of the surface oxygens, but found expansion of both $d_{12}(\text{Ti})$ (3.6%) and $d_{23}(\text{Ti})$ (5.1%). The displacement of the ions near the surface creates a static dipole moment whose polarization is estimated [1] to be about $0.17\,\text{cm}^{-2}$.

In this section we consider only the largest surface perturbations, Δw_t, the change in the electrostatic potential at the surface d-orbital site and uN_s, the average additional Coulomb repulsion at the B-ion site due to the occupation of the surface states. The effects of other surface perturbations are discussed in references [4] and [5]

As mentioned previously, because N_s is the total number of electrons per spin state in all of the surface bands we must calculate the electronic occupations of the pi(xy), pi(yz), and pi(xz) surface bands simultaneously to achieve self-consistent solutions. To begin with we assume that the surface oxygen site potential and p–d interactions are unperturbed. That is, $E_\perp = E'_\perp$, and $(pd\pi)' = (pd\pi)'' = (pd\pi)$. The only perturbation is then the change in the diagonal energy at the surface B-ion site, $\Delta w_t + uN_s$. The actual value of U for various perovskites is not known. For the Ti ion the difference between the ionization potentials for Ti^{+4} and Ti^{+3}, $\Delta I_p = 15.75\,\text{eV}$. This value, appropriate for atomic states, should be an upper bound on the possible value of U for this material. A reasonable estimate is that U is one-half to one-third of ΔI_p. For SrTiO$_3$, the band gap is 3.2 eV and $(pd\pi)$ is between 0.84 and 1.3 eV based on LCAO fits to different energy band calculations [4, 6]. The change in the Madelung potential, Δw_t, for the ideal surface is about $-2\,\text{eV}$ at a surface Ti ion. Because the actual (001) surface is puckered, the precise value is uncertain, but most likely it is negative (less repulsive than at interior ions). We will explore the self-consistent solutions as a function of Δw_t and u and for the examples here assume $(pd\pi) = 1\,\text{eV}$.

To find self-consistent solutions we need to calculate N_s as a function of κ. For the pi(yz) or pi(xz) surface bands we have the eigenvalue equation

$$(\omega_t - \omega)(\omega_\perp - \omega) - 4S_y^2 - 2 + \Delta p + \frac{1}{\Delta p} = 0, \tag{9.56}$$

with

$$\Delta p(\omega) = \kappa(\omega_\perp - \omega) + 1. \tag{9.57}$$

For doping concentrations less than or equal to about 10^{18} cm^{-3} we may use $E_F/(pd\pi) = \omega_F = \omega_t$, the bottom of the bulk π^* band. This means the occupied surface states lie in the range $\omega_t' \leq \omega \leq \omega_t$ as illustrated schematically in Fig. 9.5.

The contribution to N_s (per spin state) from pi(yz) or pi(xz) surface bands is given by

$$N_s(\alpha\beta) = \int_0^{\Omega(\omega_t)} \rho(\Omega) \, d\Omega = \frac{1}{2} - \frac{1}{\pi} \arcsin\left(1 - 2\Omega(\omega_t)\right), \tag{9.58a}$$

and from (9.56) and (9.57) we have

$$\Omega(\omega_t) = \frac{(\kappa\omega_g)^2}{4(1 - \kappa\omega_g)}. \tag{9.58b}$$

For the pi(xy) band the DOSS (per spin state) is

$$\rho_{xy}(\omega) = \frac{1}{\pi^2}(\omega - \omega_m')K(x), \tag{9.59a}$$

$$x^2 = 1 - \left(\frac{\varepsilon(\omega)}{2}\right)^2, \tag{9.59b}$$

$$\varepsilon(\omega) = \frac{1}{2}\left[(\omega - \omega_m')^2 - \left(\frac{\omega_g'}{2}\right)^2\right] - 2, \tag{9.59c}$$

where $K(x)$ is the complete elliptic integral of the first kind, $\omega_m' = \frac{1}{2}(\omega_t + \kappa + \omega_\perp)$, and $\omega_g' = \omega_g + \kappa$. The number of electrons occupying the surface band is

$$N_s(xy) = \int_{\omega_t+\kappa}^{\omega_t} \rho_{xy}(\omega) \, d\omega. \tag{9.60}$$

$N_s(xy)$ may be approximated by

$$N_s(xy) \approx \frac{1}{4\pi}(\kappa^2 - \omega_g'\kappa) + \frac{1}{16\pi}(\omega_g')^2\kappa^2 + \frac{1}{48\pi}\omega_g'\kappa^3. \tag{9.61}$$

Equation (9.61) errs less than 2% for values of $\kappa^2 < 0.2$. The total number of electrons occupying all three surface states is then,

$$N_s = N_s(xy) + 2N_s(yz) = \int_{\omega_t+\kappa}^{\omega_t} \rho_{xy}(\omega) \, d\omega + 2\left[\frac{1}{2} - \frac{1}{\pi} \arcsin\left(1 - 2\Omega(\omega_t)\right)\right]. \tag{9.62}$$

The procedure for obtaining self-consistent solutions is to first choose a value for κ, calculate N_s using (9.62), then for a given value $\Delta\omega_t$ use the expression

$$u = \frac{\kappa - \Delta\omega_t}{N_s}, \tag{9.63}$$

to obtain the self-consistent value of u.

Figure 9.6 shows the self-consistent solution of N_s versus κ. The kink in the curve at $\kappa = -2(1 + \sqrt{2})/\omega_g$ (≈ -1.5 for $SrTiO_3$) corresponds to the point at which

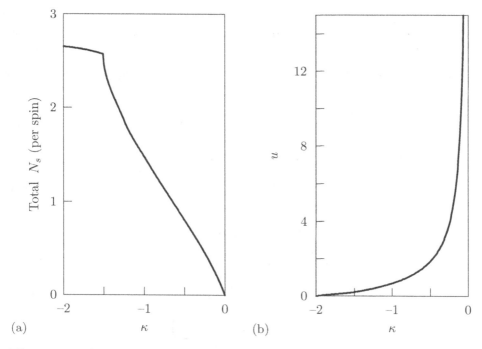

Figure 9.6. Self-consistent parameters for the surface energy bands. (a) N_s versus κ, (b) u versus κ (for these plots, $\omega_g = 3.2$ and $\Delta\omega_t = -2$).

the pi(xz) and pi(yz) bands lie entirely below the conduction-band edge so that they are completely occupied. Beyond this point the occupation of the pi(yz) and pi(xz) surface bands do not change. The dispersion of the pi(xy) band is larger than that of the pi(yz) surface band because it is two-dimensional. Therefore, it takes a larger negative value of κ to drop it entirely below the conduction-band edge. The parameters u, κ, and $\Delta\omega_t$ are in units of $(pd\pi)$. For SrTiO$_3$ with $(pd\pi) = 1$ eV, the results may be read in units of electronvolts. Figures 9.7(a) and (b) show the pi(xy) and pi(yz) surface energy bands, respectively, for $\Delta E_t = -2$ eV with $u = 0$ and $u = 6$ ($U = 0$ and 6 eV for SrTiO$_3$). As can be seen, the Coulomb repulsion forces the surface bands toward the edge of the continuum of states. For $U = 6$ eV, the pi(yz) and pi(xz) surface bands lie within 0.053 eV of the edge of the conduction band. The pi(xy) band lies lower in energy than the pi(yz) and pi(xz) bands but is still within 0.055 eV of the conduction-band edge for $U = 6$ eV. The energy displacement of the pi(xy) surface band at Γ below the bulk conduction-band edge is in general equal to $\kappa(pd\pi)$. Some results are summarized in Table 9.1. They show that N_s per surface unit cell is in the range of 10^{14} cm^{-2} for a wide variety of the parameters. The entries in the table can be used for any value of $(pd\pi)$, but the energy gap is

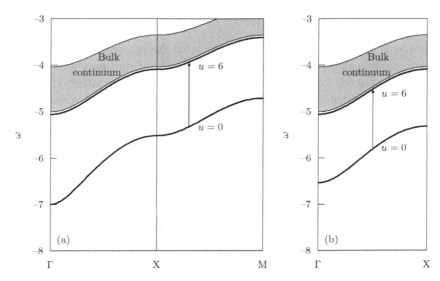

Figure 9.7. Self-consistent surface bands. (a) Pi(xy) surface band for $u = 0$ and $u = 6$ and (b) pi(yz) surface band for $u = 0$ and $u = 6$.

fixed at 3.2 eV and the electron concentration is calculated for a lattice spacing of 3.905 Å: values appropriate for SrTiO₃.

Table 9.1. *Self-consistent surface state parameters.*

U	$\Delta\omega_t$	κ	N_s(total)	$N_s(xy)$	$2N_s(yz)$	Surface concent. $(10^{14}\,\mathrm{cm}^{-2})$
15	-2.0	-0.0609	0.1293	0.0157	0.1136	0.8481
15	-1.0	-0.0294	0.0648	0.0075	0.0572	0.4249
15	-0.5	-0.0144	0.0324	0.0037	0.0287	0.2125
6	-2.0	-0.1589	0.3070	0.0415	0.2654	2.0132
6	-1.0	-0.0738	0.1545	0.0191	0.1355	1.0132
6	-0.5	-0.0354	0.0775	0.0091	0.0684	0.5082
3	-2.0	-0.3236	0.5589	0.0861	0.4729	3.6652
3	-1.0	-0.1458	0.2849	0.0381	0.2468	1.8683
3	-0.5	-0.0684	0.1441	0.0177	0.1264	0.9450

The results obtained here do not explicitly include the effects of surface charge. A high density of occupied surface states in the band gap can cause "band bending" if the bulk density of electrons is insufficient to screen these surface charges. However, the term UN_s has the same effect and therefore no explicit surface charge term needs to be added to the model.

(a) Other types of surface bands

In the previous discussion it has been assumed that ΔE_t is negative. This prejudice is based on model calculations of the electrostatic (Madelung) potential for a B ion on a ideal type I (001) surface for which ΔE_t is several eV negative. However, for non-ideal surfaces there is the possibility that ΔE_t and κ could be positive; that is, more repulsive than at an interior site. (See Appendix D for a table of Madelung potentials.) In this case, the surface theory produces truncated surface bands that lie just above the top of the π^* conduction band. Surface bands are also produced just above the π valence band, but they do not lie in the fundamental band-gap range (i.e., Γ-band-gap region between E_t and $E_\perp$). Such states would be difficult to observe optically or by photoemission. The surface bands near the top of the conduction band will be unoccupied and close to the jump in the bulk density of states at the top of the band. The surface bands split off from the valence bands would be very near to the bulk non-bonding band energies. Thus their contribution in optical or photoemission experiments could also be obscured by the high bulk density of occupied valence-band states. Figure 9.8(a) shows the surface bands for both positive and negative values of κ with $U = 0$. More details on these types of surface bands can be found in [4].

On type II surfaces calculations of the electrostatic potentials indicate that ΔE_t is nearly unchanged, but that $\Delta E_\perp$ is several eV positive (less attractive). For this perturbation nearly flat surface bands appear in the fundamental band-gap region above the π valence bands whose wavefunctions are composed primarily of p orbitals [4]. These bands are illustrated in Fig. 9.8(b). For the p-like surface bands the electron–electron repulsion is much smaller than that for the d orbitals and because the bands are flat a large DOSS results. In n-doped insulators and metallic perovskites such band states would be occupied.

For metallic perovskites such as Na_xWO_3 ($x > 0.3$) the theory of the surface bands described above applies with some modifications. First, the Fermi level is no longer pinned at the bottom of the conduction band and hence the surface bands will have much higher electron concentrations. Second, the concentration of bulk electrons is sufficient to screen the surface charge associated with the occupied surface states. Third, although the Coulomb repulsion energy is still large, the *difference* between the bulk and surface state density of electrons is much smaller. The Coulomb repulsion, UN_s is still operative for both the bulk and surface states, but it is roughly the same for both types of bands. For our empirical LCAO model this means that the correlation effects can be assumed to be incorporated into the parameters of the model and therefore the effects of self-consistency are less important when employing the theory. As a first approximation, we can calculate the surface band energies in the same manner as the bulk energy bands are calculated, as though the Coulomb repulsion parameter, U, were zero. Therefore, we expect

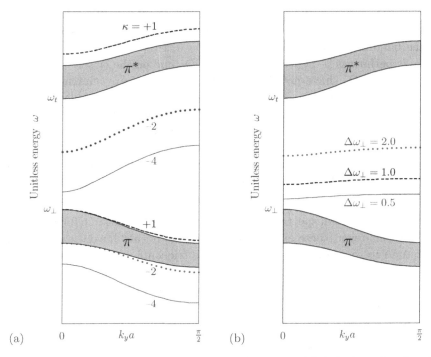

Figure 9.8. Type I (001) surface bands with $U = 0$. (a) Pi(yz) surface band for $\kappa = +1, -2$, and -4. For negative values of κ surface bands lie in the fundamental gap and below the valence band. For positive values of κ, surface bands lie above the top of the conduction band and above the valence band. (b) Band-gap surface bands for the type II (001) surface for different values of the perturbed p-orbital site potential.

that the calculated surface bands for the metallic perovskites will lie deeper in the band gap than for the insulating perovskites. For example, referring to Figs 9.7(a) and (b), the positions of the surface bands for $u = 0$ would apply to a metallic perovskite while the surface bands pushed up to the edge of the conduction band would apply for n-type doped insulating perovskites. One would expect the metallic behavior to dominate for electron concentrations greater than or about 10^{21} cm^{-3} and insulator behavior for electron concentrations less than or about 10^{18} cm^{-3}.

(b) Experimental results: SrTiO₃, TiO₂, and Na$_x$WO₃

The range of concentrations of electrons in the surface bands displayed in Table 9.1 for n-doped perovskites should be detectable in photoemission experiments.

Angle-resolved photoemission studies have been carried out for several Na$_x$WO₃ samples. Surface bands in the band gap were not found for insulating

WO_3 but a surface band was reported [6] for metallic $Na_{0.85}WO_3$. In this latter case, the surface band has a band width of about $0.9\,eV$ and dispersion similar to the surface band shown in Fig. 9.7(a) with $U = 0$. However, UPS and XPS photoemission experiments [7–9] performed on $SrTiO_3$ and on the closely related oxide TiO_2 as well as WO_3, have not detected the presence of any "intrinsic" surface states in the band-gap region. The Fermi energy appears to be pinned at or near the conduction-band edge. Whether this is pinning by the bulk conduction band or a nearby surface band is not known, but in either case the concentration of electrons must be below the detection threshold of about $10^{13}\,cm^{-2}$ for the photoemission experiments. (For n-type materials with concentrations of the order of 10^{18} electrons per cm^{-3}, the surface concentration *in the absence of band-gap surface states* is only about $4 \times 10^{10}\,cm^{-2}$.)

Inverse photoelectron spectroscopy has also been used to search for unoccupied surface states; however, these experiments probably did not have the energy resolution required to separate bulk states from the unoccupied surface band near the edge of the volume continuum. For example, the rise in the photoemission at the conduction-band edge seen in the inverse photoelectron experiments is spread out over an energy interval of $1.6\,eV$ and the quoted energy resolution [7] was $0.7\,eV$.

Why "intrinsic" band-gap surface states are not observed for $SrTiO_3$, TiO_2, or WO_3 but are observed for metallic Na_xWO_3 is unclear. As mentioned above, surface charge may play a role for the doped insulators. For WO_3 it has been suggested that band bending due to surface charge depletes the surface bands of electrons [6]. However, unless the charge is due to sources other than electrons in the surface states, band-bending is already effectively included in the Coulomb repulsion parameter, U.

Many conjectures can be put forth for why surface states are not seen in the band gap of the n-doped insulators. First, there is the question as to whether energy band theory can be applied to these materials since the correlation energy among the surface electrons is large compared to the bulk. The mean-free path in photoemission is less than $10\,\text{Å}$ and therefore photoemission samples principally the first few layers. Nevertheless, the bulk electronic structure predicted by band theory is clearly evident. Thus it is difficult to argue that the band theory applies to the bulk but not the surface. One may wonder if the LCAO model is too simplistic to correctly describe the surface electronic structure. This is certainly a possibility, particularly since the model employed here has only nearest-neighbor interactions. However, the addition of more distant interactions will not change the qualitative features of the surface bands. Furthermore LCAO models correctly describe most features of the bulk electronic structure observed in photoemission and optical experiments on insulating perovskites as well as the electronic dispersion observed in high-temperature superconducting metal oxides [10, 11].

There is the possibility that the mean-field approximation used for the electron–electron repulsion is inadequate to treat surface states. A dynamic theory may be more aggressive and force the surface bands to within a few meV of the bottom of the conduction band instead of a few hundredths of an eV. That would reduce the electron concentration in the surface band to nearly that of the bulk, a concentration that is below the threshold for detection in photoemission experiments.

Fracturing the crystal in high-vacuum conditions would be expected to produce both type I, the surface type II surfaces. Surface bands split off from the valence bands on the type II surface are also predicted to lie in the fundamental band-gap region but are also not seen experimentally for the n-doped insulators. Depending upon the size of the various patches of type I and type II surfaces it is conceivable that the long-range Madelung potentials of the two different surfaces cancel one another approximately, leaving the surface with an average potential of the bulk. In such a case surface bands would not occur.

In summary, the band-gap surface bands expected on the basis of the LCAO model have been observed in metallic $Na_x WO_3$ but not for $SrTiO_3$, TiO_2, or WO_3. The reason why surface bands are not detected for the doped insulators is not certain. Stronger electron correlation at the surface or the approximate cancelation of changes in the site potentials are possibilities.

9.4 Surface–oxygen defect states

Bombarding the surface of a perovskite with Ar^+ (argon ions) results in the removal of oxygen from the surface layer and, at high doses, from subsurface layers as well. According to an ionic model each oxygen removed donates two electrons to the material.

UPS and XPS measurements on $SrTiO_3$ (also TiO_2 and Ti_2O_3) exhibit emission from oxygen defect states in the band-gap region, centered roughly 1.0–$1.3\,eV$ below the conduction-band edge. The intensity of the emission increases with increasing Ar^+ dose until saturation occurs (sometimes accompanied by reconstruction or formation of a different phase on the surface). Since the emission from these states is reduced rapidly by exposing the surface to oxygen, the states are believed to result from oxygen vacancies. The wavefunctions for these defect states involve the d orbitals on the B ions that are adjacent to one or more of the oxygen vacancies.

When an oxygen ion is removed from the surface of a perovskite several perturbations occur: (1) electrons are released into the bulk conduction bands; (2) the two B ions adjacent to the vacancy will have unsaturated bonds and lowered symmetry; (3) the removal of the O^{2-} ion makes the electrostatic potential at the two B-ion sites much more attractive (negative); (4) but the occupation of the band-gap defect states leads to a repulsive electron–electron potential between electrons in the same orbital; (5) unless the vacancies form an ordered array, translation symmetry is destroyed and the defect states are not characterized by a wavevector.

Theories of the surface defect states are beset with difficulties because the actual surface geometry, surface defects, and surface potentials are not known. In addition, the electron–electron correlation energy for surface d ions is much larger than for an interior ions when defect states lie in the band gap.

Theoretical analysis of vacancy-induced states center around two different approaches. The first is an atomic-like picture. It is supposed that the electrons donated by the removal of an oxygen ion are retained by the d orbitals of the B ions adjacent to the vacancy site. This approach suggests the formation of different B-ion oxidation states for the surface ions. For the oxides of titanium, $SrTiO_3$, TiO_2, and Ti_2O_3 for example, the model supposes that Ti^{3+} (or even Ti^{2+}) ions are formed on the surface with energies in the band-gap region. The bulk states are assumed to be band states, but the defect states are assumed to have atomic-like wavefunctions. This picture can be conceptually useful, but treating electrons occupying surface states on a different footing from these occupying bulk states is difficult to justify theoretically.

The second approach is band theory. In this model spatially localized vacancy-induced states are derived from delocalized energy band wavefunctions. The oxygen-donated electrons are not necessarily retained by the surface ions alone, but may spread throughout the bulk conduction bands. The number of extra electrons residing in surface defect states is determined by the energy of the defect state/band relative to the Fermi energy. However, if the vacancy-induced state/band lies completely within the band gap it will be occupied in n-type, doped insulators, so that the state may begin to resemble a Ti^{3+} ion state. In this case the localized, atomic view and the band theory picture are conceptually similar.

With either the atomic approach or the band theory approach, the electron–electron repulsion energy is large for the surface ions. The difference in ionization energies of different oxidation states is large. For example, the difference in the ionization energies of Ti^{4+} and Ti^{3+} is nearly $15\,eV$ and for Ti^{4+} and Ti^{2+} it is nearly $30\,eV$. Therefore, if the vacancy-induced state is to appear in the band gap, there must be a correspondingly large decrease in the Madelung/electrostatic potential for the surface sites. While it is easy to see that such a reduction is likely to result when repulsive O^{2-} ions are removed, calculating these long-range electrostatic potentials requires detailed knowledge of the arrangement of the ions and their charges. A large surface charge can result if all of the oxygen-donated electrons reside on the surface ions. This charge can lead to band bending if it is not canceled by other local charges. For $SrTiO_3$ the excess charge may be partially neutralized by surface and subsurface Sr^{2+} vacancies. For TiO_2 it is suggested that the Ti^{4+} ions are converted to Ti^{3+} ions by forming a surface layer of Ti_2O_3.

There is currently no reliable theory for accurately treating surface defect states such as oxygen vacancy states on the surface of actual crystals. However, the qualitative features of oxygen vacancy states can be understood by studying some simple LCAO energy band models. In this section we look at two such models that relate to the oxygen vacancy states.

(a) A line of oxygen vacancies on a type I (001) surface

The planes containing d_{yz}, p_y, and p_z orbitals that are perpendicular to the surface of a cubic perovskite are uncoupled in the nearest-neighbor approximation. Therefore the surfaces states discussed in Section 9.2(d) are actually "edge" states or "line" states, and that is why they behave as one-dimensional systems. Figures 9.9(a) and (b) show schematically the geometry of a yz-layer before and after removing p_z surface orbitals along a line in the y-direction. The resulting surface line consists of only d_{yz} orbitals on B ions. Since the layers are uncoupled we can assume all the other yz-planes have their full compliment of surface p_z orbitals. Therefore the model is a single line of alternating B ions and surface oxygen vacancies on an otherwise normal (001), type I surface. For the layer with the line of vacancies the surface parameter, $(pd\pi)''$, is equal to zero. That is, $\Delta'' = -1$ in (9.15b). Clearly, the same model applies to the symmetry-equivalent xz-planes with a line of p_z vacancies along the x-direction.

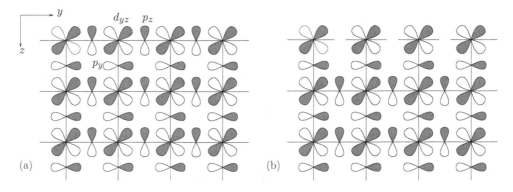

Figure 9.9. Schematic of a yz-layer (a) without vacancies and (b) with oxygen vacancies extending along the y-direction.

Using the surface state condition, $\Lambda = 0$, of (9.17) and the definitions of the surface perturbations of (9.15) with $\Delta' = 0$, $\Delta'' = -1$, $E'_\perp = E_\perp$ gives:

$$1 - \Delta p(\omega, k_y) e^{i\theta} = 0 \qquad \text{(eigenvalue condition)}, \qquad (9.64)$$

where

$$-2 \cos\theta = (\omega_t - \omega)(\omega_\perp - \omega) - 4S_y^2 - 2 \qquad (9.65)$$

$$\Delta p(\omega, k_y) = \kappa(\omega_\perp - \omega) + 1 + 4S_y^2, \qquad (9.66)$$

$$\kappa = \Delta\omega_t + uN_s.$$

For this model it is expected that the parameter, $\Delta\omega_t$ is much more negative than in the case of the ideal surface because an entire row of repulsive, O^{2-} ions has been removed. Consequently, we would expect that the "line" energy band will lie deeper in the gap than the pi(yz) surface states previously discussed (Table 9.1). Exactly where the band lies depends upon the balance between the Coulomb repulsion and increased negativity of $\Delta\omega_t$.

The DOSS for the line energy band is given by (9.41).

$$\rho_{\text{vac}}(\Omega) = \left(\frac{2a}{\pi}\right)\left(\frac{d\Omega}{dk_y}\right)^{-1} = \frac{1}{\pi}\frac{1}{\sqrt{\Omega(1-\Omega)}}. \qquad (9.67)$$

$$\Omega(\omega, k_y) = -\frac{1}{4}\left(\eta + \frac{1}{\eta + \xi}\right), \qquad (9.68)$$

$$\eta = \kappa(\omega_\perp - \omega) + 1,$$

$$\kappa = (\Delta E_t + UN_v)/(pd\pi), \qquad (9.69)$$

$$\xi = (\omega_t - \omega)(\omega_\perp - \omega) - 2. \qquad (9.70)$$

In (9.69) N_v is the number of electrons occupying the vacancy-induced surface band. The square root singularities occur at $\Omega = 0$ and $\Omega = 1$, corresponding to the bottom and top of the line band. The eigenvalue condition, $\Delta p + 1/\Delta p - 2\cos\theta = 0$, can be written as a cubic equation in the variable $F = (\omega_\perp - \omega)$:

$$\kappa F^3 + F^2[\kappa(\kappa + \omega_g) + (1 - Z)] + F[(\kappa + \omega_g)(1 - Z) - \kappa] + Z = 0, \qquad (9.71)$$

with $Z = -4S_y^2$. Solutions may be obtained using the standard formulae for the roots of a cubic equation. The three vacancy-line bands are shown in Fig. 9.10. The two vacancy bands, labeled $V1$ and $V2$ will be completely occupied with electrons, but do not enter the fundamental band-gap region between E_t and $E_\perp$.

The line band at −6 eV, labeled $C1$, has wavefunctions composed of nearly pure d orbitals and is nearly dispersionless. This oxygen-vacancy-produced band lies about 1 eV below the conduction band for $\kappa = -1.3$ eV and $(pd\pi) = 1$ eV. Because the band lies totally within the band gap it would be completely occupied by electrons in an n-type, doped insulator such as $SrTiO_3$. That is, $N_v = 2$ including both spin states. Here we have a situation of an extremely narrow d-electron

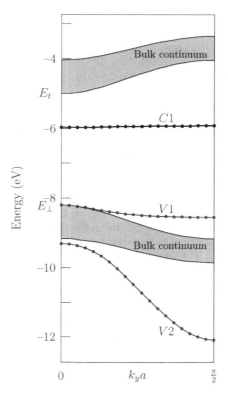

Figure 9.10. Surface bands due to a line of vacancies. The $C1$ band, derived from the d orbitals centered on the surface B ions, lies in the band-gap region. $V1$ and $V2$ are p-orbital bands.

band with large correlation energy. The surface B ions neighboring the vacancies are forming atomic-like states. Therefore, the first electron occupying the state encounters a Coulomb repulsion energy roughly equal to the $Ti^{+4} \rightarrow Ti^{+3}$ ionization energy of about $15\,eV$. Placing a second electron in the state costs $30\,eV$ relative to occupying a bulk state. Therefore, the Coulomb repulsion parameter is U for the first electron and $2U$ for the second electron. As a result the $C1$ band for double electron occupation lies well above the conduction band. Consequently, the vacancy line-band will have only a single electron per d orbital and the Ti ions are, roughly speaking, Ti^{3+} ions while the bulk Ti ions are Ti^{4+} ions.

(b) Isolated vacancy states

The previous section dealing with a line of vacancies applies when the vacancy concentration is high. At the other extreme, that of low concentration, the vacancies

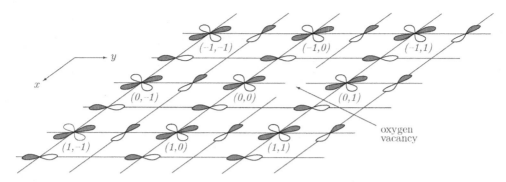

Figure 9.11. Schematic of an isolated vacancy on the xy-plane. For the pi(xy) band the vacancy is represented by the absence of a p_x orbital in the unit cell at the origin. The unit cells are indicated by the x–y coordinates in parentheses and their locations are specified by two-dimensional vectors $\vec{\rho}_{j,m}$. The values of j and m are shown in the figure.

are non-interacting and may be considered as isolated. The model for an isolated oxygen vacancy on a (001) type I surface is illustrated in Fig. 9.11. The surface unit-cell layer, parallel to the xy-plane, is uncoupled from the other unit-cell layers in the nearest-neighbor approximation. The orbitals for the pi(xy) states are shown schematically. In this case the oxygen vacancy takes the form of a missing p_x orbital. The B ions are located on the xy-plane by the set of two-dimensional vectors, $\vec{\rho}_{j,m} = 2a(j\vec{e}_x + m\vec{e}_y)$, where m and j are integers. The p_x orbitals of the O ions are located at $\vec{\rho}_{j,m} + a\vec{e}_y$ and the p_y orbitals at $\vec{\rho}_{j,m} + a\vec{e}_x$. If we assume the missing p_x orbital is in the unit cell at the origin, then, the equations for c_x, c_y, and c_{xy}, the amplitudes of the p_x, p_y, and d_{xy} orbitals respectively, are

$$(\omega_t + \kappa - \omega)\, c_{xy}(\vec{\rho}_{0,0}) - c_x(\vec{\rho}_{0,-1} + a\vec{e}_y) + c_y(\vec{\rho}_{0,0} + a\vec{e}_x)$$
$$-c_y(\vec{\rho}_{-1,0} + a\vec{e}_x) = 0, \tag{9.72a}$$

$$(\omega_t + \kappa - \omega)\, c_{xy}(\vec{\rho}_{0,1}) + c_x(\vec{\rho}_{0,1} + a\vec{e}_y) + c_y(\vec{\rho}_{0,1} + a\vec{e}_x)$$
$$-c_y(\vec{\rho}_{-1,1} + a\vec{e}_x) = 0, \tag{9.72b}$$

$$(\omega_\perp - \omega)\, c_y(\vec{\rho}_{0,0} + a\vec{e}_x) + c_{xy}(\vec{\rho}_{0,0}) - c_{xy}(\vec{\rho}_{1,0}) = 0, \tag{9.72c}$$

$$(\omega_\perp - \omega)\, c_x(\vec{\rho}_{0,1} + a\vec{e}_y) + c_{xy}(\vec{\rho}_{0,1}) - c_{xy}(\vec{\rho}_{0,2}) = 0, \tag{9.72d}$$

$$(\omega_\perp - \omega)\, c_y(\vec{\rho}_{0,1} + a\vec{e}_x) + c_{xy}(\vec{\rho}_{0,1}) - c_{xy}(\vec{\rho}_{1,1}) = 0,. \tag{9.72e}$$

$$\kappa = (\Delta E_t + U N_v)/(pd\pi). \tag{9.72f}$$

For $j \neq 0$, $m \neq 0$ or 1:

$$(\omega_t - \omega)\, c_{xy}(\vec{\rho}_{j,m}) + c_x(\vec{\rho}_{j,m} + a\vec{e}_y) - c_x(\vec{\rho}_{j,m-1} + a\vec{e}_y)$$
$$+c_y(\vec{\rho}_{j,m} + a\vec{e}_x) - c_y(\vec{\rho}_{j-1,m} + a\vec{e}_x) = 0, \tag{9.73a}$$

$$(\omega_\perp - \omega)\, c_x(\vec{\rho}_{j,m} + a\vec{e}_y) + c_{xy}(\vec{\rho}_{j,m}) - c_{xy}(\vec{\rho}_{j,m+1}) = 0, \tag{9.73b}$$

$$(\omega_\perp - \omega)\, c_y(\vec{\rho}_{j,m} + a\vec{e}_x) + c_{xy}(\vec{\rho}_{j,m}) - c_{xy}(\vec{\rho}_{j+1,m}) = 0. \tag{9.73c}$$

In (9.72f) ΔE_t is the change in the Madelung potential and N_v is the number of d electrons occupying the vacancy states. By direct substitution we can eliminate the p-orbital amplitudes and obtain equations involving only the d-orbital amplitudes,

$$\begin{aligned}
\big[(\omega_t + \kappa - \omega)(\omega_\perp - \omega) - 3\big]\, c_{xy}(\vec{\rho}_{0,0}) + c_{xy}(\vec{\rho}_{0,-1}) + c_{xy}(\vec{\rho}_{1,0}) \\
+ c_{xy}(\vec{\rho}_{-1,0}) = 0,
\end{aligned} \tag{9.74a}$$

$$\begin{aligned}
\big[(\omega_t + \kappa - \omega)(\omega_\perp - \omega) - 3\big]\, c_{xy}(\vec{\rho}_{0,1}) + c_{xy}(\vec{\rho}_{0,2}) + c_{xy}(\vec{\rho}_{1,1}) \\
+ c_{xy}(\vec{\rho}_{-1,1}) = 0.
\end{aligned} \tag{9.74b}$$

For $j \neq 0$, $m \neq 0$ or 1:

$$\begin{aligned}
\big[(\omega_t - \omega)(\omega_\perp - \omega) - 4\big]\, c_{xy}(\vec{\rho}_{j,m}) + c_{xy}(\vec{\rho}_{j+1,m}) + c_{xy}(\vec{\rho}_{j-1,m}) \\
+ c_{xy}(\vec{\rho}_{j,m+1}) + c_{xy}(\vec{\rho}_{j,m-1}) = 0.
\end{aligned} \tag{9.74c}$$

The equations represented by (9.74a)-(9.74c) can be written in matrix form:

$$[H_d(\omega) + \Delta H_d]\, \vec{C}_{xy} = 0, \tag{9.75}$$

where $H_d(\omega)$ is the effective Hamiltonian describing the interactions between the d orbitals for the unperturbed surface unit-cell layer, ΔH_d describes the perturbations due to the vacancy, and $\vec{C}_{xy}$ is a vector whose components are the d-orbital amplitudes, $c_{xy}(\vec{\rho}_{j,m})$.

Referring back to (9.74a) and (9.74b) we see that ΔH_d is a null matrix except for a 2×2 block centered on the diagonal.

$$\Delta H_d = \begin{pmatrix} \Delta H(0,0) & \Delta H(0,1) \\ \Delta H(1,0) & \Delta H(1,1) \end{pmatrix}, \tag{9.76}$$

where

$$\Delta H(0,0) = \Delta H(1,1) \equiv \Delta h = \kappa(\omega_\perp - \omega) + 1, \tag{9.77}$$
$$\Delta H(1,0) = \Delta H(0,1) = -1. \tag{9.78}$$

Equation (9.75) can be rewritten as

$$H_d[I + G_\varepsilon \Delta H_d]\, \vec{C}_{xy} = 0, \tag{9.79}$$

where I is the unit matrix and $G_\varepsilon = G_\varepsilon(\omega) = [H_d]^{-1}$ is the *d-orbital lattice Green's function*. The matrix elements of $G_\varepsilon(\omega)$ are derived and their behavior discussed in Appendix C.

The eigenvalues are given by the zeros of the determinant of (9.79),

$$\det\{H_d[I + G_\varepsilon \Delta H_d]\} = \{\det H_d\}\{\det[I + G_\varepsilon \Delta H_d]\} = 0. \tag{9.80}$$

Since the zeros of $\{\det H_d\}$ occur at the unperturbed energies it follows that the perturbed energies are given by the zeros of the 2×2 determinant, $\det[I + G_\varepsilon \Delta H_d]$. Thus the energies of the vacancy states are given by

$$\det \begin{pmatrix} 1 + \Delta h\ G(0) - G(1) & \Delta h\ G(1) - G(0) \\ \Delta h\ G(1)^* - G(0) & 1 + \Delta h\ G(0) - G(1)^* \end{pmatrix} = 0, \tag{9.81}$$

where $G(0) \equiv G_\varepsilon(\rho, \rho)$ and $G(1) \equiv G_\varepsilon(\vec{\rho}, \vec{\rho} + a\vec{e}_x) = G_\varepsilon(\vec{\rho}, \vec{\rho} + a\vec{e}_y)$. For energies in the band gap, $G(0)$ and $G(1)$ are real functions and $G(1) = G(1)^*$. Equation (9.81) yields two solutions for the vacancy states corresponding to symmetric and anti-symmetric combinations of the d orbitals adjacent to the vacancy.

$$[G(0) + G(1)](\Delta h - 1) + 1 = 0 \tag{9.82}$$

and

$$[G(0) - G(1)](\Delta h + 1) + 1 = 0. \tag{9.83}$$

In Appendix C it is shown that in the band gap

$$G(0) = \frac{1}{\pi\varepsilon} K\left(\frac{2}{\varepsilon}\right) \tag{9.84}$$

and

$$G(1) = \frac{1}{2}\left[\frac{1}{2} - \varepsilon\, G(0)\right], \tag{9.85}$$

where

$$\varepsilon = \frac{1}{2}(\omega_t - \omega)(\omega_\perp - \omega) - 2. \tag{9.86}$$

So that (9.82) and (9.83) take the form

$$\left[G(0)\left(1 - \frac{\varepsilon}{2}\right) + \frac{1}{4}\right](\Delta h - 1) + 1$$

$$= \left[\frac{1}{\pi\varepsilon} K\left(\frac{2}{\varepsilon}\right)\left(1 - \frac{\varepsilon}{2}\right) + \frac{1}{4}\right][\kappa(\omega_\perp - \omega)] + 1 = 0, \tag{9.87}$$

$$\left[G(0)\left(1 + \frac{\varepsilon}{2}\right) - \frac{1}{4}\right](\Delta h + 1) + 1$$

$$= \left[\frac{1}{\pi\varepsilon} K\left(\frac{2}{\varepsilon}\right)\left(1 + \frac{\varepsilon}{2}\right) - \frac{1}{4}\right][\kappa(\omega_\perp - \omega) + 2] + 1 = 0, \tag{9.88}$$

where K is the complete elliptic integral of the first kind.

Figure 9.12 shows the solutions of (9.87) as a function of the perturbation,

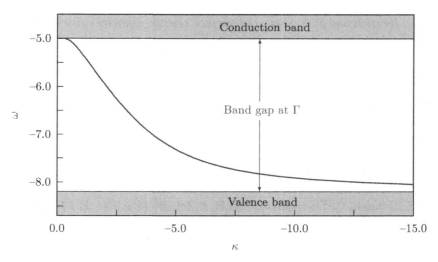

Figure 9.12. Energy of the oxygen defect state as a function of the perturbation parameter, κ. The gray areas at the top and bottom indicate the bulk continuum of states.

$\kappa = \Delta\omega_t + uN_d$. The solutions of (9.87) and (9.88) are virtually identical.[3] For zero or positive values of κ there are no vacancy states in the band-gap region. For negative κ the vacancy states move downward into the band gap with increasing negative values of κ. The energy moves down rapidly at first then approaches the top of the valence band asymptotically for very large negative values of κ. For an n-type insulator such as $SrTiO_3$ with $(pd\pi) = 1\,\mathrm{eV}$, the abscissa of Fig. 9.12 corresponds to energy in eV. As mentioned earlier, vacancy states appear about $1\,\mathrm{eV}$ below the conduction band. In Fig. 9.12 a vacancy state $1\,\mathrm{eV}$ below the conduction band corresponds to κ of about $-2.05\,\mathrm{eV}$. As in the case of the vacancy line band, N_v will be equal to 1 per unit cell since the doubly occupied state will lie above the conduction-band edge and the Fermi level. For $SrTiO_3$ each of the pair of Ti ions adjacent to the vacancy will correspond approximately to a Ti^{3+} ion while the interior ions are approximately Ti^{4+}.

The vacancy states provide coordinatively unsaturated bonds that are chemically active. These states provide d orbitals that can act as a source or sink for electrons to catalyze surface chemical reactions. A more complete discussion of the surface reactive properties can be found in [5].

---

[3] If the second-neighbor oxygen–oxygen interactions are included the two solutions will be slightly different, corresponding to symmetric and antisymmetric combinations.

References

[1] V. E. Henrich and P. A. Cox, *The surface science of metal oxides* (Cambridge, Cambridge University Press, 1996) p. 1.

[2] N. Bickel, G. Schmidt, K. Heinz, and K. Müller, *Vacuum* **41**, 46 (1990).

[3] T. Hikita, T. Hanada, M. Kudo, and M. Kawai, *Surf. Sci.* **287/288**, 377 (1993).

[4] T. Wolfram, E. A. Kraut, and F. J. Morin, *Phys. Rev.* B **7**, 1677 (1973).

[5] T. Wolfram and S. Ellialtioglu, Concepts of surface states and chemisorption on *d*-band perovskites. In *Theory of chemisorption*, ed. J. R. Smith, *Topics in current physics*, Vol. 19, Ch. 6 (Heidelberg, Springer-Verlag, 1980) pp. 149-181.

[6] H. Höchst, R. D. Bringans, and H. R. Shanks, *Phys. Rev.* B **26**, 1702 (1982).

[7] B. Reihl, J. G. Bednorz, K. A. Müller, Y. Jugnet, G. Landgren, and J. F. Morar, *Phys. Rev.* B **30**, 803 (1984).

[8] V. E. Henrich, G. Dresselhaus, and H. J. Zeiger, *Phys Rev.* B **17**, 4908 (1978).

[9] R. A. Powell and W. E. Spicer, *Phys. Rev.* B **13**, 2601 (1976).

[10] D. H. Lu, D. L. Feng, N. P. Armitage, K. M. Shen, A. Damascelli, C. Kim, F. Ronning, Z.-X. Shen, D. A. Bonn, R. Liang, W. N. Hardy, A. I. Rykov, and S. Tajima, *Phys. Rev. Lett.* **86**, 4370 (2001).

[11] M. C. Schabel, C.-H Park, A. Matsuura, Z.-H. Shen, D. A. Bonn, R. Liang, and W. N. Hardy, *Phys. Rev.* B **57**, 6090 (1998).

Problems for Chapter 9

1. For the pi(xy) surface bands with the only non-zero perturbation, $\Delta'' = -1$, (a) find the eigenvalue equation for the two surface bands. (b) Give analytic expressions for the eiqenvalues. (c) Show that the two surface bands are not truncated, that is, those surface bands that exist for all values of k_y.

2. Make graphs of the pi(xy) bulk band edge ($k_y = 0$) and the two surface bands of Problem 1 for $\omega_t = -5$ and $\omega_t = -8$.

3. Using the eigenvalue expressions of Problem 1, find an analytical expression for the DOSS, $\rho_s(\omega)$, for the surface bands and show that $\rho_s(\omega)$ has square-root singularities at the top and bottom of both surface bands.

4. Make a graph of $\rho_s(\omega)$ in Problem 3 for $\omega_t = -5$ and $\omega_\perp = -8$.

5. If the Fermi energy corresponds to $\omega_F = -4.9$, find the number of electrons per spin occupying the surface bands in Problem 4. (Hint: use $\rho_s(\Omega)$ rather than $\rho_s(\omega)$ for the calculation.) If $U = 6\,\text{eV}$, what is the surface Coulomb repulsion potential for electrons occupying the surface band?

10

Distorted perovskites

The majority of perovskites are not cubic, but many of the non-cubic structures can be derived from the cubic (aristotype) structure by small changes in the ion positions. Several types of distortions occur among the perovskites. The most important types are those involving (a) ion displacements in which, for example, the B ion or A ion (or both) moves off its site of symmetry; (b) rotations or tilting of the BO_6 octahedra; and (c) both tilting and displacements.

Departure from the cubic perovskite structure will occur whenever distortions lead to a lower total energy. The lowering of the total energy in most cases is small, typically of the order of a tenth of an eV/cell and dependent upon the temperature. The additional stabilization energy of one structure over another depends in a subtle manner upon the competition between a number of electronic factors including changes in the Coulomb interactions (Madelung potentials), changes in the degree of covalent bonding, and the number of electrons occupying the antibonding conduction bands. In many cases changes in the A–O covalent bonding is thought to play a key role in determining the distorted structure.

Clearly it is not possible to predict structures based on the simple LCAO model we have been studying. However, given a particular structure that is close to the cubic structure one can examine the changes in the electronic states with the goal of understanding why the distorted structure is more stable. In addition, there are important new electronic features that result from small changes in the structure that can be explored with the simple LCAO model.

10.1 Displacive distortions: cubic-to-tetragonal
phase transition

A number of perovskites that have the ideal cubic structure undergo a cubic-to-tetragonal phase transition as the temperature is lowered. $BaTiO_3$ and $SrTiO_3$, for example, are cubic at temperatures above their transition (or Curie) temperatures

$T_C = 408$ and $378\,\mathrm{K}$, respectively, but are tetragonal below T_C.[1] For $BaTiO_3$ the tetragonal structure is achieved by what is called a "displacive transition". The B ions and A ions are displaced from their cubic positions in one direction and the oxygen ions in the other direction as shown in Fig. 10.1. In this tetragonal phase $BaTiO_3$ is ferroelectric. The situation for $SrTiO_3$ is different. Its transition involves the rotation of alternate octahedra in opposite directions as well as displacements

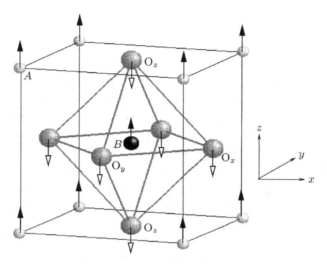

Figure 10.1. Tetragonal displacements: the displacements of the ions for the cubic-to-tetragonal phase transition. The A and B ions move upward along the z-direction while the oxygens move down. The cubic symmetry changes to tetragonal as a result of the displacements of the ions.

The spontaneous displacements (or octahedral rotations) are usually associated with the existence of a "soft" optical phonon. That is, a particular lattice vibration belonging to an optical phonon branch whose frequency tends to zero at T_C. As the phonon frequency decreases the vibrational motions increase and become anharmonic. Conceptually, one can imagine the vibrational motions of the ions in the soft mode becoming "frozen in" as T approaches T_C.

The magnitudes of the displacements in a displacive transition can be different for the different ions. For $BaTiO_3$ the magnitude of Ti-ion displacement is larger than that of the A ion and O ions [1]. Also, the O_z displacement is larger than that of the O_x and O_y ions so that the oxygen octahedron becomes slightly elongated in the z-direction. The displacements reduce the symmetry to tetragonal, but the departure from cubic symmetry is small.

---

[1] Additional structural transitions can occur at even lower temperatures. For example, $BaTiO_3$ is tetragonal in the range $285 < T < 400\,\mathrm{K}$, but undergoes a tetragonal-to-orthorhombic transition at $285\,\mathrm{K}$.

For a displacive transition the B ion of the ABO_3 structure is no longer at a site of inversion symmetry. As a result the electronic charge distribution is asymmetric and hence there is an electric dipole associated with each unit cell. In the ferroelectric state applying an external electric field can orient all the dipoles. The field of the dipoles produces long-range effects which, in fact, depend upon the macroscopic shape of the sample (or domains that form within the sample). The polarizability of the dipoles leads to a low-frequency dielectric function that is extremely large and temperature sensitive, particularly as the temperature approaches T_C. For $BaTiO_3$ the dielectric "constant" ranges from 1200 to 1600 at a frequency of 1 kHz. By contrast, in Chapter 6, we found $\varepsilon_2(\omega) \approx 6$ at the band-gap energy where $\omega \approx 5 \times 10^{12}$ kHz. Understanding the low-frequency behavior of the dielectric properties of the perovskites requires consideration of the lattice dynamics and polarizability of the dipoles and is not described by conventional band theory.

Many of the ferroelectric perovskites are also piezoelectric. If electrodes are attached to opposite faces of such crystals and a voltage applied, the electric field will induce dimensional changes. Conversely, application of an axial compressive force can produce a voltage on the electrodes. Because of these properties piezoelectric perovskites are used in electronic devices such as transducers and electrooptic modulators, and for the fine control of scanning tunneling microscope (STM) tip motion.

In Fig. 10.1 it can be seen that the B ion moves up along the z-axis toward one of the neighboring O_z ions and away from the other. As a result the $(pd\sigma)$ and $(pd\pi)$ interactions are increased for one of the B–O bonds and decreased for the other. Also, the line joining the O_x (or O_y) ions to the B ion is no longer perpendicular to the z-axis. Consequently, the wavefunctions at Γ, which are pure d orbital or pure p orbital in the cubic case, will have a small p–d mixing in the tetragonal phase. Another related effect is the splitting of the energy band degeneracies at Γ. The tetragonal displacement splits the t_{2g} states into a doubly degenerate group, d_{xz} and d_{yz}, and a non-degenerate d_{xy} state. Similarly, the doubly degenerate σ^* states are split into non-degenerate d_{x^2} and d_{z^2} states. This splitting, however, is not the same as the Jahn–Teller splitting that occurs in molecules.

For perovskites that undergo cubic-to-tetragonal phase transitions the displacements of the ions are small, usually less than a few percent of the lattice constant. For example, the calculated displacements for the Ba, Ti, O_x, O_y, and O_z ions in $BaTiO_3$ are 0.012, 0.039, 0.014, 0.014 and 0.025 Å, respectively [1]. The experimentally observed $c/2a$ ratio is 1.0086, indicating that the departure from cubic symmetry is very small [2]. The displacements are larger for $PbTiO_3$ where Ti-ion displacement is calculated to be 0.1006 Å and the experimental $c/2a$ ratio is 1.0649. The stabilizing energy per unit cell for the tetragonal phase is calculated to be about -0.4 meV for $BaTiO_3$ and -40 meV for $PbTiO_3$ [1].

To explore the effects of the displacements on the energy levels we can specify the ion locations for the tetragonal structure as:

B ions at: $\qquad 2a(n_x, n_y, n_z),$

A ions at: $\qquad 2a\left[\left(n_x + \dfrac{1}{2}\right), \left(n_y + \dfrac{1}{2}\right), \left(n_z + \dfrac{1}{2} - \delta_A\right)\right],$

O_x ions at: $\qquad 2a\left[\left(n_x + \dfrac{1}{2}\right), n_y, (n_z - \delta_{O_x})\right],$ $\qquad$ (10.1)

O_y ions at: $\qquad 2a\left[n_x, \left(n_y + \dfrac{1}{2}\right), (n_z - \delta_{O_y})\right],$

O_z ions at: $\qquad 2a\left[n_x, n_y, \left(n_z + \dfrac{1}{2} - \delta_{O_z}\right)\right],$

where $n_x, n_y,$ and n_z are positive or negative integers and

$$\begin{aligned}
\delta_A &= (d_B - d_A)/a, \\
\delta_{O_x} &= \delta_{O_y} = (d_B - d_{O_x})/a, \\
\delta_{O_z} &= (d_B - d_{O_z})/a.
\end{aligned} \qquad (10.2)$$

Here, d_B, d_A, d_{O_x}, and d_{O_z} are the displacements of the B, A, O_x, and O_z ions, respectively, and a is the B–O distance in the cubic phase. Equation (10.1) can be understood in the following way. Each type of ion can be assigned to a simple cubic sublattice, but not all of the atoms are on the lattice points. We choose a B ion as the origin, and then displace each of the sublattices relative to the B sublattice.

The Hamiltonian, H, for the cubic perovskite is given in Table 4.1. The changes in the matrix elements at Γ to first order in the displacements are:

$$\begin{aligned}
\Delta H(1,2) &= 2\Delta_\sigma \equiv d \\
\Delta H(1,11) &= \Delta H(1,14) = \delta_{O_z}[(pd\sigma) - 2\sqrt{3}(pd\pi)] \equiv e \\
\Delta H(3,11) &= -\Delta H(3,14) = -\delta_{O_x}[\sqrt{3}(pd\sigma) - 2(pd\pi)] \equiv f \\
\Delta H(9,4) &= \Delta H(5,12) = -2\delta_{O_x}[\sqrt{3}(pd\sigma) - (pd\pi)] \equiv g \\
\Delta H(9,7) &= \Delta H(8,12) = -2\delta_{O_x}(pd\pi) \equiv h \\
\Delta H(9,10) &= \Delta H(12,13) = 2\Delta_\pi \equiv u \\
\Delta H(I,J) &= \Delta H(J,I).
\end{aligned} \qquad (10.3)$$

In equation (10.3) $\Delta_\sigma \propto \delta_{O_z}$ is the increase (decrease) in $(pd\sigma)$ along the positive (negative) z-axis. Similarly, $\Delta_\pi \propto \delta_{O_x}$ is the increase (decrease) in $(pd\pi)$ along the positive (negative) z-axis. We have neglected the tetragonal perturbation on the oxygen–oxygen interaction in (10.3).

By rearranging rows and columns $(H + \Delta H)$ can be reduced to block-diagonal form. There is a 5×5 block (rows/columns 1, 2, 3, 11, 14), a 1×1 (row/column 6) and two symmetry-equivalent 4×4 blocks (rows/columns 9, 4, 7, 10 and 12, 5, 8, 13).

We label the energies as in Table 5.1. The "primed" energies of the form, $E'_{\Gamma n}$, indicate energies for the tetragonal phase and those without a superscript "prime" are for the cubic phase.

The block-diagonalized secular equation, $(H + \Delta H - E'_{\Gamma})$ takes the forms shown below.

(a) 1×1; row/column 6. This solution is the $\pi^*(xy)$ band edge.

$$E'_{\Gamma 3} = E_{\Gamma 3} = E_t \text{ (unshifted eigenvalue)} .$$ (10.4a)

(b) 4×4; rows/columns 9, 4, 7, 10. Solutions yield the shifted $\pi^*(xz)$ band edge and shifted valence-band energies.

Orbital	9	4	7	10
9	$E_t - E'_{\Gamma}$	g	h	u
4	g	$E_\| - E'_{\Gamma}$	$2c$	$2c$
7	h	$2c$	$E_\perp - E'_{\Gamma}$	p
10	u	$2c$	p	$E_\perp - E'_{\Gamma}$

(10.4b).

(c) 4×4, rows/columns 12, 5, 8, 13. Solutions yield the shifted $\pi^*(yz)$ band edge and shifted valence-band energies. These solutions are symmetry-equivalent to the solutions of (10.4b)   (10.4c)

(d) 5×5, rows/columns 1, 2, 3, 11, 14. Solutions yield the shifted σ^* band edges and shifted valence-band energies:

Orbital	1	2	3	11	14
1	$E_e - E'_{\Gamma}$	d	0	e	e
2	d	$E_\| - E'_{\Gamma}$	0	$2c$	$2c$
3	0	0	$E_e - E'_{\Gamma}$	f	$-f$
11	e	$2c$	f	$E_\perp - E'_{\Gamma}$	p
14	e	$2c$	$-f$	p	$E_\perp - E'_{\Gamma}$

(10.4d)

The matrix elements d, e, f, g, h, and u are defined in (10.3). The elements, $p \equiv 4(pp\pi)$ and $c \equiv (pp\sigma) + (pp\pi)$ are the oxygen–oxygen interactions for the cubic phase.

The solutions of $\det(H + \Delta H - E'_{\Gamma n}) = 0$ involve first-order changes in the eigenvectors and second-order changes in their eigenvalues from their cubic values.

As an example we consider the case of BaTiO$_3$ for which $\delta_{O_x} = 0.027$, $\delta_{O_z} = 0.033$. The LCAO parameters for BaTiO$_3$ are (in eV):

$$E_t = -6.70 \quad E_\perp = -10.00 \quad (pd\pi) = 1.00 \quad (pp\pi) = 0.05 \quad \Delta_\pi = 0.038$$
$$E_e = -5.40 \quad E_\parallel = -10.50 \quad (pd\sigma) = -2.00 \quad (pp\sigma) = -0.15 \quad \Delta_\sigma = -0.076.$$

To estimate the change in $(pd\pi)$ and $(pd\sigma)$ integrals (Δ_π and Δ_σ) we made use of hydrogenic orbitals to calculate the change in the overlap due to the displacements and then scaled LCAO parameters accordingly.

(a) Pi-like tetragonal states

The solutions of the 4×4 secular equation are given in Table 10.1. The splittings caused by the tetragonal distortion are shown schematically in Fig. 10.2. An important result in Table 10.1 is that to second order in the displacements there is no effect on the $E_{\Gamma 3}$ (the band edge for $\pi^*(xy)$ at Γ). This result can be seen by inspection since the angular integral of d_{xy} with any pair of neighboring p orbitals except $p_x(\vec{r} - a\vec{e}_y)$ and $p_y(\vec{r} - a\vec{e}_x)$ vanishes by symmetry independent of the displacement along the z-direction. Furthermore, the change in the interaction with the $p_x(\vec{r} - a\vec{e}_y)$ and $p_y(\vec{r} - a\vec{e}_x)$ orbitals is second order in the displacements and hence higher order in the perturbed energy.

Table 10.1. *Pi-like states at Γ. (Here $\mu = x, y, z$ and $\nu = y, z$.)*

Band at Γ (Table 5.1 notation)	Orbitals involved Cubic case	Orbitals involved Tetragonal case	Energy $E_{\Gamma n}$ (eV) Cubic case	$\Delta E_{\Gamma n}$ shift (meV)
$E'_{\Gamma 3}(\pi^*)$	d_{xy}	d_{xy}	−6.70	0.00
$E'_{\Gamma 4}(\pi^*)$	d_{xz}	$d_{xz}, p_x(\vec{r} - a\vec{e}_\mu)$	−6.70	17.63
$E'_{\Gamma 9}(\pi^0)$	$p_x(\vec{r} - a\vec{e}_\nu)$	$d_{xz}, p_x(\vec{r} - a\vec{e}_\mu)$	−10.20	−2.33
$E'_{\Gamma 10}(\pi + \sigma)$	$p_x(\vec{r} - a\vec{e}_\mu)$	$d_{xz}, p_x(\vec{r} - a\vec{e}_\mu)$	−9.70	−1.40
$E'_{\Gamma 11}(\pi + \sigma)$	$p_x(\vec{r} - a\vec{e}_\mu)$	$d_{xz}, p_x(\vec{r} - a\vec{e}_\mu)$	−10.60	−13.90

The perturbed wavefunction for the $\pi^*(xz)$ state, $E'_{\Gamma 4}$, is an admixture of (mostly) d_{xz} with a small amount of $p_x(\vec{r} - \vec{e}_\mu)$ orbitals ($\mu = x, y,$ and z). Symmetry equivalence tells us the same energy shift occurs for the d_{yz} state and mixing occurs between the d_{yz} and the oxygen $p_y(\vec{r} - \vec{e}_\mu)$ orbitals ($\mu = x, y,$ and z).

Table 10.1 shows that the triple degeneracy of the π^* band is split into an unshifted singlet $E'_{\Gamma 3}(\pi^*(xy))$ and a doublet $E'_{\Gamma 4}(\pi^*(xz))$ and the symmetry-equivalent $E'_{\Gamma 5}(\pi^*(yz))$. The $\pi^*(xz)$ and $\pi^*(yz)$ band edges are shifted upward in energy as shown schematically in Fig. 10.2. The shift increases quadratically with the relative displacements of the oxygen ions.

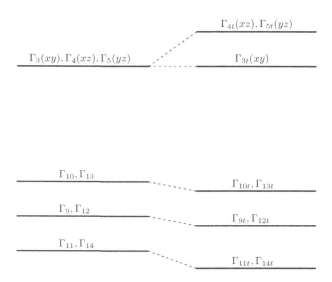

Figure 10.2. Tetragonal splitting of the pi-like states at Γ: the labeling of the levels is according to the convention of Table 5.1. The subscript 't' refers to the tetragonal phase. The threefold degeneracy of the π^* band edge is split into a doubly degenerate and a non-degenerate level. The twofold degenerate σ^* band edge is split into two non-degenerate levels.

For insulating $BaTiO_3$ (with empty π^* bands) one can see that the energy of the system is lowered because all of the filled valence states are lowered in energy. To calculate the actual stabilization energy we would need the perturbed energies over the entire Brillouin zone. The results at Γ are only suggestive and likely overestimate the effect since the perturbations vary as $\cos(k_\alpha a)$ and are therefore maximal at Γ. Nevertheless, it is interesting to look at the contributions from the states at Γ. The stabilization energy due to these occupied valence states amounts to -35.26 meV/cell $(2\times(-13.90-1.40-2.33)$ meV/cell) and results from the increased Ti–O bonding. From Table 10.1 it is also obvious that the stabilizing energy will decrease if electrons are added to the π^* bands.

The change in the degeneracy of the π^* band edge at the phase transition can produce important effects. For example, a lightly n-doped material may have its Fermi energy within a few meV of the bottom of the π^* band. A concentration of 10^{20} electrons per cubic centimeter $(6.4\times10^{-3}$ electrons per unit cell) produces a Fermi energy about 5.6 meV above the bottom of the π^* band edge at Γ (see Subsection 6.4(a)). In the cubic phase the electrons reside equally at the bottom of the three degenerate π^* bands. During the cubic-to-tetragonal transition the Fermi

energy, E_F, will change abruptly because the degeneracy of the band edge changes
from 3 (above T_C) to 1 (below T_C). Below T_C all of the electrons must now reside in
the $\pi^*(xy)$ band. To achieve the same number of occupied states the Fermi energy
must change as shown schematically in Fig. 10.3. In addition, if $E_F < E_t + \Delta E'_{\Gamma4}$,
the Fermi level will not intersect the $E'_{\Gamma4}, (\pi^*(xz))$ or the $E'_{\Gamma5}, (\pi^*(yz))$ bands so
the Fermi surface loses two of its arms. The three-dimensional "jack" described in
Chapter 6 and shown in Fig. 10.3(a) degenerates into the surface of a single circular
rod oriented along the k_z-axis as shown in Fig. 10.3(b). The radius of the rod must
increase by a factor of $\sqrt{3}$ to accommodate the same number of electrons. The above
comments would be strictly true at low temperature. However, the thermal energy,
$k_B T$, is about 35 meV at 400 K and 25 meV at 285 K so the thermal energy is larger
than the splitting throughout the temperature range of the tetragonal phase. Thus,

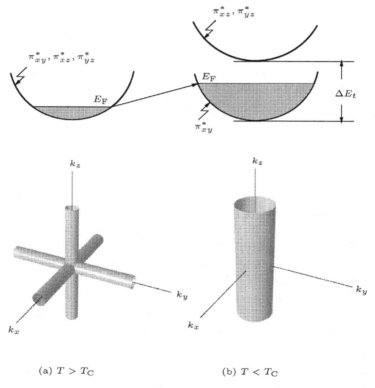

(a) $T > T_C$ (b) $T < T_C$

Figure 10.3. Splitting and the Fermi surface. (a) Above the Curie temperature, T_C. The
three π^* bands are degenerate and the Fermi surface is the "jack" described in Chapter
6. (b) Below the Curie temperature. The $\pi^*(xz)$ and $\pi^*(yz)$ energies are raised while the
$\pi^*(xy)$ is unshifted. Abrupt changes occur in the Fermi energy and the Fermi surface at
T_C that lead to abrupt changes in the electron transport properties.

one would expect thermally excited electrons in the $E'_{\Gamma 4}$, $E'_{\Gamma 5}$ bands and holes in the $E'_{\Gamma 3}$ band.

Nevertheless the abrupt change in the Fermi energy and the sudden appearance of electric dipoles will produce abrupt changes in the electron transport properties.[2] For example, one might expect to see the sudden appearance of anisotropy in the conductivity or in the Hall effect. In fact, such effects are seen experimentally in samples of BaTiO$_3$. Measurements on single-crystal BaTiO$_3$ [3] show a step-like increase of the resistivity by a factor of 2 at T_C as the crystal passes from the cubic to the tetragonal phase. In addition, measurements of the Hall-effect mobility, μ, indicate $\mu_a/\mu_c \gtrsim 10$ for electrons [4, 5] in the tetragonal phase.

(b) Sigma-like tetragonal states

The secular equation resulting from the 5×5 matrix equation can be condensed into the following form:

$$\left[\tilde{E}_e \tilde{E}^-_\perp - 2f^2\right]\left\{\tilde{E}_e(\tilde{E}_\| \tilde{E}^+_\perp - 8c^2) - d^2\tilde{E}^+_\perp - 2e^2\tilde{E}_\| + 8cde\right\} = 0, \tag{10.5}$$

where $\quad \tilde{E}^\pm_\perp = E_\perp \pm 4(pp\pi) - E'_\Gamma$,

$\qquad \tilde{E}_e = E_e - E'_\Gamma$,

$\qquad \tilde{E}_\| = E_\| - E'_\Gamma.$

The first factor in brackets yields

$$E'_{\Gamma 2,6} = \frac{1}{2}(E_e + E^-_\perp) \pm \sqrt{\left[\frac{1}{2}(E_e - E^-_\perp)\right]^2 + 2f^2} . \tag{10.6}$$

The second factor is a cubic equation in E'_Γ. The solutions are given in Table 10.2.

The energies, $E'_{\Gamma 1}$ and $E'_{\Gamma 2}$ correlate with the e_g d-orbital states, d_{z^2} and d_{x^2}, which are degenerate in the cubic case with energy E_e (equal to −5.40 eV in this example). In the tetragonal phase they are split with the d_{z^2} raised in energy by 13.64 meV and the d_{x^2} by 9.05 meV. All of the valence states are lowered in energy. The contribution to the stabilization energy from these valence states at Γ is −22.69 meV, an amount that is about two-thirds of the contribution of the pi-like states. The total stabilization energy from all the valence states at Γ is −57.95 meV/cell.

The energy shifts due to the tetragonal distortion are a result of increased B–O covalent bonding (and antibonding) They are not analogous to the Jahn–Teller effect that occurs in molecules. Here both the t_{2g} and e_g antibonding states are

---

[2] The energy band description for electron transport in BaTiO$_3$ is appropriate as recent experiments have shown [3]. Older results have been interpreted in terms of transport by small polarons [6].

Table 10.2. *Sigma-like states at* Γ. *(Here* $\mu = x, y, z$, *and* $\nu = x, y$.*)*

Band at Γ (Table 5.1 notation)	Orbitals involved Cubic case	Orbitals involved Tetragonal case	Energy $E_{\Gamma n}$ (eV) Cubic case	$\Delta E_{\Gamma n}$ shift (meV)
$E'_{\Gamma 1}(\sigma^*)$	d_{z^2}	$d_{z^2}, p_z(\vec{r} - a\vec{e}_\mu)$	−5.40	+13.64
$E'_{\Gamma 2}(\sigma^*)$	d_{x^2}	$d_{x^2}, p_z(\vec{r} - a\vec{e}_\nu)$	−5.40	+9.05
$E'_{\Gamma 6}(\pi^0)$	$p_z(\vec{r} - a\vec{e}_\nu)$	$d_{x^2}, p_z(\vec{r} - a\vec{e}_\nu)$	−10.20	−9.05
$E'_{\Gamma 7}(\pi + \sigma)$	$p_z(\vec{r} - a\vec{e}_\mu)$	$d_{z^2}, p_z(\vec{r} - a\vec{e}_\mu)$	−9.70	−4.90
$E'_{\Gamma 8}(\pi + \sigma)$	$p_z(\vec{r} - a\vec{e}_\mu)$	$d_{z^2}, p_z(\vec{r} - a\vec{e}_\mu)$	−10.60	−8.74

pushed up in energy whereas for Jahn–Teller splitting generally one of the d-orbital levels is lowered and other is raised in energy.

While it is not expected that the LCAO model can yield accurate results for the stabilization energy, it is expected that the results reflect the dominant energy-band mechanisms that act to stabilize the distorted structure.

10.2 Octahedral tilting

(a) Classification of tilting systems and space groups

Many of the cubic perovskites distort by tilting their oxygen octahedra. Phase transitions involving octahedral tilting alter the symmetry and can result in tetragonal, orthorhombic, rhombohedral, or monoclinic structures. The occurrence of a large number of different perovskites is often attributed to the fact that the octahedral structure can tilt to accommodate metal cations of widely varying sizes.

A system for classifying the tilt structures and determining their space-group symmetries has been developed by Glazer [7] and extended by others [8]. To understand this system we need to look at the aristotype (cubic) structure as an array of oxygen octahedra as depicted in Fig. 10.4(a). The oxygen ions are located at the corners of the octahedra. The B ions are located at the center of the octahedra and the A ions nestle in the spaces (interstices) between the octahedra. The tilting of an octahedron in a particular layer can be defined by specifying the angle of tilt (rotation) about each of the three pseudocubic axes. That is, the cubic axes before tilting has occurred. Since each oxygen (corner) is shared by two octahedra, it is clear that the tilting of one octahedron will require tilting of all the octahedra in a given plane. For example, rotating one of the octahedra about the z-axis will cause its neighbors to rotate oppositely, leading to an entire layer in which alternate oc-

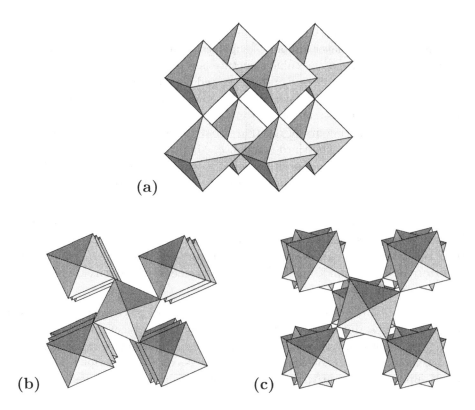

Figure 10.4. Octahedral tilting: ABO_3 cubic perovskite showing the oxygen octahedra. The oxygen ions are at the corners of the octahedra. The B ions are at the center of the octahedra and the A ions reside in the interstices between the octahedra. (a) Untilted, cubic case. (b) View looking down the $\vec{c}$-axis of the $a^0 a^0 c^+$ tilt system. The sense of rotation about the $\vec{c}$-axis is the same in adjacent layers. (c) View looking down the $\vec{c}$-axis of the $a^0 a^0 c^-$ tilt system. The sense of rotation about the $\vec{c}$-axis is opposite in adjacent layers.

tahedra are rotated in opposite senses. The tilts in the adjacent layers must also be specified. The octahedra in the next layer may be rotated in the same sense or in the opposite sense.

In Glazer's system the octahedra themselves are assumed to remain regular and tilts of a particular octahedron are described by specifying three symbols, $\vec{a}$, $\vec{b}$, $\vec{c}$ which relate to the magnitude of the rotation about the $\vec{a}$, $\vec{b}$, and $\vec{c}$ (x, y, and z) axes in a given layer. A repeated symbol indicates that the tilts are equal. Thus "aac" would indicate equal rotations about the $\vec{a}$ and $\vec{b}$ axes and a different rotation about the $\vec{c}$-axis. To specify the sense of rotation in the adjacent layer each of the symbols is assigned a superscript, "+", "−", or "0". If there is no rotation about a given axis the superscript 0 is used. A "+" superscript indicates rotation in the same sense in the adjacent layer and a "−" superscript indicates rotation in the opposite sense. With this system the cubic structure is represented by "$a^0a^0a^0$", indicating no rotation relative to $\vec{a}$, $\vec{b}$, or $\vec{c}$ in the layer and none in the adjacent layer. Consider the triad, $a^+b^-c^-$. The first symbol, a^+, indicates a rotation about the $\vec{a}$-axis in the first layer and the same rotation in the second layer. The second symbol, b^-, indicates a different amount of rotation about the $\vec{b}$-axis and the opposite rotation in the adjacent layer. The third symbol, c^-, indicates a third amount of rotation about the $\vec{c}$-axis and the opposite amount in the adjacent layer. The tilt systems $a^0a^0c^+$ and $a^0a^0c^-$ are shown in Figs 10.4(b) and (c), respectively.

Glazer's system has proven very useful for characterizing the space-group symmetry of distorted perovskites in a compact and convenient way. One should note, however, that it is assumed that the repeat of the interlayer pattern is no more than two layers of octahedra and the octahedra are assumed to remain (nearly) undistorted, or "regular". Furthermore, displacements of the ions are not described by Glazer's notation.

Others [8] have extended the system to include displacements of the B cation by adding a subscript to each of Glazer's three symbols. According to this notation, $a^0_0a^0_0c^0_+$ is an untilted tetragonal system (because all the superscripts are zero) with displacement of the B ion along the positive $\vec{c}$-axis (+ subscript). This notation would apply, for example, to the displacive phase transition we discussed for $BaTiO_3$ in Section 10.1. $SrTiO_3$ in its antiferroelectric tetragonal phase would correspond to $a^0_0a^0_0c^-_+$. This notation describes a crystal with tetragonal distortions consisting of opposite tilting about the $\vec{c}$-axis in adjacent layers and a B ion displaced along the positive $\vec{c}$-axis.

Glazer's system leads to 10 distinct tilt systems and 23 space groups. The extended notation, including displacements, yields 61 distinct space groups. Crystals representing 15 of Glazer's 23 tilt systems have been found experimentally. The most common system for the perovskites appears to be $a^-b^+a^-$ (55%) while cubic crystals with $a^0a^0a^0$ account for about 10% of the identified structures [9]. Of course, the actual structure depends upon the temperature. It is not unusual for a perovskite to go through several different octahedral-tilting transitions with increasing temperature, ending up in the cubic phase at the highest temperature. $NaNbO_3$, for example, undergoes seven transitions with increasing temperature, finally taking on the cubic structure above 923 K. A few examples of tilting systems are shown in Table 10.3.

Table 10.3. *Examples of tilt systems.* $(T_C = 110\,K$ *for* $SrTiO_3.)$

$a^0a^0a^0$	$a^0a^0c^-$	$a^-b^+a^-$	$a^-a^-a^-$
$SrTiO_3$ $(T>T_C)$	$SrTiO_3$ $(T<T_C)$	$CaTiO_3$	$LiNbO_3$
ReO_3	$NaTaO_3$	$SrZrO_3$	$NdAlO_3$
$NaWO_3$	$BaTiO_3$	$YCoO_3$	$LaCoO_3$

(b) Causes of octahedral tilting

Perovskites will have a tilted structure whenever tilting lowers the total energy of the system. Understanding exactly how the energy is reduced is not a simple matter. The physical and chemical forces (energies) that come into play are large in number and often competitive in action. A delicate balance between these various energies determines the structure with the lowest energy. This balance of energies changes with temperature, indicating that we are dealing with stabilization energies of a few tens of meV/cell at most.

The physics and chemistry of octahedral tilting in perovskites is a topic receiving a great deal of attention. Woodward [10] has given an overview of the factors that influence the tilt systems. Approaches to understanding the phenomenon are varied and include: (1) empirical correlations focused on ion sizes, coordination spheres, and various geometrical considerations; (2) empirical model calculations of total lattice energy; and (3) energy band calculations using various approximations. Each of these different approaches has its strengths and weaknesses and each offers a different perspective on the problem.

(c) Geometric considerations

From a geometrical point of view, the largest effect of octahedral tilting is the change in the immediate environment of the A ion. Tilting can substantially alter the A–O distances and the coordination spheres of the A ion, while the B-ion coordination sphere remains roughly octahedral. Figure 10.5 shows the oxygen coordination spheres for the cubic, $a^0a^0a^0$ and the $a^0b^-b^-$ tilt systems. For the ideal cubic system the A ion is surrounded by 12 equidistant oxygen ions. For the $a^0b^-b^-$ tilt system there are seven nearest and next nearest-neighbor oxygens (Fig. 10.5(b)). The nearest five enclose the A ion in a pyramidal cage and the remaining two lie further out. The nearest-neighbor A–O distances are less than for the cubic case, suggesting increased orbital interaction between the A and O ions. Such interactions tend to depress the energy of the non-bonding oxygen valence states and therefore lower the total energy of the system if the antibonding states are unoccupied.

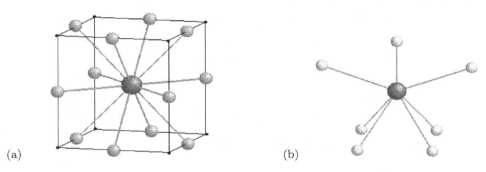

(a)   (b)

Figure 10.5. A-ion coordination spheres. (a) A cubic perovskite showing the 12 oxygen ions that form the coordinations sphere of the A ion. (b) The first and second nearest-neighbor oxygen ions for the $a^0b^-b^-$ tilt structure. The first coordination sphere consists of five oxygens that enclose A in a tetrahedral cage. The A–O distance for these nearest neighbors is less than in the cubic case. Two oxygen ions at a greater distance constitute the second nearest oxygens.

Some theories focus on the size of the A ion as the key factor in tilting. The geometrical "fit" of the A ion in the structure is measured by the Goldschmidt tolerance factor, "t", which is defined in terms of the ionic radii R_A, R_B, and R_O of the constituent ions:

$$t = \frac{(R_A + R_O)}{\sqrt{2}(R_B + R_O)}. \tag{10.7}$$

The ideal fit occurs for $t = 1$ and significant departures from unity suggest the crystal will tilt to accommodate the mismatch. Most of the known perovskites have $0.78 < t < 1.05$ [8]. Most cubic crystals are in the range $0.986 < t < 1.049$, but many tilted structures also lie in this range. The tolerance factor is suggestive of when tilting might be expected but gives no information about what tilt system should occur. Thomas [8] has proposed an empirical, geometric approach that yields a predictive relationship between the tilt angles and the polyhedral volumes:

$$\frac{V_A}{V_B} = 6 \cos^2 \theta_m \cos \theta_z - 1, \tag{10.8}$$

where V_A is the volume of the A–O_{12} coordination polyhedron and V_B is the volume of the BO_6 polyhedron, θ_m is the average of the tilt angles in the xy-plane and θ_z is the rotation angle about the z-axis. According to equation (10.8) V_A/V_B is maximized for untilted structures where $V_A/V_B = 5$, and decreases with increasing tilt angles. For a number of perovskite structures good agreement is found between the predicted volume ratio and measured tilt angles. For example, for $NaNbO_3$ ($293 < T < 773\,K$) the measured tilt angles are $\theta_m = 3.617°$ and $\theta_z = 4.209°$ and (10.8) yields $V_A/V_B = 4.964$. The ratio calculated from the actual structure is 4.956.

Thomas defines the degree of tilt by the parameter $\Phi = 1 - \cos^2 \theta_m \cos \theta_z$, which is zero for untilted systems. The experimental data points for most of the perovskites examined lie on the straight-line plot of V_A/V_B versus Φ. Those data which do not fall on the line (e.g., $NaTaO_3$) correspond to structures believed to have significantly distorted octahedra.

(d) Lattice energy calculations

Models that use empirically derived atomic potentials to represent Coulombic and short-range forces have been employed to examine the stability of various tilting systems [11, 12]. Woodward [10] used this approach to obtain results for idealized $YAlO_3$ subject to various normalizing constraints in order to compare the repulsive, attractive, and total lattice energy of different tilt systems. The repulsive energy is minimized by the cubic structure, but the attractive potential is maximized for the $a^+b^-b^-$ system. Table 10.4, second column, shows the stabilization energies relative to the cubic structure for several of the tilt systems [10].

Table 10.4. *Total energies of tilt systems relative to the cubic phase [10].*

Tilt system	Empirical lattice energy $YAlO_3$ (eV)	Extended Hückel $YAlO_3$ (eV)	Extended Hückel AlO_3 (eV)
$a^+b^-b^-$	-0.72	-1.76	$+0.08$
$a^-a^-a^-$	-0.61	-1.60	$+0.07$
$a^0b^-b^-$	-0.55	-1.62	$+0.05$
$a^0a^0c^-$	-0.20	-1.49	$+0.19$
$a^0a^0c^+$	-0.19	-1.44	$+0.19$
$a^0a^0a^0$	0.00	0.00	0.00

The total lattice energy difference between different tilt systems ranges from 0.19 to 0.72 eV. The Al–O–Al angles were found to vary from $145.4°$ (for $a^0b^-b^-$ and $a^0a^0c^-$) to $180°$ for the cubic case. Woodward suggested that oversized A cations ($t > 1$) are best accommodated by the cubic structure because that structure minimizes the repulsive energy. When the A ion is small ($t < 0.975$) the tilt rotation angle becomes large and the orthorhombic $a^+b^-b^-$ tilt system is favored according to Woodward [10].

(e) Electronic structure considerations

For many of the tilted structures the BO_6 octahedron remains nearly undistorted. The energy band structure will differ from the cubic structure for a number of reasons: (1) the interaction of the A-ion orbitals with the oxygen orbitals may be substantially increased and (2) the changes of the ionic site potentials shift the diagonal energies, (3) the angles of the O–B–O and O–O interactions change, and (4) splitting of degeneracies occur because of lower symmetry. All of these effects change the energy band structure from that of the cubic crystal. Most of the changes are minor in terms of the overall band structure but are of vital importance for properties such as ferroelectricity or magnetic order, metal–insulator and structural phase transitions. The decreased A–O distance can lead to admixing the A orbitals into the wavefunctions for the p–d valence and conduction bands and the p–p non-bonding bands. If the A-orbital state is at an energy higher than the d-orbital diagonal energy, as is usually the case, the effect of A–O interactions is to lower the energy of the valence bands and raise the energy of the A-orbital antibonding bands. In extreme cases where the A orbitals are strongly interacting even in the untilted phase, the basic structure of the energy bands is substantially altered. For example, for $BaBiO_3$ the $6s$ Bi orbitals hybridize with the oxygen sigma orbitals to form an antibonding band that is partially occupied. Similarly, for $PbTiO_3$ the Pb $6s$ and $6p$ orbitals are involved in the primary electronic structure. In these cases, tilting further enhances the roles of the A-ion orbitals.

Calculations of the total energy using the extended Hückel band structure model [10] have also been carried out for $YAlO_3$. The stabilization energy relative to the cubic phase is shown in the third column of Table 10.4. The results are quite different from those of the empirical potential calculations. In order to isolate the contribution of the Y ion, similar calculations were carried out for AlO_3^{3-}. The results are shown in the fourth column of Table 10.4. Omitting the Y (A ion) resulted in much smaller differences in the energies of the various tilt systems. More importantly, the results indicate cubic structure is the most stable system. Comparing columns three and four of Table 10.4 indicates that the majority of the tilt stabilization energy for $YAlO_3$ is derived from the presence of the Y ion. These results might suggest that tilting is mainly due to the A ion, but that is certainly not true. On the contrary, the perovskite, WO_3, has no A-site ion yet undergoes four phase transitions between 0 and 950 K. Furthermore, displacement of the W ion from its center of symmetry appears to occur in all of the phases [13–15]. Therefore, electronic effects alone are sufficient to cause tilting. For example, theoretical energy band calculations using density functional theory [14] for WO_3 as a function of electron doping produce all four of the tilting phases observed for $NaWO_3$ despite the fact than the A ion (Na) is omitted from the calculation. In this case adding electrons to the conduction bands drives the phase transitions.

As the number of electrons is theoretically increased from zero to one, the charge goes into the W $5d$-like antibonding bands. This reduces the stabilization energy of the filled valence bands and allows tilting structures to compete. Doping above $\frac{1}{2}$ electron per unit cell leads to displacement of the W ion along the z-axis. The $5d_{xz}$ and $5d_{yz}$ bands are then preferentially filled in a manner similar to that discussed in Section 10.1 for tetragonal $BaTiO_3$.

In summary, octahedral tilting occurs in most perovskites. Tilting provides the perovskite structure a way to accommodate a wide range of cation sizes and to lower its total energy. Typical stabilization energies are small, usually in the range 0.01–0.1 eV per unit cell. Despite the small energies involved, tilting phase transitions can lead to dramatic changes in the physical properties of the perovskite (e.g., ferroelectricity, magnetic order, superconductivity). Theoretical prediction of which tilting system will occur at a given temperature is a formidable task because there are a large number of different, competing mechanisms at work. Calculations of energy differences must be accurate to better than 0.01 eV to identify the lowest-energy tilt system with certainty.

References

[1] T. Nishimatsu, T. Hashimoto, H. Mizuseki, Y. Kawazoe, A. Sasaki, and Y. Ikeda, http://xxx.lanl.gov/PS_cache/cond-mat/pdf/0403/0403603.pdf

[2] Y. Kuroiwa, S. Aoyagi, A. Sawada, J. Harada, E. Nishibori, M. Takata, and M. Sakata, *Phys. Rev. Lett.* **87**, 217601 (2001).

[3] T. Kolodiazhnyi, A. Petric, M. Niewczas, C. Bridges, A. Safa-Sefat, and J. E. Greedan, *Phys. Rev.* B **68**, 085205 (2003).

[4] C. N. Berglund and W. S. Bear, *Phys. Rev.* **157**, 358 (1967).

[5] P. Bernasconi, I. Biaggio, M. Zgonik, and P. Günter, *Phys. Rev. Lett.* **78**, 106 (1997).

[6] E. Iguchi, N. Kubota, T. Nakamori, N. Yamamoto, and K. J. Lee, *Phys. Rev.* B **43**, 8646 (1991).

[7] A. M. Glazer, *Acta Cryst.* B **28**, 3384 (1972).

[8] N. W. Thomas, *Acta Cryst.* B **52**, 16, (1996); H. T. Stokes, E. H. Kisi, D. M. Hatch, and C. J. Howard, *Acta Cryst.* B **58**, 934 (2002).

[9] M. W. Lufaso and P. M. Woodward, *Acta. Cryst.* B **57**, 725 (2001).

[10] P. M. Woodward, *Acta Cryst.* B **53**, 44 (1997).

[11] T. S. Bush, J. D. Gale, R. A. Catlow, and P. D. Battle, *J. Mater. Chem.* **4**, 831 (1994).

[12] T. S. Bush, C. R. A. Catlow, A. V. Chadwick, M. Cole, R. M. Geatches, G. N. Greaves, and S. M. Tomlinson, *J. Mater. Chem.* **2**, 309 (1992).

[13] P. M. Woodward, A. W. Sleight, and T. Vogt, *J. Phys. Chem. Solids* **56**, 1305 (1995).

[14] A. D. Walkingshaw, N. A. Spaldin, and E. Artacho, *Phys. Rev.* B **70**, 165110 (2004).

[15] F. Cora, M. G. Stachiotti, C. R. A. Catlow, and C. O. Rodriguez, *J. Phys. Chem.* B **101**, 3945 (1997); M. G. Stachiotti, F. Cora, C. R. A. Catlow, and C. O. Rodriguez, *Phys. Rev.* B **55**, 7508 (1997).

Problems for Chapter 10

1. Calculate the magnitude of the spontaneous polarization (dipole moment/unit volume) in C/m^2 for $BaTiO_3$ using the displacements given in this chapter (see Fig. 10.1) and assuming charges of $+2e$, $+4e$, and $-2e$ for the Ba, Ti and O ions, respectively ($a = 3.992$ Å, $c = 4.036$ Å).

2. The perturbation matrix element $\Delta H(1,14)$ is the change in the LCAO interaction between the $d_{z^2}(\vec{r})$ orbital and the two orbitals $p_z(\vec{r} + a\vec{e}_y)$ and $p_z(\vec{r} - a\vec{e}_y)$ due to the relative oxygen-ion displacement. Show that $\Delta H(1,14) = \delta O_z[(pd\sigma) - 2\sqrt{3}(pd\pi)]$ to first order in δO_z.

3. Show that the unitary transformation, U, block-diagonalizes the 5×5 matrix of (10.4d) into a 2×2 and a 3×3 block, where

$$U = \frac{1}{\sqrt{2}} \begin{pmatrix} \sqrt{2} & 0 & 0 & 0 & 0 \\ 0 & \sqrt{2} & 0 & 0 & 0 \\ 0 & 0 & \sqrt{2} & 0 & 0 \\ 0 & 0 & 0 & 1 & 1 \\ 0 & 0 & 0 & -1 & 1 \end{pmatrix}.$$

Using this transformation derive the eigenvalue equations and find the eigenvectors (in terms of the amplitudes of the orbitals) for the 2×2 block.

4. Describe the tilting and ion displacements of a crystal with the $a_0^+ b_0^0 c_-^+$ structure.

5. Consider the interactions between an s orbital on the A ions and the d orbitals on a B ion. The B ion at $a(0,0,0)$ has eight nearest-neighbor A-ion s orbitals located at $(\pm a, \pm a, \pm a)$. Find the Slater–Koster parameters for the s–d interactions for the five types of d orbitals. What effect would you expect this to have on the (a) π and π^* bands at Γ and (b) the σ and σ^* bands at Γ?

6. Suppose the B ion in Problem 5 is displaced by a small amount, δa, along the positive z-axis. Find the new Slater–Koster s–d parameters to first order in δ. What effect would you expect this to have on the (a) π and π^* bands at Γ and (b) the σ and σ^* bands at Γ?

11

High-temperature superconductors

11.1 Background

In 1986 Bednorz and Müller [1] made the surprising discovery that the insulating, ceramic compound, La_2CuO_4, was superconducting at low temperature when suitably doped with divalent ions. In fact, all of the members of the class of copper oxides, $La_{2-x}M_xCuO_4$ (where M is Ba^{2+}, Sr^{2+}, or Ca^{2+} ions), were found to be superconducting for x in the range, $0.1 \lesssim x \lesssim 0.25$. At optimal doping of $x \approx 0.15$, the superconducting transition temperature, T_c, of $La_{2-x}Sr_xCuO_4$ was about 38 K.

Since 1986 there has been an enormous scientific effort focused on copper oxides with similar structures. Over 18 000 research papers were published in four years following Bednorz and Müller's report. As work progressed around the world, new compounds were discovered with higher T_c's. In 1987 doped samples of $YBa_2Cu_3O_7$ ("YBCO") were found to be superconducting at 92 K, thus becoming the first superconductors with T_c higher than the boiling point of liquid nitrogen (77 K). Recent studies [2] on mercuric cuprates report T_c in excess of 165 K.

A very brief list of some of the most studied high-temperature superconductors (HTSC) and their transition temperatures is given in Table 11.1. In column two "alias" refers to the frequently used name of the undoped, "parent" composition. For example, Tl1223 refers to: one Tl, two Ba, two Ca, and three Cu atoms. The tetragonal structure for the La_2CuO_4, is shown in Fig. 11.1(a). The copper ions are located at the centers of the octahedra and the oxygen ions occupy the corners of the octahedra. The La ions fill the spaces between the octahedra. The figure also shows the details of one of the octahedra. It is similar to the perovskite metal–oxygen octahedron, but elongated in the z-direction. The Cu–O_{ip} (in-plane oxygen) distance is about 1.9 Å, while the Cu–O_z distance is significantly greater at around 2.4 Å. As a result the O_z ions interact only weakly with the central copper ion compared with the O_{ip} ions. The strongly interacting Cu-O_{ip}'s of the octahedral arrays form planar sheets. Each sheet consists of a square copper-oxygen lattice whose unit cell contains one Cu and two oxygen ions. The Cu–O_2 sheets are

Table 11.1. *Some high-T_c superconductors and their T_c values. The formula for the undoped compound and its alias in the scientific literature are given along with the number of Cu–O$_2$ layers, n, in close proximity.*

Compound formula	Alias	n	T_c (K)
La_2CuO_4	La214	1	38
$TlBa_2CuO_5$	Tl1201	1	50
$Tl_2Ba_2CuO_6$	Tl2201	1	80
$Bi_2Sr_2CaCu_2O_8$	BiSCO	2	85
$YBa_2Cu_3O_7$	YBCO	2	92
$TlBa_2Ca_2Cu_3O_9$	Tl1223	3	110
$TlBa_2Ca_3Cu_4O_{11}$	Tl1234	4	122
$HgBa_2Ca_2Cu_3O_8$	Hg1223	3	166

separated by a large distance, about 6.6 Å. Figure 11.1(b) shows the structure of La_2CuO_4 projected onto the ac-plane.. The Cu–O$_2$ superconducting layers are in the ab-plane separated from each other by O$_z$ and La ions. Other high-T_c cuprates have two or more Cu–O$_2$ sheets in close proximity. For example, Y123 ($YBa_2Cu_3O_7$, $T_c = 92$ K) has two copper-oxygen layers about 3.2 Å apart. This pair of adjacent layers is separated (8.2 Å) from the next pair by three Y–Ba–O isolation layers. Generally, it is found that T_c increases with, n, the number of adjacent Cu–oxygen layers. The schematic phase diagram, T versus acceptor (hole) concentration, for $La_{2-x}Sr_xCuO_2$ shown in Fig. 11.2 is typical for the cuprates. The undoped parent compound from which the HTSC material is derived by acceptor doping[1] with Ba, Sr, or Ca is antiferromagnetic with the nearest-neighbor Cu spins aligned in opposite directions. The Neel temperature, T_N, for undoped La_2CuO_4 is 340 K, but decreases sharply with hole-doping. The magnetic phase exists up to about $x = 0.04$. There is a small region between the antiferromagnetic phase and the superconducting phase in which La_2CuO_4 is a non-magnetic insulator. In the antiferromagnetic and insulating phases the carrier dynamics are described by a Hubbard model or the so-called "t–J model" (discussed later). The mobile holes tend to hop between sites on the same spin sublattice and have large effective masses due to the antiferromagnetic correlations. Increased doping begins to interfere with the long-range antiferromagnetic order resulting in some cases in the formation of a spin glass. Further doping produces a poor metallic material that becomes superconducting at low temperatures. In the under-doped region the material is no longer antiferromagnetic but the spin correlations persist. The coexistence of antiferromagnetic

---

[1] Not all HTSC materials are hole-doped. The superconducting compound $Nd_{2-x}Ce_xCuO_4$ is electron-doped. Similar to the hole-doped cuprates this compound is antiferromagnetic, but structurally it lacks the apical oxygen and superconductivity exists only over a narrow range of doping.

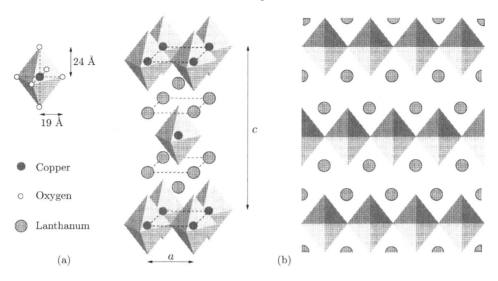

Figure 11.1. (a) Structure of La_2CuO_4. The Cu ions are located at the center and the oxygen ions occupy the corners of the tetragonally distorted octahedra. The O_z (apical) oxygens are 2.4 Å above or below the central Cu, but the in-plane oxygens are at a distance of only 1.9 Å. The La and apical oxygens act as isolation layers separating the Cu-in-plane-oxygen layers. (b) Projection of the structure onto the (001) plane showing the layers end-on.

correlations and superconductivity appears to be common in all the HTSCs and both phenomena may arise from the Coulomb repulsion effects at the Cu site.

As the doping increases, T_c increases in the region, $0.1 \lesssim x < 0.15$. At the "optimal doping" concentration, $x \simeq 0.15$, T_c is maximized. Beyond that point further doping results in a decrease in T_c ("over-doped" region). At even higher doping concentrations ($x > 0.25$ or so) the material is no longer superconducting. The "normal" metal phase is characterized by extreme anisotropy in the transport properties. The conductivity parallel to the $Cu–O_2$ planes is orders of magnitude greater than that perpendicular to the layers. While the resistivity of conventional metals typically varies as $A + BT^5$ at low temperatures due to phonon scattering, HTSC materials display a nearly linear decrease with temperature of the form $C + DT$ (or in some cases $C/T + DT$) down to the temperature at which superconductivity sets in. The in-plane conductivity is unaffected by the addition of some impurities, but modified by others. Dopants that increase resistivity and reduce T_c are those which substitute for Cu and form localized spins in the $Cu–O_2$ layers. Also shown in Fig. 11.2 is a structural phase boundary (dashed line) which separates the tetragonal phase from the orthorhombic phase. The occurrence of such structural transitions in the HTSC materials is typical.

Although the scientific literature dealing with the electronic properties of

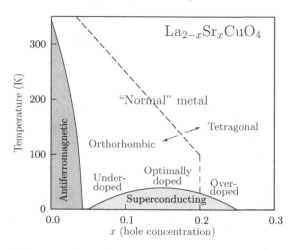

Figure 11.2. Schematic phase diagram for $La_{2-x}Sr_xCuO_4$. The dashed line running diagonally across the sketch separates the tetragonal from the orthorhombic structures. Three regions in the superconducting phase are identified as "under-doped", "optimally doped", and "over-doped". There is a small insulating region between the antiferromagnetic and superconducting phases at small doping.

HTSCs is large and rapidly growing, there is as yet no single, complete, and widely accepted quantitative theory. Theoretical understanding of how the electronic structure evolves from an antiferromagnetic insulator to a superconductor and then to a "normal" metal as a function of the doping concentration is still incomplete. It is generally agreed that electron pairs are responsible for the superconductivity, but the mechanism of pairing is still uncertain. BCS (Bardeen–Cooper–Schrieffer) theory, which has been so successful for "conventional" (lower-temperature) superconductors, does not appear to adequately describe the behavior of HTSC materials. It is also generally agreed that the superconducting energy gap is anisotropic in $\vec{k}$-space and usually has d-wave symmetry. Theoretical explanations of the electronic properties of the "normal" state are still controversial. A variety of models for transport in the normal state have been proposed including the total absence of coherent states and collective modes in which charge and spin are separated and propagate at different velocities [3].

The development of a quantitative theory for the HTSCs has been difficult because not only are the systems in the strong electron-correlation regime, but also there are a large number of competing phenomena with small but comparable stabilization energies to be considered. These phenomena include magnetic order, superconductivity, charge density waves, spin density waves, formation of charge and spin stripe domains, as well as structural phase transitions. In addition, the number of different HTSC compounds discovered is growing rapidly. Perhaps the most remarkable thing is not the differences displayed by this highly diverse collection of compounds, but rather the striking similarities.

A review of the field of HTSC in general, and the various current candidate theories in particular, is beyond the scope of this introductory text. Instead, in this chapter we shall concentrate on discussions of some of the qualitative electronic models that can be related to experimental results, particularly results obtained from angle-resolved photoemission. For this purpose we will relate our results mainly to the $La_{2-x}Sr_xCuO_4$ (LSCO) system because it has a relatively simple crystal structure, has single, isolated Cu–oxygen layers, and can be reliably doped up to a hole concentration of $x = 0.35$.

11.2 Band theory and quasiparticles

The application of conventional energy band theory to the electronic structure of the Cu–oxygen layer leads to the prediction that the parent cuprate materials should be metals with half-filled, d-like conduction bands. This is not the case. Band theory fails because the electron correlation energy associated with two electrons occupying the same copper site is large.

A simple ionic model of La_2CuO_4, for example, would have La^{3+}, Cu^{2+}, and O^{2-} ions. Thus, the copper ions would have the electronic configuration $[Ar]\ 3d^9$ while the La^{3+} and O^{2-} ions have closed-shell or subshell configurations. In the solid the d orbitals are split into the t_{2g} and e_g states that can accommodate six and four electrons, respectively (including spin states). The tetragonal symmetry further splits the levels. The t_{2g} states split into a singlet, d_{xy}, and a doublet d_{xz} and d_{yz}. The e_g states split into d_{z^2} and $d_{x^2-y^2}$ levels. According to energy band theory these states are broadened and covalently admixed with oxygen p orbitals to form one-electron pi and sigma bands. The valence-band states below the p–d gap are filled and the nine outer electrons of the copper ion are distributed among the antibonding σ^* and π^* bands. Six of the outer nine d electrons will occupy the π^* bands, two will occupy the lowest σ^* band, and the remaining electron will reside in the upper σ^* band. Since each σ^* band can hold two electrons per unit cell, the highest occupied band is half-filled and therefore the material should be metallic. However, a transition metal oxide with a half-filled d band usually forms a Mott–Hubbard antiferromagnet insulator as a result of the strong correlation effects. In the antiferromagnetic regime the Hubbard model for strong correlations [4, 5] leads to splitting (of what would have been the upper conduction band) into two bands separated by a gap of several electronvolts.[2] The lower Hubbard band is

---

[2] Depending on the relative magnitudes of the band gap, repulsive correlation energy, and band width, the filled, non-bonding bands may hybridize with the upper Hubbard band. This contributes additional structure to the filled bands [6]. Nevertheless, the upper Hubbard band remains empty and is still separated from the filled states by a sizable energy gap.

fully occupied and the upper band is empty so the material is an insulator and the states are localized rather than extended as in band theory. In this situation the p–d interaction causes virtual "hopping" of electrons between the metal and oxygen ions rather than forming delocalized energy band states. Acceptor dopants such as Sr^{2+} that substitute for the trivalent ions introduce holes in the lower Hubbard band (as for example in LSCO $\doteq La_{2-x}Sr_xCuO_4$). The holes reduce the effect of the correlation energy. As x increases, a point is reached where delocalized electron states are competitive with the localized Hubbard states. Further acceptor doping results in superconductivity at low temperatures. This, however, does not mean that conventional band theory applies. Strong correlations still operate and must be taken into account in any description of the electronic states. Experimental results, however, suggest that a quasiparticle-like description of the low-energy electronic excitations is appropriate in the superconducting phase.

The details of the low-energy excitations and the Fermi surface (FS) of the HTSCs have been investigated extensively using angle-resolved photoemission spectroscopy (ARPES) [7]. These experiments show that in the doped materials a FS or at least portions of a FS and quasiparticle peaks can be identified that are related to states derived from the d- and p-orbital bands. The quasiparticle states are similar to the one-electron states of energy band theory, but their energies are much smaller than what would be expected from band theory. These quasiparticle states include electron correlation effects that spread the spectral weight of a state over a range of energies and $\vec{k}$-vectors and introduce lifetime effects. Such quasiparticles are described by Landau's Fermi liquid theory, which, for example, provides a justification for mean-field theories of electronic structure and a starting point for the BCS theory of superconductivity [8].

The character of Fermi-liquid quasiparticles can be explained heuristically by extending the density of states (DOS) concept. In one-electron energy band theory the states have sharp energies in $\vec{k}$-space. The DOS for a single band state is

$$\rho(\omega, \vec{k}) = -\frac{1}{\pi}\Im\left\{\frac{1}{\omega - E(\vec{k}) + i0^+}\right\} = \delta\big(\omega - E(\vec{k})\big). \tag{11.1}$$

The effects of electron–electron interactions lead to a complex, proper self-energy correction, $E(\vec{k}) \rightarrow E(\vec{k}) + \Sigma(\vec{k}, \omega)$ with

$$\Sigma(\vec{k}, \omega) = \Sigma'(\vec{k}, \omega) + i\Sigma''(\vec{k}, \omega). \tag{11.2}$$

The DOS takes the form

$$\rho(\vec{k}, \omega) = -\frac{1}{\pi}\Im\left\{\frac{1}{\omega - \big(E(\vec{k}) + \Sigma'(\vec{k}, \omega) + i\Sigma''(\vec{k}, \omega)\big)}\right\}$$

$$= \frac{1}{\pi} \frac{\Sigma''(\vec{k}, \omega)}{\left[\omega - \left(E(\vec{k}) + \Sigma'(\vec{k}, \omega)\right)\right]^2 + \left[\Sigma''(\vec{k}, \omega)\right]^2} . \tag{11.3}$$

The result in (11.3) is the *one-particle spectral function*, usually denoted by $A(\vec{k}, \omega)$. It includes the effects of electron–electron interactions, which broaden the energy and introduce lifetime effects. The spectral weight described by $A(\vec{k}, \omega)$ is distributed according to the $\vec{k}$ and ω dependences of $\Sigma'(\vec{k}, \omega)$ and $\Sigma''(\vec{k}, \omega)$. The Fermi liquid quasiparticle state is not an eigenstate of the system. Instead it is coherent superposition of eigenstates with a narrow spread of energies and momenta. As a result, the quasiparticle has a finite lifetime that is related to $\Sigma''(\vec{k}, \omega)^{-1}$. Nevertheless, there remains a FS similar to that for non-interacting electrons, but in this case the sharp edges of the Fermi distribution function are smeared out, even at $T = 0\,\mathrm{K}$. The "line shape" described by (11.3) consists of a Lorentzian peak and a broad, relatively smooth background. In fact, $A(\vec{k}, \omega)$ can be separated into two parts [9, 10]:

$$A(\vec{k}, \omega) = A(\vec{k}, \omega)_{\text{coherent}} + A(\vec{k}, \omega)_{\text{incoherent}} , \tag{11.4}$$

representing the "coherent" and "incoherent" parts of the spectral function. The coherent part is given by

$$A(\vec{k}, \omega)_{\text{coherent}} = Z(\vec{k}) \frac{\Gamma(\vec{k})/\pi}{\left(\omega - \epsilon(\vec{k})\right)^2 + \Gamma(\vec{k})^2} \tag{11.5}$$

$$Z(\vec{k}) = \left(1 - \frac{\partial \Sigma'(\vec{k}, \omega)}{\partial \omega}\right)^{-1} \bigg|_{\omega = \epsilon(\vec{k})} , \tag{11.6}$$

$$\Gamma(\vec{k}) = Z(\vec{k}) |\Sigma''(\vec{k}, \omega)| \big|_{\omega = \epsilon(\vec{k})} , \tag{11.7}$$

$$\epsilon(\vec{k}) = Z(\vec{k}) \left[E(\vec{k}) + \Sigma'(\vec{k}, \omega)\right] \big|_{\omega = \epsilon(\vec{k})} . \tag{11.8}$$

For non-interacting electrons the occupation number of a state, $n(E(\vec{k}))$, has a discontinuity of 1 as $E(\vec{k})$ passes through the Fermi energy. For quasiparticles the discontinuity is reduced due to electron–electron interactions. $Z(\vec{k})$ is a measure of the discontinuity in $n(\vec{k})$ across the FS surface and of the validity of the quasiparticle representation. Small $Z(\vec{k})$ indicates that the quasiparticles undergo strong interactions and have short lifetimes, while $Z(\vec{k}) = 1$ indicates a sharp Fermi distribution, sharp energies, and infinite lifetimes. Equation (11.8) shows that the "quasiparticle" energy, $\epsilon(\vec{k})$, is smaller than the bare energy, $E(\vec{k})$, when $Z(\vec{k}) < 1$.

Fermi liquid theory is rigorously applicable only when the effects of correlation are weak. This does not seem to be the case for the HTSCs at low temperatures. The electron–electron repulsion on the Cu ions, U_d, is comparable to or greater than the p–d interaction parameters. On the other hand, ARPES experiments clearly show a well-defined FS and electronic excitations with the characteristics of quasiparticles exist for the HTSC materials in the superconducting phase.

11.3 Effective Hamiltonians for low-energy excitations

The shortcoming of band theory is that the total wavefunction contains states that have finite probabilities of two electrons of opposite spin simultaneously occupying the same Cu orbital, but the one-electron Hamiltonian either ignores this situation or includes a repulsive energy through the introduction of a mean-field potential. The LDA or local density approximation method, often used to calculate the electronic structures, assumes that the exchange and correlation can be represented by a potential that is a function of the average local density of charge. The potential is obtained self-consistently by iterative calculations. Unfortunately, the LDA does not capture the full effect of strong electron correlations.

A different approach is to exclude the double occupation states from the system's wavefunction (e.g., by using a reduced Hilbert space for the possible solutions). The problem may be treated by a variety of methods including "renormalization" schemes and so-called "slave-boson" formulations. The slave-boson scheme uses two separate particle operators to describe the spin (spinon) and charge (holon) degrees of freedom of the electrons [11–14]. In many cases the Hilbert space used excludes the doubly occupied d orbitals. Both renormalization and slave-boson approaches lead to the same type of effective Hamiltonian for the case when the repulsion parameter U_d is large (usually taken to be infinitely large). The "extended Hubbard model" Hamiltonian, H_{eH}, obtained in this manner has the form:

$$H_{eH} = \sum_{ij\sigma} h_{ij}\, a^\dagger_{i\sigma} a_{j\sigma} + \frac{1}{2} \sum_{ij\sigma\sigma'} U_{ij}\, a^\dagger_{i\sigma} a_{i\sigma}\, a^\dagger_{j\sigma'} a_{j\sigma'}, \qquad (11.9)$$

where h_{ij} are the "renormalized", metal–oxygen interactions for $i \neq j$ and the "renormalized" diagonal site energies for $i = j$. U_{ij} are the electron–electron Coulomb repulsion parameters of which $U_{ii} = U_d$ or U_p. In (11.9), the subscript, σ, indicates the spin state. Equation (11.9) looks relatively simple, but the solutions are extremely complex even with a myriad of approximations. Exact solutions have been found for one-dimensional systems but not otherwise. Because of the complexity of finding meaningful approximate solutions, usually only a single band is considered. Approximate effective Hamiltonians can be obtained from (11.9) by retaining nearest- (and sometime next nearest) neighbor interactions. An approximate effective Hamiltonian called the "Rice model" is also often employed. This Hamiltonian is of the form

$$H_R = P \sum_{ij\sigma} t_{ij}\, a^\dagger_{i\sigma} a_{j\sigma}\, P + J \sum_{ij\sigma} \left[\vec{s}_i \cdot \vec{s}_j - \frac{1}{4} n_{i\sigma} n_{j,-\sigma} \right]. \qquad (11.10)$$

The operator, P, projects out the zero and single occupancy states and J is the effective exchange energy. The parameters, t_{ij} are the renormalized tight-binding (LCAO-like) parameters and $\vec{s}_i$ is the electron spin operator. For a half-filled band the second term of the Hamiltonian is dominant and the low-lying quasiparticles are antiferromagnetic spin excitations. For small doping, mobile holes become possible. In this regime the holes acquire a sizable effective mass due principally to the "resistance" of the ordered spin system through which they move. At higher hole concentration the first term of (11.10) becomes competitive with the second term, resulting in the loss of long-range spin order.

Effective Hamiltonians (essentially equivalent to the Rice model) called the "t–J" model (nearest neighbors) and the "tt′–J" model (nearest- and next-nearest-neighbor interactions) are also commonly used. The Rice or "t–J" model recasts the calculation of the electronic/spin structure and low-energy excited states into a form that is similar to conventional tight-binding (LCAO) band theory. However, the effective parameters such as E_g, $(pd\sigma)$ and $(pd\pi)$, are typically 5–10 times smaller than those of conventional band theory [6, 15, 16]. On the other hand, it has been argued [17, 18] that the O–O parameters, $(pp\sigma)$ and $(pp\pi)$, which govern the hopping between oxygen orbitals are much less affected than the p–d parameters. In our previous discussions of the band structure of the perovskites these O–O interactions were of minor importance since they were roughly an order of magnitude smaller than the $(pd\sigma)$ and $(pd\pi)$ parameters. For the cuprates, after renormalization, it is likely that the effective $(pp\sigma)$ (oxygen–oxygen, second-neighbor parameter) is comparable to $(pd\sigma)$ and therefore the O–O interactions must be considered in calculating the quasiparticle band states.

In general, it is found that the energy scale for the effective Hamiltonian is of the order of 0.1 eV when the band is approximately half-filled and the scale for the superconducting gap is of the order of a few times $k_B T_c$, a few hundredths of an eV. These scales are much smaller than typical band structure parameters that tend to be of the order of 0.5 eV to several electronvolts.

11.4 Angle-resolved photoemission

Angle-resolved photoemission spectroscopy (ARPES) is one of the most important tools for investigating the electronic properties of the HTSC materials. The process of photoemission and its relation to the energy band states was described in Chapter 8. In that discussion we concentrated on the emitted electron intensity, $I(E)$, versus the energy of the initial state. In practice a beam of monochromatic ultraviolet photons of energy, $\hbar\omega$ (usually less than 100 eV), is incident on the surface of the sample. Photoelectrons are emitted from the surface in all directions with a distribution of kinetic energies. For a fixed kinetic energy the photoelectron intensity is measured as a function of the polar angle, θ, relative to the surface normal of the sample. Using this information the initial state energy and wavevector parallel to the surface can in principle be inferred if the momentum perpendicular to the surface is assumed to be conserved,

$$E_{\text{kin}} = \hbar\omega - \Phi - E(\vec{k}), \tag{11.11}$$

$$\hbar\vec{k}_\| = \sqrt{2mE_{\text{kin}}} \sin\theta. \tag{11.12}$$

If E_{kin} is fixed and θ varied, a peak in the photoelectron intensity is expected when $\vec{k}_\|$ corresponds to an initially occupied state on the energy-band dispersion curve defined by (11.11). Conversely, if $\vec{k}_\|$ is fixed and E_{kin} varied, a peak in the intensity is expected when the energy corresponds to an initially occupied state on the energy-band dispersion curve. For a typical solid with three-dimensional energy bands, such peaks are broadened because the component of momentum perpendicular to the surface, $k_\perp$ is *not* conserved. Therefore the photoemitted electrons arriving at the detector are coming from a range of states with different $k_\perp$ values. On the other hand, for the HTSCs the CuO_2-layer energy bands are two-dimensional and $E(\vec{k}_\|)$ is the same for all values of $k_\perp$. As a result, the two-dimensional dispersion curve $E(\vec{k}) = E(\vec{k}_\|)$ for states below the Fermi energy can be mapped with reasonable accuracy.

In the BCS theory of superconductivity the formation of Cooper pairs opens a small gap, 2Δ, in the quasiparticle density of states near the Fermi energy. The DOS is given by

$$\rho_{\text{BCS}}(E) = \rho(E_{\text{F}}) \frac{E - E_{\text{F}}}{\sqrt{(E - E_{\text{F}})^2 - \Delta^2}} \tag{11.13}$$

where $\rho(E_{\text{F}})$ is the DOS at E_{F} in the normal state. The DOS has a square-root singularity at $E = E_{\text{F}} \pm \Delta$ that is clearly observed in electron tunneling experiments. HTSC material may not be described by BCS theory, but in the superconducting phase they possess a similar energy gap that can be measured by ARPES experiments.

Figure 11.3 illustrates the type of data obtained from ARPES experiments. Figure 11.3(a) shows energy distribution curves for a typical HTSC for $T > T_{\text{c}}$ and $T < T_{\text{c}}$. The wavevector $\vec{k}_\|$ is fixed at a point on the FS and the energy scanned. The midpoint of the leading edge of the intensity curve (indicated by horizontal tick marks) is taken as the position of the quasiparticle energy. For $T > T_{\text{c}}$, the midpoint of the intensity edge corresponds to the state $E = E_{\text{F}}$. For $T < T_{\text{c}}$, the midpoint moves away from E_{F} to a lower energy leaving an energy gap, Δ, in which no quasiparticle states exist. By choosing different wavevectors on the FS the dependence of Δ on $\vec{k}_\|$ can be determined. In many cases $\Delta(\vec{k}_\|)$ possesses

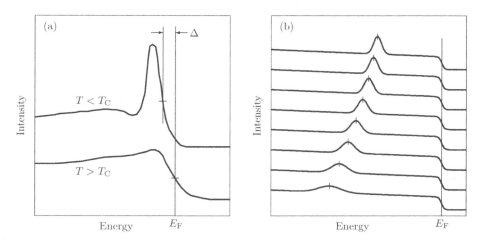

Figure 11.3. Typical ARPES data for a typical HTSC. (a) Photoemission intensity versus energy for $T > T_c$ and $T < T_c$, showing the formation of a superconducting energy gap, Δ, near the Fermi surface. (b) Energy distribution curves. Each curve is data taken along a line in the Brillouin zone that cuts across the Fermi surface. The locus of the peaks defines the energy dispersion, $E(\vec{k}_\parallel)$.

approximate d-wave symmetry given by the expression, $\Delta(\vec{k}_\parallel) = \text{constant} \cdot |\cos k_x a - \cos k_y a|$.

Figure 11.3(b) is an illustration of a series of ARPES energy distribution curves. Each curve corresponds to a particular $\vec{k}_\parallel$ along a line segment in the Brillouin zone. The intensity, $I(E)$, shows two features. On the right-hand side of the figure it can be seen that $I(E)$ drops to zero as the energy scans across the Fermi energy. In the middle of the figure are a series of quasiparticle peaks, one for each energy distribution curve. Therefore the locus of these peaks is the energy dispersion of the quasiparticles, $E(\vec{k}_\parallel)$.

It should be mentioned that the vector $\vec{k}_\parallel$, determined by E_{kin} and θ in (11.11) and (11.12), often lies in the second Brillouin zone, but the results can be projected back to the first Brillouin zone by symmetry considerations.

It is evident that tracking the quasiparticle dispersion up to the Fermi energy along different line segments in the Brillouin zone provides a means of determining the shape of the FS. Thus, ARPES experiments are extremely important because they provide direct measurements of both the quasiparticle dispersion and the FS.

Finally, we mention that photoemission is a many-body process. The process may be equated to the operation of the initial-state destruction operator $\vec{a}_{\vec{k}\ \text{initial}}(0)$ acting on the N-electron state followed by the final-state creation operator, $a^{\dagger}_{\vec{k}\,\text{final}}(t)$, acting on the $(N-1)$-electron state. Ignoring final-state effects, the shape of the quasiparticle peak is directly related to the spectral intensity function, $A(\vec{k}, \omega)$ discussed above. Approximate results for real and imaginary parts of the self-energy function have been derived from ARPES energy distribution curves [16].

11.5 Energy bands of the Cu–O₂ layers

As mentioned above, an empirical LCAO model with effective parameters that approximately include correlation effects can be employed to investigate the nature of the Fermi surface and density of states when the high-T_c material is sufficiently doped so that the FS and quasiparticle peaks are identifiable experimentally in ARPES data.

In this section we use our LCAO model to describe the energy bands of the Cu–O₂ layers. The results can not be applied to the parent (undoped) cuprates because of the strong correlation effects discussed above. They can, however, be used to describe the cuprates when they are suitably doped. That is, when the materials are in the superconducting phase.

We begin by considering the square, planar array formed by the Cu and in-plane oxygen ions. The LCAO Hamiltonian can be derived directly from the Hamiltonian for the perovskites given in Chapter 4 by setting to zero the matrix elements associated with the O_z ions. That is, by simply deleting the rows and columns labeled 2, 10, and 13 in Table 4.1. This eliminates the orbitals $p_\alpha(\vec{r} - a\vec{e}_z)$ ($\alpha = x, y$, and z) and reduces the matrix to that appropriate for the square, copper–oxygen layer whose unit cell is CuO₂. The result is the 11×11 matrix in Table 11.2. We have introduced the notation E_z and E_x for the diagonal energies of the d_{z^2} and $d_{x^2-y^2}$ orbitals in place of the cubic parameter, E_e. Due to Jahn–Teller and tetragonal ligand-field effects, E_z is expected to be more negative than E_x and therefore the uppermost, half-filled band is derived from $d_{x^2-y^2}$. For convenience we show in Table 11.2 the orbitals involved as well as retaining the labeling used in Table 4.1. (We remind the reader that we use "a" as the metal–oxygen distance, a distance that is half of the lattice spacing, a_0. Discussions in the scientific literature use "a" as the lattice spacing. As a result the high symmetry points such as X and M are designated by wavevector components apparently larger by a factor of 2. Also it should be mentioned that most papers seem to neglect the factor of a_0 altogether. Thus one sees the wavevector for X written as $k = (\pi, 0)$ instead of $ka_0 = (\pi, 0)$. In our notation the X point is $ka = (\pi/2, 0)$.)

To begin with we consider the solutions neglecting the oxygen–oxygen interactions. Then, as can be seen from Table 11.2, the Hamiltonian reduces to a block-diagonal form. There is (a) a 4×4 sigma block involving the σ-type p orbitals,

Table 11.2. The 11×11 matrix $(\mathbb{H} - E_{\vec k}\mathbb{I})$, with vanishing determinant, for $Cu\text{–}O_2$ bands.

| | d_{z^2} | $d_{x^2-y^2}$ | p_{xx} | p_{yy} | d_{xy} | p_{xy} | p_{yx} | d_{xz} | p_{zx} | d_{yz} | p_{zy} |
	1	3	4	5	6	7	8	9	11	12	14
1	$E_z-E_{\vec k}$	0	$-i(pd\sigma)S_x$	$-i(pd\sigma)S_y$	0	0	0	0	0	0	0
3		$E_x-E_{\vec k}$	$\sqrt{3}i(pd\sigma)S_x$	$-\sqrt{3}i(pd\sigma)S_y$	0	0	0	0	0	0	0
4			$E_\parallel-E_{\vec k}$	$-2bS_xS_y$	0	$2cC_xC_y$	0	0	0	0	0
5				$E_\parallel-E_{\vec k}$	0	0	$2cC_xC_y$	0	0	0	0
6					$E_t-E_{\vec k}$	$2i(pd\pi)S_y$	$2i(pd\pi)S_x$	0	0	0	0
7						$E_\perp-E_{\vec k}$	$-2bS_xS_y$	0	0	0	0
8							$E_\perp-E_{\vec k}$	0	0	0	0
9								$E_t-E_{\vec k}$	$2i(pd\pi)S_x$	0	0
11									$E_\perp-E_{\vec k}$	0	$4(pp\pi)C_xC_y$
12										$E_t-E_{\vec k}$	$2i(pd\pi)S_y$
14											$E_\perp-E_{\vec k}$

$H_{ij} = H_{ji}^*$

Parameters used are $b \equiv (pp\sigma) - (pp\pi)$, $c \equiv (pp\sigma) + (pp\pi)$, $S_\alpha \equiv \sin k_\alpha a$, and $C_\alpha \equiv \cos k_\alpha a$.

$p_{\alpha\beta} \equiv p_\alpha(\vec r - a\vec e_\beta)$, $E_z = d_{z^2}$ site energy, $E_x = d_{x^2-y^2}$ site energy, $E_x - E_z =$ tetragonal splitting of e_g levels.

Table 11.3. *Energy bands for the* Cu–O_2 *plane in the nearest-neighbor approximation. AB, B, and NB stand for antibonding, bonding, and non-bonding, respectively.*

Band #	Band	Type	Energy band $E_{\vec{k}}$	E_Γ
Pi-type bands				
6	$\pi^*(xy)$	AB	$E_{m\pi} + \sqrt{(\frac{1}{2}E_{g\pi})^2 + 4(pd\pi)^2(S_x^2 + S_y^2)}$	E_t
7	$\pi(xy)$	B	$E_{m\pi} - \sqrt{(\frac{1}{2}E_{g\pi})^2 + 4(pd\pi)^2(S_x^2 + S_y^2)}$	$E_\perp$
8	$\pi_{\pi^0}(xy)$	NB	$E_\perp$	$E_\perp$
9	$\pi^*(x)$	AB	$E_{m\pi} + \sqrt{(\frac{1}{2}E_{g\pi})^2 + 4(pd\pi)^2 S_x^2}$	E_t
11	$\pi(x)$	B	$E_{m\pi} - \sqrt{(\frac{1}{2}E_{g\pi})^2 + 4(pd\pi)^2 S_x^2}$	$E_\perp$
12	$\pi^*(y)$	AB	$E_{m\pi} + \sqrt{(\frac{1}{2}E_{g\pi})^2 + 4(pd\pi)^2 S_y^2}$	E_t
14	$\pi(y)$	B	$E_{m\pi} - \sqrt{(\frac{1}{2}E_{g\pi})^2 + 4(pd\pi)^2 S_y^2}$	$E_\perp$
Sigma-type bands				
1	σ^*	AB	$E_0 + E_1 \cos(\varphi/3)$	E_x
3	$\sigma^0_{z^2}$	NB	$E_0 + E_1 \cos(\varphi/3 - 2\pi/3)$	E_z
4	σ	B	$E_0 + E_1 \cos(\varphi/3 + 2\pi/3)$	$E_\|$
5	σ^0	NB	$E_\|$	$E_\|$

$p = -(E_x + E_\| + E_z), \qquad q = E_z E_\| + E_x E_z + E_x E_\| - 4(pd\sigma)^2\Omega, \qquad s = (p^2 - 3q)/9, \qquad E_0 = -p/3.$

$r = -E_x E_z E_\| + (3E_z + E_x)(pd\sigma)^2\Omega, \qquad t = (9pq - 2q^3 - 27r)/54, \qquad \cos(\varphi) = t/\sqrt{s^3}, \qquad E_1 = 2\sqrt{s}.$

and the d_{z^2} and $d_{x^2-y^2}$ orbitals, (b) a 3×3 block involving the oxygen π orbitals, $p_x(\vec{r} - a\vec{e}_y)$, $p_y(\vec{r} - a\vec{e}_x)$, and d_{xy}, (c) a 2×2 block involving the oxygen π orbital, $p_z(\vec{r} - a\vec{e}_x)$, and d_{xz}, (d) a 2×2 block involving the oxygen π orbital, $p_z(\vec{r} - a\vec{e}_y)$, and d_{yz}. The eigenvalues of the block-diagonalized matrix are easily obtained.

(a) Filled valence bands: the pi bands

The pi-band energies are given in Table 11.3. There are several interesting features. The $\pi^*(xy)$ and $\pi(xy)$ bands have the same type of energy dependence as the perovskite pi bands and hence the DOS is the same as that given by equation (6.28). Therefore, we expect two dimensional, logarithmic singularities in the $\pi^*(xy)$ and $\pi(xy)$ density of states. The $\pi^*(\alpha)$ and $\pi(\alpha)$ ($\alpha = x$ or y) however, are one-dimensional (depend on only one component of the wavevector) and therefore their DOSs will have square-root van Hove singularities at the top and bottom of the band. The DOS is given by

$$\rho(\varepsilon) = \frac{1}{\pi^2} \int_{-\pi/2}^{\pi/2} \frac{dx}{\varepsilon - S_x^2 + i0+} = \frac{1}{\pi} \frac{\Theta(\varepsilon)\Theta(\varepsilon - 1)}{\sqrt{\varepsilon(\varepsilon - 1)}} \tag{11.14}$$

$$\varepsilon(E) = \frac{(E - E_{m\pi})^2 - (\frac{1}{2}E_{g\pi})^2}{4(pd\pi)^2}, \tag{11.15}$$

$$\rho(E) = \rho(\varepsilon) \frac{d\varepsilon}{dE}, \tag{11.16}$$

where $E_{m\pi} = \frac{1}{2}(E_t + E_\perp)$ and $E_{g\pi} = (E_t - E_\perp)$.

The ideal logarithmic, square-root and flat band singularities in the DOS will be broadened in actual data, but should be observable in photoemission from the filled valence bands below the Fermi level [19].

(b) Sigma bands for the Cu–O$_2$ layer

The secular equation for the four sigma bands yields one flat band with $E = E_\parallel$ and three bands whose energies are given by the solution of the cubic equation,

$$(E_z - E)\left[(E_x - E)(E_\parallel - E) - 3(pd\sigma)^2\Omega\right] - (E_x - E)(pd\sigma)^2\Omega = 0, \tag{11.17}$$

$$\Omega = \sin^2(k_x a) + \sin^2(k_y a). \tag{11.18}$$

Analytical expressions for the three roots are given in Table 11.3.

The band or bands that intersect the Fermi energy determine the topology of the FS in the cuprates; that is, the highest partially occupied bands. In the case of the Cu–O$_2$ layer, this is usually the antibonding σ^* band. We shall refer to the four sigma bands in Table 11.3 above as the *four-band* model. This model is generally *not* the model used to discuss the electronic properties of the cuprates. Instead, it has been customary to ignore the role of the d_{z^2} orbitals. A three-band model is employed that considers only the $d_{x^2-y^2}$, $p_x(\vec{r} - a\vec{e}_x)$ and $p_y(\vec{r} - a\vec{e}_y)$ orbitals and the interactions among these three orbitals. The model is defined by the matrix elements shown in Table 11.4, which includes the second-neighbor interactions between the two oxygen orbitals. It should be noted, however, that the model neglects other comparable oxygen–oxygen interactions such as the interaction between $p_x(\vec{r} - a\vec{e}_y)$ and $p_y(\vec{r} - a\vec{e}_x)$. The justification for neglecting these other oxygen–oxygen interactions is that the orbitals $p_x(\vec{r} - a\vec{e}_y)$ and $p_y(\vec{r} - a\vec{e}_x)$ have no direct interaction with the $d_{x^2-y^2}$ orbital. As a result, they have only a very minor effect on the σ^* band (see Problem 11.7). In Table 11.4 (and Table 11.3) we use the simplified notation $p_{xx} = p_x(\vec{r} - a\vec{e}_x)$ and $p_{yy} = p_y(\vec{r} - a\vec{e}_y)$.

Again, ignoring the oxygen–oxygen interaction (i.e., setting $b = 0$) the resulting

Table 11.4. *Matrix for the three-band model, with vanishing determinant.*

	$d(x^2)$	p_{xx}	p_{yy}
$d(x^2)$	$E_x - E(\vec{k})$	$\sqrt{3}i(pd\sigma)S_x$	$-\sqrt{3}i(pd\sigma)S_y$
p_{xx}	$-\sqrt{3}i(pd\sigma)S_x$	$E_{\parallel} - E(\vec{k})$	$-2bS_xS_y$
p_{yy}	$\sqrt{3}i(pd\sigma)S_y$	$-2bS_xS_y$	$E_{\parallel} - E(\vec{k})$

solutions include one non-bonding band with $E = E_{\parallel}$, together with an antibonding band and a bonding band:

$$E_{\sigma^0}(\vec{k}) = E_{\parallel}, \tag{11.19}$$

$$E_{\sigma^*}(\vec{k}) = E_{m\sigma} + \sqrt{\left(\frac{1}{2}E_{g\sigma}\right)^2 + 3(pd\sigma)^2(S_x^2 + S_y^2)}, \tag{11.20}$$

$$E_{\sigma}(\vec{k}) = E_{m\sigma} - \sqrt{\left(\frac{1}{2}E_{g\sigma}\right)^2 + 3(pd\sigma)^2(S_x^2 + S_y^2)}. \tag{11.21}$$

The parameters in (11.20) and (11.21) are defined as $E_{m\sigma} = \frac{1}{2}(E_x + E_{\parallel})$ and $E_{g\sigma} = (E_x - E_{\parallel})$. In Fig. 11.4 the energy bands for the four-band model are compared with those of the three-band model using the same parameters. For the figure we plot the dimensionless energy, $\epsilon \equiv (E - E_x)/|(pd\sigma)|$. As can be seen the bonding and antibonding bands are similar for both the three- and four-band models with the four-band model having somewhat larger dispersion. The non-bonding band labeled σ^0 corresponds to the state with $E(\vec{k}) = E_{\parallel}$ (that is, at $\epsilon = (E_{\parallel} - E_x)/|(pd\sigma)| = -2.3333$). It is the same for both models. However, the nearly flat band derived from d_{z^2} and labeled $\sigma_{z^2}^0$ does not occur in the three-band model. This band has weak dispersion, indicating weak covalent bonding to the oxygen ions, but it is not entirely dispersionless. The Fermi energy for the undoped material, ϵ_F, is at the center of the σ^* band, the energy at the X-point,

$$\epsilon_F = \frac{E(\vec{k}_F) - E_x}{|(pd\sigma)|}, \qquad \vec{k}_F = (\pi/2a, 0).$$

For the cuprates all of the bands are filled except for the σ^* band, which for the parent material is half-filled. Since the electronic and superconducting properties are determined by the states near the Fermi level the only band that is significant is the σ^* antibonding band. Therefore it can be argued that the three-band model is sufficient. *This argument will be valid provided the $\sigma_{z^2}^0$ band is well below the σ^* band. If not, hybridization of the two bands will change the composition of the wavefunctions.* We shall return to this point later in this chapter.

Much of the theoretical work in the field is based on the three-band model

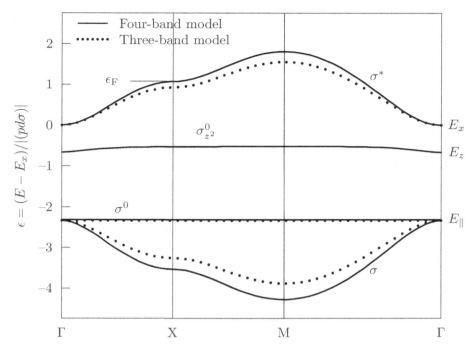

Figure 11.4. Comparison of the energy bands of the three-band and four-band models. The dotted curves are for the three-band model. The solid curves are for the four-band model. The flat, non-bonding, σ^0 band occurs in both models. The nearly flat band $\sigma^0_{z^2}$, derived from the d_{z^2} orbitals occurs only in the four-band model. The parameters used for the calculations of the curves are $E_{g\sigma}/|(pd\sigma)|=2.3333$, $E_x/|(pd\sigma)|=-1$, and $E_z/|(pd\sigma)|=-1.6667$.

because it is simpler to work with than the four-band model. The energy bands of the three-band model given by (11.20) and (11.21) are of the same analytical form as those encountered for the pseudo-two-dimensional pi bands of the cubic perovskites. Therefore we can use (4.41) and (6.28) to immediately obtain the DOS for the σ^* band by the replacement $(pd\pi) \rightarrow \frac{\sqrt{3}}{2}(pd\sigma)$. This gives

$$\rho_{\sigma^*}(E) = \frac{1}{\pi^2} \left| \frac{E - \frac{1}{2}E_{m\sigma}}{\frac{3}{4}(pd\sigma)^2} \right| K\left(\sqrt{1 - \left(\frac{\varepsilon(E)}{2}\right)^2}\right), \quad \text{for } \left(\frac{\varepsilon(E)}{2}\right)^2 \leq 1, \quad (11.22)$$

$$\varepsilon(E) = \frac{[E - E_{m\sigma}]^2 - [E_{g\sigma}/2]^2}{\frac{3}{2}(pd\sigma)^2} - 2, \quad (11.23)$$

where K is the complete elliptic integral of the first kind. *It follows that the DOS for the σ^* band of the three-band model has the same van Hove singularities as the DOS for the perovskite π^* band including the jump discontinuities at the band edges and the logarithmic singularity at the center of the band. There is, however,* a critical difference. For the perovskites the three, symmetry-equivalent π^* bands can accommodate six electrons so the possibility of a half-filled conduction band is not realized for any known perovskite oxide. For example, to half-fill the π^*-conduction bands of WO_3 would require doping of 3 electrons per unit cell. On the other hand, for the cuprates, the σ^* band is a single band that can accommodate only two electrons and hence the parent compounds invariably have a half-filled Cu–O_2 conduction band. *Therefore, if energy band theory were applicable to the parent materials it would predict that the Fermi energy is always precisely at the logarithmic singularity in the DOS.* This common feature is undoubtedly an important factor in the electronic properties of the high-T_c superconductive materials.

(c) Effects of the oxygen–oxygen interactions

Now we consider the effect of the second nearest neighbor, oxygen–oxygen interactions on the three-band model. Using the full matrix of Table 11.4 gives the secular equation

$$(E_x - E)\left[(E_\| - E)^2 - 4b^2 S_x^2 S_y^2\right] - 3(pd\sigma)^2(E_\| - E)(S_x^2 + S_y^2) + 12b(pd\sigma)^2 S_x^2 S_y^2 = 0. \tag{11.24}$$

Figure 11.5 shows the results for the three-band model for different values of b. Since the oxygen–oxygen interaction parameters b and b^2 in (11.24) are multiplied by $S_x^2 S_y^2$ the dispersion along the $\Gamma \rightarrow X$ (along the k_x or k_y directions) in the Brillouin zone is unchanged from the case with $b = 0$. In the interior of the Brillouin zone the dispersion is larger for $b > 0$ than for $b = 0$, while for $b < 0$ the dispersion is reduced. The parameter $b = (pp\sigma) - (pp\pi)$ enters (11.24) linearly through the last term and hence the results are sensitive to the sign of b. Negative b leads to a fairly flat dispersion as can be seen in the figure.

DOS plots are shown in Fig. 11.6. Several features are evident. First, the peak position always occurs at the same energy regardless of the value of b. That is because the logarithmic singularity in the DOS occurs at a saddle point in the two-dimensional energy dispersion. The saddle points are the X-points where the energy is independent of the value of b. The proof of this is not difficult. In two dimensions a saddle point occurs at E_0 when the energy dispersion has the form $E(\vec{k}) \rightarrow E_0 + E_1\alpha^2 - E_2\beta^2$, where α and β are small, orthogonal components of the wavevector and E_1 and E_2 are positive constants. Writing a power-series expansion of (11.24) near the X point with $k_x a = \pi/2 + \alpha$, and $k_y a = \beta$, we obtain

$$\left[(E_\| - E)(E_x - E) - 3(pd\sigma)^2\right](E_\| - E) + 3\alpha^2(pd\sigma)^2(E_\| - E)$$
$$- \beta^2\left\{\left[3(E_\| - E) - 12b\right](pd\sigma)^2 + 4b^2(E_x - E)\right\} \cong 0. \tag{11.25}$$

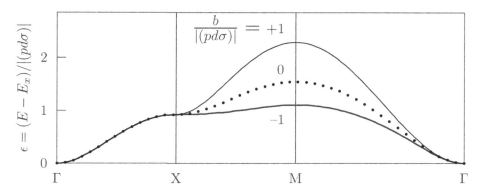

Figure 11.5. Energy bands for the Cu–oxygen layer (three-band model) showing the effect of the oxygen–oxygen interaction parameter, b. The parameters used are $E_{g\sigma}/|(pd\sigma)| = 2.3333$, $E_x/|(pd\sigma)| = -1$, and $E_z/|(pd\sigma)| = -1.6667$.

Next, write $E = E_0 + \Delta E$, where E_0 is the energy at X and satisfies the equation

$$(E_x - E_0)(E_\parallel - E_0) - 3(pd\sigma)^2 = 0 \qquad \text{at X.} \qquad (11.26)$$

Using (11.25) and (11.26) we find that $\Delta E = E_1 \alpha^2 - E_2 \beta^2$, where E_1 and E_2 are positive coefficients. Therefore, if $(pd\sigma) \neq 0$, X is a saddle point. Furthermore, the saddle-point energy at X is unaffected by the value of b since $S_x S_y = 0$ at the X points. We can conclude that (i) there is a logarithmic singularity at X and (ii) it occurs at E_0 independent of the value of b. This does not mean that the Fermi energy, ϵ_F, is the same for different values of b. On the contrary, ϵ_F depends on the magnitude and sign of b. It is clear from Fig. 11.6 that for a fixed number of electrons in the band, ϵ_F moves to higher energy for $b > 0$ and to lower energy for $b < 0$. The effect of b on the shape of the FS is illustrated in Fig. 11.7. The straight line is the edge of the FS for $b = 0$ and $\epsilon_F = \epsilon_0 = \epsilon(\pi/2a, 0)$ corresponding to the van Hove singularity and exactly one electron. The curves for $b/|(pd\sigma)| = 1$ and -1 are calculated for the same Fermi energy, $\epsilon_F = \epsilon_0$. For $b/|(pd\sigma)| = 1$ the curve is bowed inward (concave) compared to the $b = 0$ curve. For $b/|(pd\sigma)| = -1$ the curve is bowed outward (convex).

The Fermi level will also shift with the number of holes. Clearly, as the hole concentration increases ϵ_F must decrease, therefore qualitatively hole-doping counteracts the effect of $b > 0$ interactions. Arguments have been advanced that the optimal hole-doping occurs when the effect of the holes just offsets the effect of the oxygen–oxygen interactions leaving the Fermi energy at the logarithmic singularity. The positions of the Fermi energies are indicated in Fig. 11.6 for a hole concentration of 15%, an amount that is usually near optimal hole-doping. The results of Figs 11.6 and 11.7 assume that the LCAO parameters are unchanged by the doping. That is unlikely to be the case. In Chapter 6, Subsection 6.4(b) we

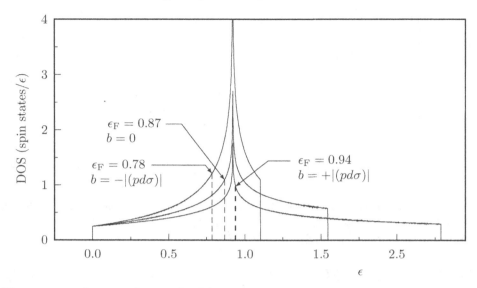

Figure 11.6. Density of states for different values of the oxygen–oxygen interaction parameter, $b=0$, 1, and -1 (in units of $|(pd\sigma)|$). The logarithmic singularity occurs at the same value of ϵ for all values of b. For $b > 0$ the band is expanded. For $b < 0$ the band is contracted. The positions of the Fermi energy, ϵ_F, are shown for a hole concentration of 0.15, corresponding to optimal doping.

discussed the x dependence of the LCAO parameters for $NaWO_3$ and found substantial changes, particularly in the energy gap and the p–d interaction parameter [20]. Band calculations [21] for the cuprates indicate that the parameters change, but that ϵ_F remains near the logarithmic singularity as the hole-doping varies. Theoretical studies indicate that $E_{g\sigma}/b$ and $(pd\sigma)/b$ decrease by a factor of about 6 and 3, respectively, as the hole concentration varies from 0.3 to 0.05 in LSCO [17]. This dramatic reduction presumably results because the correlation effects increase as the band approaches a half-filled condition.

(d) Extended singularity in the $\dot{D}OS$

Returning to the energy bands of Fig. 11.5, for the case of $b < 0$ it can be seen that the dispersion curve is flattened. In fact, it turns out that there is a value of b for which the energy band is mathematically flat along the entire line $X \to M$ with energy fixed at E_0. Consider (11.24) with $k_x a = \pi/2$ and $E = E_0$. For this choice of parameters the secular equation reduces to

$$\left\{ \left[12b - 3(E_\parallel - E_0) \right] (pd\sigma)^2 - 4b^2 (E_x - E_0) \right\} S_y^2 = 0. \tag{11.27}$$

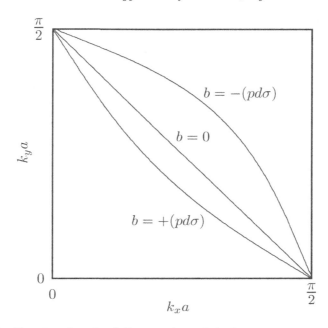

Figure 11.7. Fermi surface for different values of, b, the oxygen–oxygen interaction parameter. The upper right-hand quadrant of the Brillouin zone shows the effect of b on the shape of the Fermi surface with the energy fixed at its value ($\epsilon_F = 0.9217$) at the X-point. For $b < 0$ the surface bows outward, while for $b > 0$ it bows inward. The parameters used are $E_{g\sigma}/|(pd\sigma)| = 2.3333$, $E_x/|(pd\sigma)| = -1$, and $E_z/|(pd\sigma)| = -1.6667$.

If there is a value of b that satisfies the equation then $E = E_0$ independent of the value of k_y. That is, the band is flat along X to M. The solutions to the quadratic equation for b are doubly degenerate with the root, b_{crit}, given by

$$b_{\text{crit}} = \frac{\frac{3}{2}(pd\sigma)^2}{(E_x - E_0)}. \tag{11.28}$$

This critical value of b is necessarily negative because E_0, the energy at X, is always greater then E_x, the energy at Γ. For the parameters used for Fig. 11.5 this corresponds to $b_{\text{crit}} = -1.6275$ in units of $|(pd\sigma)|$. For this special value of b, the energy is independent of k_y along X → M. Furthermore, according to (11.25), near the X → M line, the energy varies as $E_0 - 3\alpha^2(pd\sigma)^2/|(E_\parallel - E_x)|$ for small values of α. That means the energy behaves as a one-dimensional system along the line from X to M. Therefore the DOS singularity changes from logarithmic to a square-root singularity as E approaches E_0 and has the form

$$\rho(E) \propto |(E - E_0)|^{-1/2}, \quad \text{as} \quad E \to E_0. \tag{11.29}$$

This type of singularity is called an "extended singularity" [25, 26] because the

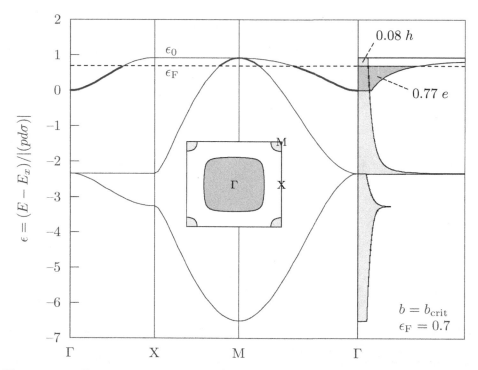

Figure 11.8. Energy bands showing the flat antibonding band that occurs for the critical value, $b = -1.6275$ (in units of $|(pd\sigma)|$). The flat portion of the upper band along X to M leads to an extended singularity of the inverse square-root type. The other parameters used are $E_{g\sigma}/|(pd\sigma)| = 2.3333$, $E_x/|(pd\sigma)| = -1$, and $E_z/|(pd\sigma)| = -1.6667$. Inset: Fermi surface for $\epsilon_F = 0.7$, below the top of the upper band, $\epsilon_0 = 0.9217$.

singularity is extended from a point at X to a line passing through X. Extended singularities have been reported for a number of HTSCs including Bi2212, Bi2201, Y123, and Y124. *However, it seems that these features are not consistent with the extended singularity of the three-band model. The problem is discussed in the following sections.* The energy bands for $b = b_{\mathrm{crit}}$ are shown in Fig. 11.8. As is evident the upper band is flat along $X \rightarrow M$. Another new feature is that the previously flat non-bonding band develops dispersion along $X \rightarrow M$ and touches the upper band at M. An interesting result is that for $E_F = E_0$ the FS for $b = b_{\mathrm{crit}}$ is the entire first Brillouin zone and includes exactly two electrons while for $b = 0$ the same energy corresponds to exactly one electron. Because of the dispersion of the second lowest band, the FS for fewer than two electrons consists of two areas; one centered at Γ due to the highest band and a second centered at M due to the lower band. The FS is illustrated in the inset in Fig. 11.8 using an extended zone scheme to show the symmetry.

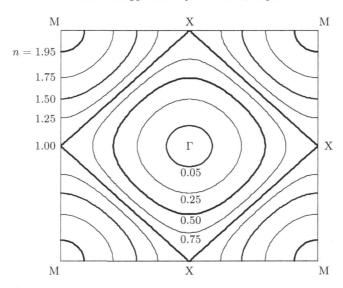

Figure 11.9. Equi-energy contours (Fermi surfaces) when the oxygen–oxygen interaction, b, is zero. The rectangle in the center occurs when the σ^* band is half-filled (one electron). Contours external to the rectangle correspond to increasing electron concentrations for which the Fermi surface consists of four empty pockets centered on the M points in the Brillouin zone. This is called a "hole-like" Fermi surface. Contours internal to the rectangle correspond to decreasing electron concentrations for which the Fermi surface consists of a single filled pocket centered on Γ called an "electron-like" Fermi surface.

(e) Comparison of the three-band results with Fermi surface experimental data

The shape or topology of the FS of the HTSC materials can be determined experimentally by careful ARPES measurements. It is important to know if empirical tight-binding (LCAO) models are capable of reproducing the FS topologies in order to establish whether the models are qualitatively useful in understanding the electronic properties of the cuprates. Before discussing the experimental results it is worthwhile recalling some of the important features of the FS. First, in the absence of oxygen–oxygen interactions the density of states, $\rho(\varepsilon)$, for the Cu–O₂ layer is a universal function just as it is for the perovskites. This means that the FS topology is independent of the empirical parameters. The shape of the FS is determined entirely by the number of electrons occupying the band. Figure 11.9 shows the universal FSs for $b = 0$ for different numbers of electrons occupying the σ^* band. The figure is an extended zone scheme with the constant-energy curves continued into the adjacent Brillouin zones. The square curve is the FS for exactly one electron in the σ^* band. Moving inward toward the center starting from the side of the square corresponds to decreasing the number of electrons. Moving outward from the square

Table 11.5. $LCAO$ parameters for $La_{2-x}Sr_xCuO_4$.

x	$E_{g\sigma}/b$	$(pd\sigma)/b$	p/d ratio at X	p/d ratio at M
0.05	0.33	0.33	0.53	0.93
0.10	0.50	0.50	0.66	1.11
0.15	1.00	0.84	0.64	1.18
0.22	1.50	1.00	0.75	1.78
0.30	2.33	1.00	0.75	2.45

toward the edges of the diagram corresponds to increasing the number of electrons in the σ^* band. Thus if the number of electrons in the band is increased beyond one, the FS is bounded externally by four areas each centered at an M point. This is often referred to as a "hole-like" Fermi surface. On the other hand, starting from one electron, the addition of holes leads to a FS, that is a single area centered on Γ. This is called an electron-like FS.

When the oxygen–oxygen interaction, b, is added to the model, the universality of the FS is lost. The FS shape becomes dependent upon both the size and the sign of b. For the three-band model, the spectrum of FS topologies are determined by only two parameters. We shall take the two parameters to be the ratio, $E_{g\sigma}/b$, and the ratio, $|(pd\sigma)|/b$. For $b > 0$ the hole-doped FS is concave, bowing inward toward Γ. For $b < 0$ it is convex, bowing out toward M.

Some experimental results for the FS of $La_{2-x}Sr_xCuO_4$ (LSCO) are compared with theoretical results in Fig. 11.10. The solid lines and shaded areas are theoretical FSs and the tick marks indicate the experimental data and its uncertainty. It is apparent that the FSs bow inward and therefore the parameter $b > 0$. The empirical parameters used to fit the data are shown in Table 11.5. Similar results have been reported by Mrkonjic and Barisic [17, 18] for both $La_{2-x}Sr_xCuO_4$ and $YBa_2Cu_3O_{6.95}$. Within the accuracy of the experimental data the LCAO parameters are not precisely determined and variations of 10%–15% in the parameters can lead to fits to the data that are equally "good". The first column of Table 11.5 lists the hole concentration per CuO_2 unit cell. The second and third columns give the ratio of the energy gap and p–d interaction parameter to the oxygen–oxygen interaction parameter, b. These ratios show a reduction in $(pd\sigma)$ and the energy gap as the doping decreases which is consistent with the idea that correlation increases as a half-filled band is approached.

The last two columns of Table 11.5 show the ratio of the p-orbital to d-orbital composition of the wavefunction at X and M. The ratios are defined as $\sqrt{|a_{p_{xx}}^2 + a_{p_{yy}}^2|/|a_d^2|}$, where the wavefunction is $\Psi = a_{p_{xx}}\, p_{xx} + a_{p_{yy}}\, p_{yy} + a_d\, d_{x^2}$. The values obtained indicate bonding similar to that of the d-band perovskites with a trend toward less covalency as x decreases.

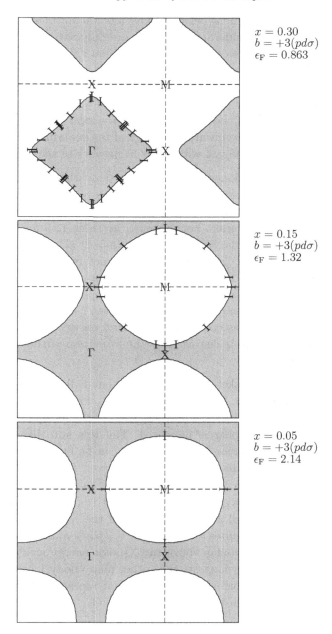

Figure 11.10. Fermi surfaces (extended into the adjacent Brillouin zones) versus the hole concentration, x, for $La_{2-x}Sr_xCuO_4$. The shaded areas are the filled states. The shapes are calculated from the three-band theory with oxygen–oxygen interaction, b, included. The parameters employed in the theoretical calculations are summarized in Table 11.5. The "I-shaped" tick marks are ARPES data from [22, 23]. (a) $x = 0.30$ (over-doped region) showing an isolated "electron-like" Fermi surface. (b) $x = 0.15$ (optimal doping) showing the Fermi surface evolved into a large "hole-like" surface. (c) $x = 0.05$ (under-doped) showing a "hole-like" Fermi surface.

In Fig. 11.10, the agreement between experimental and theoretical FSs is quite encouraging, especially considering that there are only two free parameters. However, there are other aspects of the experimental data that do not agree with the three-band model. One important disagreement is that experiment shows a definite flattening of the dispersion curve near X for low hole concentrations. This implies an extended singularity. The problem is, as discussed above, that b is required to be negative in order to achieve the flat-band condition near X. On the other hand, b must be positive to match the inward curvature of the experimental FS curves. This contradiction can not be resolved within the framework of the three-band model.

It should also be mentioned that obtaining good theoretical agreement with the measured FSs does not guarantee that the theoretical energy bands are also in agreement. Choosing parameters to fit the FS only fixes certain ratios and does not fix the absolute energy scale. For example, in Table 11.5 for $x = 0.30$ two sets of LCAO parameters with the same ratios are $\{b = 1,\ E_{g\sigma} = 2.33,\ (pd\sigma) = 1\}$ and $\{b = 3,\ E_{g\sigma} = 7,\ (pd\sigma) = 3\}$. The two sets of parameters give σ^* band widths that are very different, but yield the same FS. Therefore, the parameters should be determined from the experimental energy band dispersion curves when they are available. Unfortunately, complete dispersion curves can not be obtained experimentally because the data is cut off at the Fermi energy. In addition, very near the FS the dispersion curve is distorted by the formation of the energy gap in the superconducting state.

Figure 11.11 shows a sketch of the energy band dispersion curves obtained by ARPES experiments [22, 23] on $La_{2-x}Sr_xCuO_4$ for values of the hole-doping ranging from $x = 0.05$ to 0.30. As should be expected the Fermi energy moves down the dispersion curve as doping is increased. The FSs and dispersion curves for $x \geq 0.15$ (metallic and superconducting phases) can be fitted by the simple three-band model we have been discussing above. However, for $x \leq 0.10$ the dispersion is flat near the X point. For the three-band model this flat behavior can only be achieved with the oxygen–oxygen interaction negative ($b < 0$). However, it is clear that the Fermi energy is well above the energy of the X-point energy for $x = 0.10$ and 0.05 and this feature requires $b > 0$. In fact, $b/|(pd\sigma)|$ must be large and positive to position the Fermi level as far above the X-point energy as shown for $x = 0.05$ in Fig. 11.10(c). This failure suggests that other many-body mechanisms [27] such as antiferromagnetic spin correlations that are not represented by the three-band model are operative in the insulating phase of LSCO.

(f) Possible role of the d_{z^2} non-bonding band

As mentioned earlier in this chapter the validity of the three-band model rests on the assumption that the non-bonding d_{z^2} band is sufficiently far removed from the

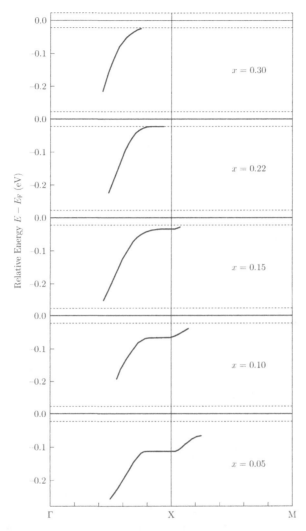

Figure 11.11. A sketch of ARPES data for the energy band dispersion as a function of the hole concentration, x. As x decreases E_F moves up the quasiparticle dispersion curve. A flat band (extended singularity) occurs for $x \leq 0.15$. The energy band parameters change as x changes (Table 11.5). The experimental results are from [22, 23].

σ^* band to be neglected. If E_z lies within the energy range of the σ^* band the two bands will hybridize and repel one another. Figure 11.12 illustrates the results of the nearest-neighbor four-band model for $E_z > E_x$, at an energy that intersects the σ^* band. The wavefunctions for the two bands are mixed. Starting from Γ on the left, the d-orbital composition of the lower band is mostly $x^2 - y^2$ in character while the upper band is mostly d_{z^2}. Proceeding left to right, toward X, the d-

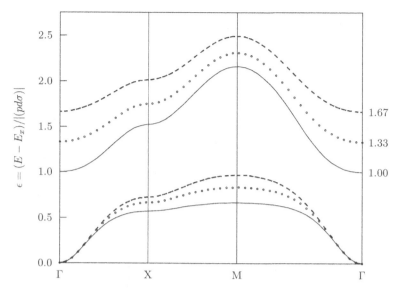

Figure 11.12. Hybridization of the d_{z^2} and $d_{x^2-y^2}$ bands in the four-band model. The non-bonding band (lower band) hybridizes with the antibonding band (upper band). From Γ to X the upper band d-orbital composition is predominantly $d_{x^2-y^2}$ and the lower band mostly d_{z^2}. Along X to M the d-orbital character reverses, then reverses again along M to Γ. The Fermi energy for hole-doped material will lie below the energy at X and therefore the FS states are predominantly of $d_{x^2-y^2}$ character. The values of the dimensionless parameters used for the calculation of the three examples are $E_x/|(pd\sigma)| = -1$, $(E_z - E_x)/|(pd\sigma)| = 1.0000$, 1.3333, and 1.6667, $E_{g\sigma}/|(pd\sigma)| = 2.3333$.

orbital composition of the lower band evolves into mostly d_{z^2} and vice versa for the upper band. Near M the wavefunctions again reverse their d-orbital characters. In the absence of the oxygen–oxygen interaction this hybridization does not alter the shape of the FS. The lower band will be fully occupied and the DOS as well as the FS of the upper band are still universal curves, independent of the parameters. The addition of the oxygen–oxygen interactions will have the same effect as described in the previous section. Because there is always a sizable gap between the upper and lower curves, ϵ_F always intersects the upper band and never intersects the lower band.

Although the shape of the FS is not significantly altered by the hybridization with the d_{z^2} non-bonding band, the matrix elements that govern the ARPES intensity as well as resolution effects can be significantly modified. However, if ϵ_F lies reasonably near the energy of the logarithmic singularity (energy at X) the portion of the hybridized σ^* band that determines the FS will be predominantly of $x^2 - y^2$ character. Therefore it appears that the qualitative FS results based on the three-band model are not invalidated by hybridization with the d_{z^2}, non-bonding band.

Theoretical justification for why E_z may lie above E_x has been given by Perry *et al.* [24] based on including certain "static correlations" omitted in local-density-approximation band calculations. These authors propose a model in which the z^2 band possesses modest dispersion in the k_z-direction and intersects the $x^2 - y^2$ band at the Fermi energy. In their model the interaction is weak and the resulting hybridization gap is very narrow. This model does not seem to be compatible with the wide gap that occurs naturally in the four-band model when $E_z > E_x$.

11.6 Chains in $YBa_2Cu_3O_{6.95}$

The high-temperature superconductor, $YBa_2Cu_3O_{6.95}$, is slightly orthorhombic with $(b/2a) = 1.025$ ($2a =$ lattice spacing). It has a pair of closely spaced, Cu–O_2 planes parallel to the a–b plane separated by Y ions. In addition to these super-conducting planes it also has copper–oxygen chains oriented along the b-axis. A schematic of the structure is shown in Fig. 11.13(a). Figure 11.13(b) and (c) illustrate the Brillouin zone and one of the chains, respectively. It is evident that the unit cell has one copper and three oxygen ions. Using the LCAO model including

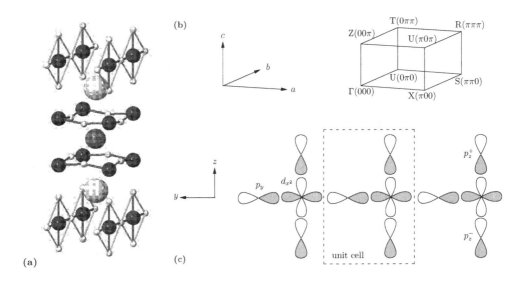

Figure 11.13. $YBa_2Cu_3O_{6.95}$. (a) Crystal structure. (b) Unit vectors and the Brillouin zone. (c) Cu–O_3 chain showing the orbitals.

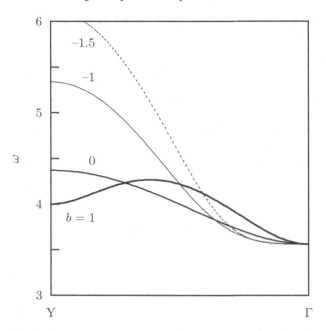

Figure 11.14. Chain dispersion curves for different values of the oxygen–oxygen interaction, b. Other parameters used are $\omega_x = 3$ and $\omega_\| = 0$.

oxygen–oxygen interactions give the following matrix equation for the chain bands:

$$\begin{pmatrix} (\omega_x - \omega) & -\beta & 1 & -1 \\ -\beta^* & (\omega_\| - \omega) & b\beta & -b\beta \\ 1 & b\beta^* & (\omega_\| - \omega) & 0 \\ -1 & -b\beta^* & 0 & (\omega_\| - \omega) \end{pmatrix} \begin{pmatrix} C_{x^2} \\ C_{p_y} \\ C_{p_z^+} \\ C_{p_z^-} \end{pmatrix} = 0 \qquad (11.30)$$

where $\omega_x = E_{d_{x^2-y^2}}/\sigma$, $\omega_\| = E_\|/\sigma$, $\beta = 1 - \exp(2ik_y a)$, $b = \frac{1}{2}[(pp\sigma) - (pp\pi)]/\sigma$, and $\sigma = \frac{\sqrt{3}}{2}(pd\sigma)$. In (11.30) C_{x^2}, C_{p_y}, $C_{p_z^+}$, and $C_{p_z^-}$ are the amplitudes of the $d_{x^2-y^2}$, p_y, p_z^+ and p_z^- orbitals for the chain wavefunction. One of the eigenvalues is $\omega = \omega_\|$ corresponding to a non-bonding band for the combination $\frac{1}{\sqrt{2}}(p_z^+ + p_z^-)$. The secular equation determining the three remaining eigenvalues is

$$(\omega_x - \omega)(\omega_\| - \omega)^2 - 2(\omega_\| - \omega) - \{(\omega_\| - \omega) - 4b + 2b^2(\omega_x - \omega)\}4S_y^2 - 32bS_y^4 = 0. \qquad (11.31)$$

The roots of (11.31) can be obtained by use of standard formulae for the roots of a cubic equation. The root of principal interest is the partially occupied antibonding chain energy band. Its dependence on the oxygen–oxygen interaction parameter, b, is illustrated in Fig. 11.14. Positive values of b depress the dispersion relative to $b = 0$. Negative b results in a curve with an initial flat portion and then a very sharp transition to a rapid rise that eventually results in an enhancement of the

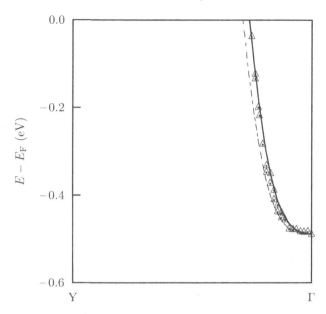

Figure 11.15. Comparison of experimental and theoretical dispersion curves. The triangles represent the data [28]. The solid curve is for $b = -1.5$ and the dashed curve is for $b = -1.3$. Other parameters used are $E_F = 2.175\,\sigma$ (eV), $w_x = 0.5$, and $w_\parallel = 0$.

dispersion compared to $b = 0$. The antibonding energy band of the Cu–O$_3$ chain is one-dimensional and the dispersion along $\Gamma \to Y$ (and $\Gamma \to S$) in the Brillouin zone have been observed experimentally [28, 29]. The bottom of the chain band lies about $0.5\,\text{eV}$ below the Fermi level and appears to be rather flat as it approaches Γ, suggesting that the oxygen–oxygen interaction, b, is negative. Figure 11.15 gives a comparison of the LCAO chain band with the experimentally observed band. The theoretical curve for $b = -1.5$ yields an excellent fit to the experimental data.

11.7 Summary

Since the discovery of HTCS in LSCO by Bednorz and Müller in 1986 a large number of similar Cu–oxygen layered compounds have been discovered. With time the record superconducting transition temperature has increased from $30\,\text{K}$ to over $165\,\text{K}$ for the mercuric cuprates. Essentially all of these compounds are antiferromagnetic in the undoped condition and become superconductors when the doping concentration is greater than about 0.10 holes per unit cell. Optimal doping at around 0.15 maximizes T_c and above this concentration T_c decreases. For hole-doping in excess of about 0.25 the material is no longer a superconductor. The superconducting energy gap is anisotropic and usually has d-wave symmetry.

Tunneling and other experiments indicate that electron pairs are responsible for the superconductivity. The mechanism(s) for electron pair formation is uncertain but likely results from electron–electron correlations instead of (or in addition to) phonon-mediated pairing. The HTSC materials are unstable to the formation of charge density waves, spin density waves, and charge and spin segregation into stripes [20].

Conventional energy band theory does not describe the HTSCs because of strong electron correlations. The antiferromagnetic and insulating phases are described by the Mott–Hubbard theory and renormalized effective Hamiltonians that include correlation effects approximately. At temperatures above T_c the HTSCs are poor metals with highly anisotropic transport properties and a nearly linear dependence of the resistivity on temperature. Conductivity parallel to the Cu–oxygen layers is several orders of magnitude greater than that perpendicular to the layers. Despite strong electron–electron correlations, Fermi-liquid-like quasiparticles are seen in ARPES experiments that are reasonably well defined in the superconducting phase. In addition, Fermi surfaces have been mapped out by ARPES experiments and fitted to tight-binding models based on the three-band energy band model with renormalized parameters. In the antiferromagnetic and insulating regions the dispersion curves for the electron excitations display a flat region in the neighborhood of the X-point in the Brillouin zone that can not be explained by the three-band model. The utility of the quasiparticle description is questionable in the "normal" metal phase and it has been suggested that in that phase there are no coherent excitations and that collective modes, holons, and spinons, involving the separation of the spin and charge are operative.

The field of HTSC is moving at a rapid pace both experimentally and theoretically. New materials are being discovered and explored. Experiments are becoming more precise and able to distinguish between different proposed theories. As a result one can expect that a better understanding of the physics and chemistry of the HTSC materials will emerge in the near future.

References

[1] J. G. Bednorz and A. Z. Müller, *Z. Physik* B **64**, 189 (1986).

[2] D. A. Pavlov, *Doctoral Thesis*, Academic Archive On-Line. Oai:DiVa.se:su-246 (2004); M. Monteverde, M. Núñez-Regueiro, C. Acha, K. A. Lokshin, D. A. Pavlov, S. N. Putilin, and E. V. Antipov, *Physica* C **408–410**, 23 (2004).

[3] P. W. Anderson, *The theory of superconductivity in the high-T_c cuprates, Princeton Series in Physics*, (Princeton, NJ, Princeton University Press, 1999).

[4] P. W. Anderson, *Science* **235**, 1196 (1987).

[5] J. Hubbard, *Proc. Roy. Soc. London* A **277**, 237 (1964).

[6] F. C. Zhang and T. M. Rice, *Phys. Rev.* B **37**, 3759 (1988).

[7] A. Damascelli, Z. Hussain, and Z.-X. Shen, *Rev. Mod. Phys.* **75**, 473 (2003), (A. Damascelli, Z.-X. Shen, and Z. Hussain, arXiv:cond-mat/0208504 v1 27 Aug 2002).

[8] J. Bardeen, L. N. Cooper, and J. R. Schrieffer, *Phys. Rev.* **108**, 1175 (1957).

[9] D. Pines and P. Nozières, *The theory of quantum liquids*, Vol. 1 (New York, Benjamin, 1966).

[10] P. Nozières, *The theory of interacting Fermi systems* (New York, Benjamin, 1964).

[11] G. Kotliar and A. E. Ruckenstein, *Phys. Rev. Lett.* **57**, 1362 (1986).

[12] H. Hasegawa, *J. Phys. Soc. Jpn.* **66**, 1391 (1997).

[13] J. E. Hirsch, arXiv:cond-mat/0111294 v1 16 Nov 2001.

[14] R. Fresard and M. Lamboley, arXiv:cond-mat/0109543 v1 28 Sep 2001.

[15] E. A. Dagotto, A Nazarenko, and M. Boninsegni, *Phys. Rev. Lett.* **73**, 728 (1994).

[16] V. J. Emery, *Phys. Rev. Lett.* **58**, 2794 (1987).

[17] I. Mrkonjic and S. Barisic, arXiv:cond-mat/0103057 v2 5 Mar 2001.

[18] I. Mrkonjic and S. Barisic, arXiv:cond-mat/0206032 v1 4 Jun 2002.

[19] K. M. Shen, F. Ronning, D. H. Lu, W. E. Lee, N. J. C. Ingle, W. Meevasana, F. Baumberger, A. Damascelli, N. P. Armitage, L. L. Miller, Y. Kohsaka, M. Azuma, M. Takano, Hi. Takagi, and Z.-X. Shen, *Phys Rev. Lett.* **93**, 267002 (2004), (arXiv:cond-mat/0407002 v2 6 Jul 2004).

[20] B. G. Levi, *Phys. Today* **57**, 24 (December 2004).

[21] D. M. Newns, P. C. Pattnaik, and C. C. Tsuei, *Phys. Rev.* B **43**, 3075 (1991).

[22] A. Ino, C. Kim, M. Nakamura, T. Yoshida, T. Mizokawa, A. Fujimori, Z.-X. Shen, T. Kakeshita, H. Eisaki, and S. Uchida, *Phys. Rev.* B **65**, 094504 (2002); A. Ino, C. Kim, T. Mizokawa, Z. X. Shen, A. Fujimori, M. Takaba, K. Tamasaku, H. Eisaki, and S. Uchida, arXiv:cond-mat/9809311 v3 27 Mar 2000.

[23] T. Yoshida, X. J. Zhou, M. Nakamura, S. A. Kellar, P. V. Bogdanov, E. D. Lu, A. Lanzara, Z. Hussain, A. Ino, T. Mizokawa, A. Fujimori, H. Eisaki, C. Kim, Z.-X. Shen, T. Kakeshita, and S. Uchida, *Phys. Rev.* B **63**, 220501 (2001).

[24] J. K. Perry, *J. Phys.* A **104**, 2438 (2000), (arXiv:cond-mat/9903088 v2 2 Sep 1999).

[25] A. A. Abrikosov, J. C. Campuzano, and K. Gofron, *Physica* C **214**, 73 (1993).

[26] K. Gofron, J. C. Campuzano, A. A. Abrikosov, M. Lindroos, A. Bansil, H. Ding, D. Koelling, and B. Dabrowski, *Phys. Rev. Lett.* **73**, 3302 (1994).

[27] N. Bulut, D. J. Scalapino, and S. R. White, *Phys. Rev.* B **50**, 7215 (1994).

[28] D. H. Lu, D. L. Feng, N. P. Armitage, K. M. Shen, A. Damascelli, C. Kim, F. Ronning, Z.-X. Shen, D. A. Bonn, R. Liang, W. N. Hardy, A. I. Rykov, and S. Tajima, *Phys Rev. Lett.* **86**, 4370 (2001).

[29] M. C. Schabel, C.-H. Park, A. Matsuura, Z.-X. Shen, D. A. Bonn, R. Liang, and W. N. Hardy, *Phys. Rev.* B **57**, 6090 (1998).

Problems for Chapter 11

1. The electron creation and destruction operators, $a_{i\sigma}^\dagger$ and $a_{i\sigma}$, anticommute:

$$a_{i\sigma}\, a_{k\sigma'}^\dagger + a_{k\sigma'}^\dagger\, a_{i\sigma} = \delta_{ik}\,\delta\sigma\sigma',$$

$$a_{i\sigma}^\dagger\, a_{k\sigma'}^\dagger + a_{k\sigma'}^\dagger\, a_{i\sigma}^\dagger = a_{i\sigma}\, a_{k\sigma'} + a_{k\sigma'}\, a_{i\sigma} = 0,$$

the latter equations require that $a_{i\sigma}a_{i\sigma} = -a_{i\sigma}a_{i\sigma}$, and $a_{i\sigma}^+ a_{i\sigma}^+ = -a_{i\sigma}^+ a_{i\sigma}^+$ and therefore $a_{i\sigma}a_{i\sigma} = a_{i\sigma}^+ a_{i\sigma}^+ = 0$. If $[A, B] \equiv AB - BA$, show that

$$\left[a_{i\sigma}, \sum_{j,k,\sigma'} H_{jk}\, a_{j\sigma'}^\dagger\, a_{k\sigma'} \right] = \sum_k H_{ik}\, a_{k\sigma}.$$

2. The time derivative of an operator A is given by $i\hbar\, dA/dt = [A, H_{\mathrm{op}}]$, where H_{op} is the Hamiltonian in operator form. Find the time derivative of $a_{i\sigma}$ for the Hamiltonian

$$H_{\mathrm{op}} = \sum_{j,\sigma'} E_j\, a_{j\sigma'}^\dagger\, a_{j\sigma'} + \sum_{j,k\neq j,\sigma'} H_{jk}\, a_{j\sigma'}^\dagger\, a_{k\sigma'}.$$

3. The operator form of the three-band model for Cu–O$_2$ can be expressed as

$$H_{three-band} = \sum_{\alpha,i} E_{\alpha,i}\, a_{\alpha,i}^+\, a_{\alpha,i} + \sum_{\alpha,i;\beta,j(\mathrm{nn})} H_{\alpha,i;\beta,j}\, a_{\alpha,i}^+\, a_{\beta,j}$$

$$+ \sum_{\alpha,i;\beta,j(\mathrm{nnn})} H_{\alpha,i;\beta,j}\, a_{\alpha,i}^+\, a_{\beta,j}$$

where α, i denotes an α-type (d_{x2}, p_{xx}, and p_{yy}) orbital centered at $\vec{R}_i$. The notation $\alpha, i; \beta, j(\mathrm{nn})$ indicates a sum over (β, j)th orbitals that are nearest neighbors of the (α, i)th orbital. Similarly, the notation $\alpha, i; \beta, j(\mathrm{nnn})$ indicates a sum over (β, j)th orbitals that are the next-nearest neighbors of the (α, i)th orbital. The Hamiltonian components, $H_{\alpha,i;\beta,j}$ are the LCAO two-center interactions between the α-type orbital centered at $\vec{R}_i$ and the β-type orbital centered at $\vec{R}_j$. (a) Find the equations for the time derivatives of the three $a_{\alpha,i}$ ($\alpha = d_{x2}$, p_{xx}, and p_{yy}). (b) Assume $a_{\alpha,i}(t) = c_\alpha \exp(-i\omega t + \vec{k} \cdot \vec{R}_i)$ and show that the equations are equivalent to the matrix equation of Table 11.4.

4. Consider the three-band model of Table 11.4 and suppose $(pd\sigma) = 0$.
   (a) Find expressions for the three energy bands and make a graph of them for $\vec{k}$ along the lines $\Gamma \to X \to M \to \Gamma$ for $b = 1$, $E_x = -3$, and $E_\parallel = -7$.
   (b) Show that the density of states for the two dispersive energy bands, is given by $\rho(\xi) = (4/\pi^2)\, K(\sqrt{1 - \xi^2})$, where $\xi = (E - E_\parallel)/(2b)$.

5. Make a graph of the curves of constant energy (Fermi surfaces) in k_x–k_y space for the two dispersive energy bands in Problem 4 for $\xi = 0, 0.01, 0.25, 0.5, 0.75, 0.99$, and 1.

6. For the three-band model, assume $b = -3$, $E_x = -3$, $E_\parallel = -7$, and $(pd\sigma) = 0.707\,\mathrm{eV}$.
   (a) Make a graph of the three energy bands for $\vec{k}$ along the lines $\Gamma \to X \to M \to \Gamma$.

(b) Discuss the nature of the wavefunction of the upper band along $\Gamma \rightarrow X \rightarrow M$.

7. Using the five unit-cell basis orbitals, $d_{z^2-y^2}(\vec{r})$, $p_x(\vec{r} - a\vec{e}_x)$, $p_y(\vec{r} - a\vec{e}_y)$, $p_x(\vec{r} - a\vec{e}_y)$, and $p_y(\vec{r} - a\vec{e}_x)$ construct the $5{\times}5$ matrix eigenvalue equation. Find the eigenvalues at Γ, X, and M in the Brillouin zone. Show that the energies for the σ^* band are identical to those of the three-band model at Γ, X, and M.

Appendix A

Physical constants and the complete elliptic integral of the first kind

A.1 Selected physical constants

Symbol	Constant	Value	Unit
c	Speed of light in vacuum	2.9979×10^8	m/s
m_e	Electron rest mass	9.1094×10^{-31}	kg
m_p	Proton rest mass	1.6726×10^{-27}	kg
m_n	Neutron mass	1.6749×10^{-27}	kg
m_p/m_e	Proton mass/electron mass	1836.1527	
e	Electron charge	1.6022×10^{-19}	C
eV	electronvolt	1.6022×10^{-19}	J
e/m_e	Electron charge/mass ratio	-1.7588×10^{11}	C/kg
h	Plank constant	6.6261×10^{-34}	J s
$\hbar$	Reduced Plank constant	1.0546×10^{-34}	J s
		6.5821×10^{-16}	eV s
μ_e	Electron magnetic moment	-9.2848×10^{-24}	J/T
μ_B	Bohr magneton	9.2740×10^{-24}	J/T
		5.7884×10^{-5}	eV/T
a_0	Bohr radius	5.2918×10^{-11}	m
Å	Angström	1×10^{-10}	m
N_A	Avogadro constant	6.0221×10^{23}	mol^{-1}
k_B	Boltzmann constant	1.3807×10^{-23}	J/K
		8.6173×10^{-5}	eV/K
Ry	Rydberg	2.1799×10^{-18}	J
		13.6057	eV
Cal	Mean Calorie	4.1868	J

A.2 The complete elliptic integral of the first kind

$$K(x) \equiv \int_0^{\pi/2} \frac{d\phi}{\sqrt{1 - x^2 \sin^2 \phi}} \tag{A.1}$$

$$K'(x) = K(x'), \qquad x' \equiv \sqrt{1 - x^2} \tag{A.2}$$

$$K(-x) = K(x) \tag{A.3}$$

$$K(0) = \frac{\pi}{2} \tag{A.4}$$

$$K(x) \to \ln\left(\frac{4}{x'}\right) \to \infty \quad \text{as} \quad x \to 1 \tag{A.5}$$

$$\int_0^1 K(x)\, dx = 2G, \qquad G = \frac{1}{1^2} - \frac{1}{3^2} + \frac{1}{5^2} - \cdots \approx 0.915\ 9656 \tag{A.6}$$

$$\int_0^1 K'(x)\, dx = \frac{\pi^2}{4} \tag{A.7}$$

$$\frac{1}{x} K\left(\frac{1}{x}\right) = K(x) + iK'(x) \tag{A.8}$$

$$K(ix) = \frac{1}{\sqrt{x^2 + 1}} K\left(\sqrt{\frac{x^2}{x^2 + 1}}\right). \tag{A.9}$$

Table A.1. *Numerical table of $K(x)$.*

x^2	$K(x)$	$K(x')$	x'^2	x^2	$K(x)$	$K(x')$	x'^2
0.00	1.570 796	∞	1.00	0.26	1.691 208	2.138 970	0.74
0.01	1.574 746	3.695 637	0.99	0.27	1.696 749	2.122 132	0.73
0.02	1.578 740	3.354 141	0.98	0.28	1.702 374	2.105 948	0.72
0.03	1.582 780	3.155 875	0.97	0.29	1.708 087	2.090 373	0.71
0.04	1.586 868	3.016 112	0.96	0.30	1.713 889	2.075 363	0.70
0.05	1.591 003	2.908 337	0.95	0.31	1.719 785	2.060 882	0.69
0.06	1.595 188	2.820 752	0.94	0.32	1.725 776	2.046 894	0.68
0.07	1.599 423	2.747 073	0.93	0.33	1.731 865	2.033 369	0.67
0.08	1.603 710	2.683 551	0.92	0.34	1.738 055	2.020 279	0.66
0.09	1.608 049	2.627 773	0.91	0.35	1.744 351	2.007 598	0.65
0.10	1.612 441	2.578 092	0.90	0.36	1.750 754	1.995 303	0.64
0.11	1.616 889	2.533 335	0.89	0.37	1.757 269	1.983 371	0.63
0.12	1.621 393	2.492 635	0.88	0.38	1.763 898	1.971 783	0.62
0.13	1.625 955	2.455 338	0.87	0.39	1.770 647	1.960 521	0.61
0.14	1.630 576	2.420 933	0.86	0.40	1.777 519	1.949 568	0.60
0.15	1.635 257	2.389 016	0.85	0.41	1.784 519	1.938 908	0.59
0.16	1.640 000	2.359 264	0.84	0.42	1.791 650	1.928 526	0.58
0.17	1.644 806	2.331 409	0.83	0.43	1.798 918	1.918 410	0.57
0.18	1.649 678	2.305 232	0.82	0.44	1.806 328	1.908 547	0.56
0.19	1.654 617	2.280 549	0.81	0.45	1.813 884	1.898 925	0.55
0.20	1.659 624	2.257 205	0.80	0.46	1.821 593	1.889 533	0.54
0.21	1.664 701	2.235 068	0.79	0.47	1.829 460	1.880 361	0.53
0.22	1.669 850	2.214 022	0.78	0.48	1.837 491	1.871 400	0.52
0.23	1.675 073	2.193 971	0.77	0.49	1.845 694	1.862 641	0.51
0.24	1.680 373	2.174 827	0.76	0.50	1.854 075	1.854 075	0.50
0.25	1.685 750	2.156 516	0.75	0.50	1.854 075	1.854 075	0.50
x'^2	$K(x')$	$K(x)$	x^2	x'^2	$K(x')$	$K(x)$	x^2

Appendix B

The delta function

As we have seen in Chapter 6 the δ function (sometimes called the Dirac delta function) is a useful mathematical tool. In this Appendix we derive formulae for the representation of the delta functions employed in Chapter 6.

The δ function is defined by its properties:

$$\delta(x - x_0) = 0 \qquad \text{for } x \neq x_0, \tag{B.1}$$

$$\int \delta(x - x_0) f(x) \, dx = f(x_0), \tag{B.2}$$

where $f(x)$ and its derivative are continuous, single-valued functions and the integral is over any range containing x_0. The result, $\int \delta(x - x_0) \, dx = 1$, follows from (B.2) for $f(x) = 1$. Another property, $\delta(x - x_0) \to \infty$ as $x \to x_0$, is implied by (B.1) and (B.2). Clearly if (B.1) holds, the δ function must be arbitrarily large at x_0 if (B.2) is valid.

There are numerous analytical representations of the delta function. We shall use a frequently employed representation wherein $\delta(x - x_0)$ is the limit of a particular function:

$$\delta(x - x_0) = -\frac{1}{\pi} \, \Im \, \frac{1}{x - x_0 + i\lambda}, \qquad \text{in the limit as } \lambda \to 0 \tag{B.3}$$

where "$\Im$" indicates the imaginary part of the quantity and λ is a small *positive* number. In using this representation there is an implied order of doing things. The limiting process $\lambda \to 0$ ($\lambda > 0$) is to be performed last. This means one must calculate the imaginary part first, then take the limit as $\lambda \to 0$. This limiting process is often indicated by using the symbol 0^+ as we did in Chapter 6.

The imaginary part of (B.3) is

$$\delta(x - x_0) = \lim_{\lambda \to 0} \frac{\lambda}{\pi} \frac{1}{(x - x_0)^2 + \lambda^2}. \tag{B.4}$$

288

It is easy to show that the delta function defined by (B.4) satisfies the equations (B.1) and (B.2). For $x \neq x_0$, in the limit as $\lambda \to 0$ the right-hand side of (B.4) tends to zero as the first power of λ and hence $\delta(x - x_0) = 0$. However, when $x = x_0$, then the right-hand side tends to infinity as $1/\lambda$. To show that (B.2) holds we use (B.4) and write

$$\int \delta(x - x_0)\, f(x)\, dx = \lim_{\lambda \to 0} \left(\frac{\lambda}{\pi}\right) \int \frac{f(x)\, dx}{(x - x_0)^2 + \lambda^2}\,. \tag{B.5}$$

The integration range in (B.5) may be taken as a small interval around x_0, say from $x_0 - a$ to $x_0 + a$. For our purposes we may assume that $f(x)$ possesses a convergent power-series expansion near x_0,

$$f(x) = \sum_{n=0}^{\infty} \frac{1}{n!}\, f^{(n)}(x_0)(x - x_0)^n\,, \tag{B.6}$$

where the constant, $f^{(n)}(x_0)$, is the nth derivative of $f(x)$ with respect to x evaluated at x_0. Inserting (B.6) into (B.5) and changing the integration variable to $z = (x - x_0)/\lambda$, we obtain

$$\lim_{\lambda \to 0} \frac{1}{\pi} \sum_{n=0}^{\infty} \frac{\lambda^n}{n!} \int_{-a/\lambda}^{+a/\lambda} \frac{f^{(n)}(x_0)\, z^n\, dz}{z^2 + 1}\,. \tag{B.7}$$

The nth term of the sum vanishes as λ^n and therefore only the $n = 0$ term will survive in the limit as $\lambda \to 0$. Thus, (B.7) becomes

$$\frac{1}{\pi}\, f(x_0) \int_{-\infty}^{+\infty} \frac{dz}{z^2 + 1} = \frac{1}{\pi}\, f(x_0)\, \arctan(z)\Big|_{-\infty}^{+\infty} = f(x_0)\,. \tag{B.8}$$

This result shows that the delta function defined by (B.3) satisfies condition (B.2) that $\int \delta(x - x_0)\, f(x)\, dx = f(x_0)$ and that $\int \delta(x - x_0)\, dx = 1$ (by choosing $f(x){=}1$).

As mentioned in Chapter 6 it is often convenient to work with a function, $f(x)$, rather than x. If $(df(x)/dx)\big|_{x_0} \equiv f^{(1)}(x_0) \neq 0$, we can define $\delta[f(x) - f(x_0)]$ in the same way as $\delta(x - x_0)$,

$$\delta[f(x) - f(x_0)] = \lim_{\lambda \to 0} \left\{ -\frac{1}{\pi}\, \mathfrak{Im}\, \frac{1}{f(x) - f(x_0) + i\lambda} \right\}$$

$$= \lim_{\lambda \to 0} \frac{1}{\pi} \left\{ \frac{\lambda}{\left[\sum_{n=1}^{\infty} \frac{1}{n!} f^{(n)}(x_0)(x - x_0)^n \right]^2 + \lambda^2} \right\}\,. \tag{B.9}$$

For $x \neq x_0$, $\delta[f(x) - f(x_0)] \to 0$ in the limit $\lambda \to 0$. As $x \to x_0$ we have two limiting processes, but as mentioned earlier, the limit $\lambda \to 0$ is to be performed last. Therefore, for x sufficiently close to x_0 only the linear term of the power-series expansion of $f(x)$ need be retained. Thus,

$$\delta[f(x) - f(x_0)] = \lim_{\lambda \to 0} \frac{1}{\pi} \left\{ \frac{\lambda}{\left[f^{(1)}(x_0)(x - x_0) \right]^2 + \lambda^2} \right\}$$

$$= \lim_{\lambda \to 0} \left\{ -\frac{1}{\pi} \Im \frac{1}{f^{(1)}(x_0)(x - x_0) + i\lambda} \right\}. \qquad (B.10)$$

Equation (B.10) can be rewritten as

$$\delta[f(x) - f(x_0)] = -\frac{1}{\pi} \frac{1}{f^{(1)}(x_0)} \lim_{\lambda \to 0} \Im \frac{1}{(x - x_0) + i\lambda'} \qquad (B.11)$$

where $\lambda' \equiv \lambda / f^{(1)}(x_0)$. We note that when $f^{(1)}(x_0) < 0$, λ' will approach zero from the negative side as $\lambda \to 0$ from the positive side. As a result the imaginary part will also change sign. This change in sign is canceled by the change in sign of the multiplying factor, $1/[\pi f^{(1)}(x_0)]$. Thus, the final result can be written for either case by using the absolute value of $f^{(1)}(x_0)$:

$$\delta[f(x) - f(x_0)] = -\frac{1}{\pi} \frac{1}{|f^{(1)}(x_0)|} \lim_{\lambda \to 0} \Im \frac{1}{(x - x_0) + i|\lambda'|}$$

$$= -\frac{1}{\pi} \frac{1}{|f^{(1)}(x_0)|} \Im \frac{1}{(x - x_0) + i0^+} \qquad (B.12)$$

where $i0^+$ is shorthand for the limiting process. The relationship between δ functions is then,

$$\delta[f(x) - f(x_0)] = \frac{1}{|f^{(1)}(x_0)|} \delta(x - x_0). \qquad (B.13)$$

Appendix C

Lattice Green's function

We first consider the lattice Green's function for the d orbitals alone. In Section C.3 we calculate the total Green's function including the p-orbital functions. The lattice Green's function, $\mathbb{G}$, for the pi(xy) states of the cubic perovskite is defined here as the inverse of the matrix, $\mathbb{H}_d$, that describes the interactions between the d orbitals for a unit-cell layer parallel to the xy-plane. Within the nearest-neighbor approximation different unit-cell layers are uncoupled and may therefore be treated as two-dimensional systems. The B ions are located on the xy-plane by the set of two-dimensional vectors, $\vec{\rho}_{j,m} = 2a(j\vec{e}_x + m\vec{e}_y)$, where j and m are integers. The p_x orbitals of the O ions are located at $\vec{\rho}_{j,m} + a\vec{e}_y$ and the p_y orbitals at $\vec{\rho}_{j,m} + a\vec{e}_x$. For the pi($xy$) states the equations for c_x, c_y, and c_{xy}, the amplitudes of the p_x, p_y, and d_{xy} orbitals, respectively, are

$$(\omega_t - \omega)\, c_{xy}(\vec{\rho}_{j,m}) + c_x(\vec{\rho}_{j,m} + a\vec{e}_y) - c_x(\vec{\rho}_{j,m-1} + a\vec{e}_y)$$
$$+ c_y(\vec{\rho}_{j,m} + a\vec{e}_x) - c_y(\vec{\rho}_{j-1,m} + a\vec{e}_x) = 0, \qquad (C.1)$$
$$(\omega_\perp - \omega)\, c_x(\vec{\rho}_{j,m} + a\vec{e}_y) + c_{xy}(\vec{\rho}_{j,m}) - c_{xy}(\vec{\rho}_{j,m+1}) = 0, \qquad (C.2)$$
$$(\omega_\perp - \omega)\, c_y(\vec{\rho}_{j,m} + a\vec{e}_x) + c_{xy}(\vec{\rho}_{j,m}) - c_{xy}(\vec{\rho}_{j+1,m}) = 0. \qquad (C.3)$$

Using (C.2) and (C.3) to eliminate the p-orbital amplitudes from (C.1) we obtain an equation involving only the d-orbital amplitudes.

$$\big[(\omega_t - \omega)(\omega_\perp - \omega) - 4\big]\, c_{xy}(\vec{\rho}_{j,m}) + c_{xy}(\vec{\rho}_{j+1,m}) + c_{xy}(\vec{\rho}_{j-1,m})$$
$$+ c_{xy}(\vec{\rho}_{j,m+1}) + c_{xy}(\vec{\rho}_{j,m-1}) = 0. \qquad (C.4)$$

The equations represented by (C.4) can be written in matrix form:

$$H_d(\omega)\, \vec{C}_{xy} = 0, \qquad (C.5)$$

where $\mathbb{H}_d(\omega)$ is the effective Hamiltonian describing the interactions between the d orbitals and $\vec{C}_{xy}$ is a vector whose components are the d-orbital amplitudes,

$c_{xy}(\vec{\rho}_{j,m})$. The matrix elements of $\mathbb{H}_d(\omega)$ are given by

$$H_d[\vec{\rho}_{j,m}; \vec{\rho}_{j',m'}] = (\omega_t - \omega)(\omega_\perp - \omega)\delta_{j,j'}\delta_{m,m'} + \delta_{j+1,j'}\delta_{m,m'}$$
$$+ \delta_{j-1,j'}\delta_{m,m'} + \delta_{j,j'}\delta_{m+1,m'} + \delta_{j,j'}\delta_{m-1,m'}. \qquad (C.6)$$

Using a more compact notation (C.6) can be expressed as

$$H_d(\vec{\rho}, \vec{\rho}') = (\omega_t - \omega)(\omega_\perp - \omega)\delta_{\vec{\rho},\vec{\rho}'}$$
$$+ \delta_{\vec{\rho},(\vec{\rho}'+a\vec{e}_x)} + \delta_{\vec{\rho},(\vec{\rho}'-a\vec{e}_x)} + \delta_{\vec{\rho},(\vec{\rho}'+a\vec{e}_y)} + \delta_{\vec{\rho},(\vec{\rho}'-a\vec{e}_y)} \qquad (C.7)$$

In this notation eigenvectors, $\vec{C}^k$, are characterized by a two-dimensional wavevector, $\vec{k}$. The components of $\vec{C}^k$ are

$$c_{xy}^k(\vec{\rho}) = \frac{1}{\sqrt{N}} e^{-i\vec{k}\cdot\vec{\rho}}. \qquad (C.8)$$

Our aim here is to construct the matrix elements of the function $[\mathbb{H}_d]^{-1}$. To do this we note that $H_d(\vec{k}, \vec{k}')$ is diagonal in $\vec{k}$-space and hence its inverse also diagonal. Then, transforming $H_d^{-1}(\vec{k}, \vec{k}')$ back to lattice space we obtain the desired matrix. Using (C.8) we have,

$$H_d(\vec{k}, \vec{k}') = \frac{1}{N} \sum_{\vec{\rho},\vec{\rho}'} e^{-i\vec{k}\cdot\vec{\rho}} H_d(\vec{\rho}, \vec{\rho}') e^{-i\vec{k}'\cdot\vec{\rho}'}$$
$$= \{(\omega_t - \omega)(\omega_\perp - \omega) - 4 + 2C_{2x} + 2C_{2y}\}\delta_{\vec{k},\vec{k}'}. \qquad (C.9)$$

Therefore for the matrix elements of the inverse:

$$H_d^{-1}(\vec{k}, \vec{k}') = \{(\omega_t - \omega)(\omega_\perp - \omega) - 4 + 2C_{2x} + 2C_{2y}\}^{-1}\delta_{\vec{k},\vec{k}'}, \qquad (C.10)$$

where $C_{2\alpha} = \cos(2k_\alpha a)$ and a is the B–O distance. Since we know the matrix elements of $\mathbb{H}_d$ in $\vec{k}$-space, we can easily obtain the desired lattice space function, $\mathbb{G}$. We transform the matrix from the $\vec{k}$-space representation of (C.10) to lattice-space representation using the eigenvectors.

$$G_\varepsilon[\vec{\rho}, \vec{\rho}'] = \frac{1}{\sqrt{N}} \sum_{\vec{k}} \sum_{\vec{k}'} \frac{e^{i\vec{k}\cdot\vec{\rho}_{j,m}} e^{-i\vec{k}'\cdot\vec{\rho}_{j',m'}} \delta_{\vec{k},\vec{k}'}}{(\omega_t - \omega)(\omega_\perp - \omega) - 4 + 2C_{2x} + 2C_{2y}} \qquad (C.11)$$

$$= \frac{1}{N} \sum_{\vec{k}} \frac{e^{i\vec{k}\cdot(\vec{\rho}-\vec{\rho}')}}{2\varepsilon + 2C_{2x} + 2C_{2y}}, \qquad (C.12)$$

where ε has its usual definition $2\varepsilon = (\omega - \omega_m)^2 - (\omega_g/2)^2 - 4$. Because of translational symmetry, $G_\varepsilon(\vec{\rho}, \vec{\rho}')$ depends only on the difference $\vec{\rho} - \vec{\rho}'$. Since H_d is a hermitian matrix so is G_ε and therefore $G_\varepsilon(\vec{\rho}, \vec{\rho}') = G_\varepsilon(\vec{\rho}', \vec{\rho})^*$. In the limit of large N, the sum can be converted to an integral,

$$G_\varepsilon(\vec{\rho}, \vec{\rho}') = \frac{(2a)^2}{(2\pi)^2} \int \frac{e^{i\vec{k}\cdot(\vec{\rho}-\vec{\rho}')}}{2\varepsilon + i0^+ + 2C_{2x} + 2C_{2y}} d\vec{k}, \qquad (C.13)$$

where the integration is over the two-dimensional Brillouin zone. We now make use of an integral representation of the Bessel function

$$J_p(t) = \frac{i^{-p}}{2\pi} \int_{-\pi}^{\pi} dx \; e^{it\cos(x)} \, e^{ipx} \tag{C.14}$$

and

$$\frac{1}{\lambda} = -i \int_0^\infty dt \; e^{i(\lambda+i0^+)t} \,, \tag{C.15}$$

to arrive at the result

$$G_\varepsilon(\vec{\rho}_{j,m}, \vec{\rho}\,'_{j',m'}) = i^{(q+r-1)} \int_0^\infty e^{i(\varepsilon t)} \, J_q(t) \, J_r(t) \, dt \tag{C.16}$$

$$= i^{(q+r-1)} \int_0^\infty \cos(\varepsilon t) \, J_q(t) \, J_r(t) \, dt$$

$$+ \; i^{(q+r)} \int_0^\infty \sin(\varepsilon t) \, J_q(t) \, J_r(t) \, dt \tag{C.17}$$

where $q = j - j'$ and $r = m - m'$.

C.1 Function $G_\varepsilon(0)$

For $\vec{\rho} = \vec{\rho}\,'$ in (C.13) we obtain the diagonal d-orbital, lattice Green's function,

$$G_\varepsilon(0) = \frac{1}{(2\pi)^2} \int_{-\pi}^{\pi} \int_{-\pi}^{\pi} \frac{dx \, dy}{2\varepsilon + i0^+ + 2\cos(x) + 2\cos(y)} \tag{C.18}$$

$$= \frac{2}{(2\pi)^2} \int_0^\pi \int_0^\pi \frac{dx \, dy}{2\varepsilon + i0^+ + 2\cos(x) + 2\cos(y)} \,. \tag{C.19}$$

Inside the pi(xy) band, $|\varepsilon| \leq 2$, the imaginary part of $G_\varepsilon(0)$ is related to the DOS function. In fact, from (6.21) we see that,

$$-\frac{1}{\pi} \Im G_\varepsilon(0) = 2\rho(\varepsilon) = \rho(2\varepsilon) \,. \tag{C.20}$$

Making use of (C.16) and (C.17) we have

$$G_\varepsilon(0) = -\frac{i}{2} \int_0^\infty dt \, e^{i\varepsilon t} \, J_0(t)^2$$

$$= -\frac{i}{2} \int_0^\infty dt \, \cos(\varepsilon t) \, J_0(t)^2 + \frac{1}{2} \int_0^\infty dt \, \sin(\varepsilon t) \, J_0(t)^2. \tag{C.21}$$

The result may also be written in terms of the complete elliptic integrals:

$$G_\varepsilon(0) = \frac{1}{2\pi} \begin{cases} (k_1)\, K(k_1) & |\varepsilon| > 2, \\ \text{sign}(\varepsilon)\, K(k) - iK(k') & |\varepsilon| < 2, \end{cases} \tag{C.22}$$

with $k = \varepsilon/2$, $k' = \sqrt{1 - k^2}$, and $k_1 = 1/k$. The imaginary part of $G_\varepsilon(0)$ vanishes for $|\varepsilon| > 2$. Inside the band, the relation between the imaginary part of $G_\varepsilon(0)$ and the DOS is

$$-\frac{1}{\pi} \Im G_\varepsilon(0) = \rho(\varepsilon) = \frac{1}{2}\, \rho(2\varepsilon). \tag{C.23}$$

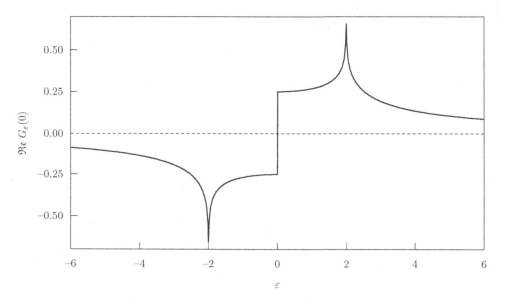

Figure C.1. Real part of Green's function $G_\varepsilon(0)$.

Figure C.1 shows a graph of the real part of $G_\varepsilon(0)$. The function possesses logarithmic singularities at the band edges ($\varepsilon = \pm 2$) and a jump discontinuity of magnitude $1/2$ at $\varepsilon = 0$. For large values of ε the function decays as $1/(2\varepsilon)$. It is antisymmetric about $\varepsilon = 0$.

C.2 Function $G_\varepsilon(1)$

We define

$$G_\varepsilon(1) = G_\varepsilon(\vec{\rho}, \vec{\rho} + 2a\vec{e}_\alpha), \qquad (\alpha = x \text{ or } y), \tag{C.24}$$

and obtain the result:

$$G_\varepsilon(1) = \frac{1}{2(2\pi)^2} \int_{-\pi}^{\pi} \int_{-\pi}^{\pi} \frac{e^{ix}\, dx\, dy}{\varepsilon + i0^+ + \cos(x) + \cos(y)}, \tag{C.25}$$

$$= \int \cos(\varepsilon t)\, J_0(t)\, J_1(t)\, dt + i \int \sin(\varepsilon t)\, J_0(t)\, J_1(t)\, dt. \tag{C.26}$$

The imaginary part of $G_\varepsilon(1)$ vanishes for ε outside the pi energy band, that is for $|\varepsilon| > 2$. From (C.25) it follows that

$$G_\varepsilon(1) + G_\varepsilon(-1) = \frac{1}{2} - \varepsilon\, G_\varepsilon(0), \quad \text{or}$$

$$\Re\{G_\varepsilon(1)\} = \Re\{G_\varepsilon(-1)\} = \frac{1}{2}\left[\frac{1}{2} - \varepsilon\, G_\varepsilon(0)\right]. \tag{C.27}$$

Using (C.22) we can write

$$\Re\{G_\varepsilon(1)\} = \frac{1}{2\pi} \begin{cases} \pi/2 - K(k_1) & |\varepsilon| > 2, \\ \pi/2 - |k|\, K(k) & |\varepsilon| < 2. \end{cases} \tag{C.28}$$

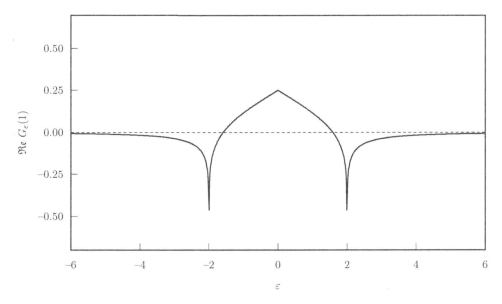

Figure C.2. Real part of Green's function $G_\varepsilon(1)$.

Figure C.2 shows a graph of real part of $G_\varepsilon(1)$. The function possesses logarithmic singularities at the band edges and a cusp at $\varepsilon = 0$. For large ε it tends to zero as $-1/(4\varepsilon^2)$. Unlike $\Re\, G_\varepsilon(0)$, $\Re\, G_\varepsilon(1)$ is symmetric about $\varepsilon = 0$.

C.3 Lattice Green's function for the pi bands

In the previous sections we obtained the Green's function for the d orbitals. In this section we calculate the Green's function for the pi bands including the Green's functions for the p orbitals. From these results we calculate the "partial" DOS functions associated with the square of the amplitudes of each of the orbitals involved in the pi bands. The partial DOS functions provide a way of quantifying the degree of covalent mixing for the pi bands and are employed in obtaining the XPS photoelectron cross-sections in Chapter 8.

(a) The pi-band lattice Green's function

According to Chapter 4, equation (4.43), the matrix equation that determines the pi($\alpha\beta$) bands, with $\alpha\beta = xy$, xz, or yz, is

$$\left[\hat{\mathbf{h}}(\vec{k}) - E\right] C(\vec{k}) = 0, \tag{C.29}$$

where the 3×3 matrix $\left[\hat{\mathbf{h}}(\vec{k}) - E\right]$ is given by

$$\begin{pmatrix} E_t - E & 2iS_\beta(pd\pi) & 2iS_\alpha(pd\pi) \\ -2iS_\beta(pd\pi) & E_\perp - E & 0 \\ -2iS_\alpha(pd\pi) & 0 & E_\perp - E \end{pmatrix} \tag{C.30}$$

with $S_\alpha = \sin(k_\alpha a)$. The components of the vector, $C(\vec{k})$, are the amplitudes of the $d_{\alpha\beta}$, p_α, and p_β orbitals. The total Hamiltonian $\left[\mathbb{H} - E\right]$ is a $3N \times 3N$ matrix given by

$$\left[\mathbb{H} - E\right]_{\vec{k},\vec{k}';r,s} = \left[\hat{\mathbf{h}}(\vec{k}) - E\right]_{r,s} \delta_{\vec{k},\vec{k}'}. \tag{C.31}$$

We define the lattice Green's function by the equation,

$$\left[\mathbb{H} - E\right] \mathbb{G}(E) = \mathbb{I} \qquad \text{or} \qquad \mathbb{G}(E) = \left[\mathbb{H} - E\right]^{-1}, \tag{C.32}$$

where $\mathbb{I}$ is a $3N \times 3N$ unit matrix. In $\vec{k}$-space the matrix elements of the inverse are given by

$$\left[\mathbb{H} - E\right]^{-1}_{\vec{k},\vec{k}';r,s} = \left[\hat{\mathbf{h}}(\vec{k}) - E\right]^{-1}_{r,s} \delta_{\vec{k},\vec{k}'}. \tag{C.33}$$

The matrix elements of the inverse,

$$\left[\hat{\mathbf{h}}(\vec{k}) - E\right]^{-1}_{r,s} = \frac{1}{D} M_{s,r}, \tag{C.34}$$

where

$$D = \det(\vec{k}, E) = (E_\perp - E)\left[(E_\perp - E)(E_t - E) - 4(pd\pi)^2(S_\alpha^2 + S_\beta^2)\right] \tag{C.35}$$

is the determinant and $M_{s,r}$ is the (s, r) element of the matrix of the minors, given by

$$
\begin{pmatrix}
(E_\perp - E)^2 & -2i(E_\perp - E)S_\beta(pd\pi) & -2i(E_\perp - E)S_\alpha(pd\pi) \\
2i(E_\perp - E)S_\beta(pd\pi) & [(E_\perp - E)(E_t - E) - 4S_\alpha^2(pd\pi)^2] & 4S_\alpha S_\beta(pd\pi)^2 \\
2i(E_\perp - E)S_\alpha(pd\pi) & 4S_\alpha S_\beta(pd\pi)^2 & [(E_\perp - E)(E_t - E) - 4S_\beta^2(pd\pi)^2]
\end{pmatrix}.
$$

$$(C.36)$$

The lattice-space Green's function is obtained by transforming $G(E)_{\vec{k},\vec{k}';r,s}$ to lattice-space by means of the unitary transformation

$$
G(E; \vec{R}_m, \vec{R}_n)_{r,s} = \frac{1}{N} \sum_{\vec{k}, \vec{k}'} e^{i(\vec{k} \cdot \vec{R}_m - \vec{k}' \cdot \vec{R}_n)} \, G(E)_{\vec{k},\vec{k}';r,s}.
\tag{C.37}
$$

The d-orbital Green's function discussed earlier in this Appendix is related to the $(1, 1)$ elements of the full lattice Green's function in (C.37) as follows:

$$
G_\varepsilon(\vec{\rho}, \vec{\rho}') = (pd\pi)^2 \, G(E; \vec{R}_m, \vec{R}_n)_{11} \qquad (\vec{R}_m - \vec{R}_n = \vec{\rho} - \vec{\rho}').
\tag{C.38}
$$

(b) Relation to the density of sates

An important result is the relationship between the DOS and the trace of the $3N \times 3N$ Green's function matrix. Consider

$$
-\frac{1}{\pi} \Im\left\{ \mathfrak{Tr} \, \mathbb{G}(E + i0^+) \right\} = -\frac{1}{\pi} \Im\left\{ \frac{1}{N} \sum_{\vec{k}} \mathfrak{Tr}[\hat{h}(\vec{k}) - E + i0^+]^{-1} \right\}
\tag{C.39}
$$

where $\mathfrak{Tr}$ indicates the trace. Since the trace of a matrix is invariant under unitary transformation it follows that we may evaluate the trace of $[\hat{h}(\vec{k}) - E + i0^+]^{-1}$ in its diagonalized form. This gives

$$
\mathfrak{Tr}[\hat{h}(\vec{k}) - E + i0^+]^{-1} = \sum_j [E_j(\vec{k}) - E + i0^+]^{-1}
\tag{C.40}
$$

where $E_j(\vec{k})$ are the pi-band energies for the π^0, π^* and π,

$$
E_1(\vec{k}) = E_{\pi^0} = E_\perp,
\tag{C.41}
$$

$$
E_2(\vec{k}) = E_{\pi^*}(\vec{k}) = E_m + \sqrt{(E_g/2)^2 + 4(pd\pi)^2(S_x^2 + S_y^2)},
\tag{C.42}
$$

$$
E_3(\vec{k}) = E_\pi(\vec{k}) = E_m - \sqrt{(E_g/2)^2 + 4(pd\pi)^2(S_x^2 + S_y^2)}.
\tag{C.43}
$$

With these results, equation (C.39) takes the form

$$
\frac{1}{\pi} \Im\left\{ \left(\frac{2a}{2\pi}\right)^3 \sum_j \int d\vec{k} \, [E_j(\vec{k}) - E + i0^+]^{-1} \right\} = \sum_j \rho_j(E),
\tag{C.44}
$$

where the integration is over the first Brillouin zone. Thus, $-\frac{1}{\pi} \Im \{ \mathfrak{Tr} \, \mathbb{G}(E + i0^+) \}$ is the sum of the density of states functions for each of the three bands and therefore it is the total density of states for the pi($\alpha\beta$) bands.

(c) Partial density of states

The "partial" DOS functions associated with each type of orbital involved in the pi bands can be obtained from the lattice Green's function. From (C.36) and (C.37) we have

$$G(E; \vec{R}_m, \vec{R}_m)_{1,1} \equiv \frac{1}{N} \sum_{\vec{k}} [\hat{\mathbf{h}}(\vec{k}) - E]_{11}^{-1} = \left(\frac{2a}{2\pi}\right)^3 \int d\vec{k} \, \frac{(E_\perp - E)^2}{D(\vec{k}, E)}, \qquad (C.45)$$

$$G(E; \vec{R}_m, \vec{R}_m)_{2,2} = \left(\frac{2a}{2\pi}\right)^3 \int d\vec{k} \, \frac{[(E_\perp - E)(E_t - E) - 4(pd\pi)^2 S_\alpha^2]}{D(\vec{k}, E)}, \qquad (C.46)$$

$$G(E; \vec{R}_m, \vec{R}_m)_{3,3} = \left(\frac{2a}{2\pi}\right)^3 \int d\vec{k} \, \frac{[(E_\perp - E)(E_t - E) - 4(pd\pi)^2 S_\beta^2]}{D(\vec{k}, E)}. \qquad (C.47)$$

The partial DOS functions are given by

$$\rho_{\alpha\beta}(E) = -\frac{1}{\pi} \Im \left\{ \mathbb{G}(E + i0^+)_{\vec{R}_m, \vec{R}_m; 1,1} \right\}, \qquad (C.48)$$

$$\rho_\alpha(E) = -\frac{1}{\pi} \Im \left\{ \mathbb{G}(E + i0^+)_{\vec{R}_m, \vec{R}_m; 2,2} \right\}, \qquad (C.49)$$

$$\rho_\beta(E) = -\frac{1}{\pi} \Im \left\{ \mathbb{G}(E + i0^+)_{\vec{R}_m, \vec{R}_m; 3,3} \right\}, \qquad (C.50)$$

where $\rho_{\alpha\beta}(E)$ is the part of the total DOS that the d orbitals participate in. Similarly $\rho_\alpha(E)$ and $\rho_\beta(E)$ are the parts of the total DOS that the p_α orbitals and p_β orbitals participate in. Since $\rho_{\alpha\beta}(E) + \rho_\alpha(E) + \rho_\beta(E) = -\frac{1}{\pi} \Im\{\mathfrak{Tr} \, [\mathbb{G}(E + i0^+)]\}$ it follows that $\rho_\pi(E) = \rho_{\alpha\beta}(E) + \rho_\alpha(E) + \rho_\beta(E)$, where $\rho_\pi(E)$ is the total pi-band DOS for all three bands together. From (C.35) and (C.36),

$$\rho_{\alpha\beta}(E) + \rho_\alpha(E) + \rho_\beta(E) = -\frac{1}{\pi} \Im \left\{ \frac{1}{(E_\perp - E + i0^+)} \right.$$
$$\left. + \left(\frac{2a}{2\pi}\right)^3 \int_{-\pi/2}^{\pi/2} \int_{-\pi/2}^{\pi/2} \frac{dk_x \, dk_y \, [(E_\perp - E) + (E_t - E)]}{[(E_\perp - E + i0^+)(E_t - E + i0^+) - 4(pd\pi)^2(S_\alpha^2 + S_\beta^2)]} \right\}.$$
$$(C.51)$$

The first term on the right-hand side of (C.51) is $\delta(E_\perp - E)$, the DOS for the flat, non-bonding, π^0 band. The second term on the right can be evaluated from (6.28) as

$$\rho_\pi(E) = \frac{|E - E_m|}{(pd\pi)^2} \rho_\pi(\varepsilon(E)). \tag{C.52}$$

The function in (C.52), discussed in Chapter 6, is the DOS for the π and the π^* bands. It is given by

$$\rho_\pi(\varepsilon(E)) = \frac{1}{\pi^2} K\left(\sqrt{1 - \left(\frac{\varepsilon(E)}{2}\right)^2}\right), \tag{C.53}$$

$$\varepsilon(E) = \frac{(E - E_m)^2 - (E_g/2)^2}{(pd\pi)^2} - 2. \tag{C.54}$$

For $E < E_m$, $\rho_\pi(E)$ gives the DOS of the π band and for $E > E_m$ it gives the DOS of the π^* band. This proves that $\rho_\pi(E) = \rho_{\alpha\beta}(E) + \rho_\alpha(E) + \rho_\beta(E) = \rho_{\pi^0}(E) + \rho_\pi(E) + \rho_{\pi^*}(E)$.

(d) d-Orbital and p-orbital partial DOS functions

The d-orbital partial DOS, $\rho_{\alpha\beta}(E)$, is given by (C.45). The integral can be evaluated immediately in terms of $\rho_\pi(E)$. We have

$$-\frac{1}{\pi} \Im G(E; \vec{R}_m, \vec{R}_m)_{1,1} = -\frac{1}{\pi} \Im\left\{\left(\frac{2a}{2\pi}\right)^3 \int d\vec{k} \, \frac{(E_\perp - E)^2}{D(\vec{k}, E)}\right\}$$

$$= -\frac{1}{\pi} \Im\left\{\left(\frac{2a}{2\pi}\right)^3 \int_{-\frac{\pi}{2}}^{\frac{\pi}{2}} \int_{-\frac{\pi}{2}}^{\frac{\pi}{2}} \frac{dk_x \, dk_y \, (E_\perp - E)}{[(E_\perp - E + i0^+)(E_t - E + i0^+) - 4(pd\pi)^2(S_\alpha^2 + S_\beta^2)]}\right\}$$

$$= \frac{|E - E_\perp|}{2(pd\pi)^2} \rho_\pi(\varepsilon(E)). \tag{C.55}$$

The final result of (C.55) for the d-orbital partial DOS gives the contribution of the square of the $d_{\alpha\beta}$-orbital amplitude to the states in the range E to $E + dE$. For convenience we shall name this function $\rho_{\pi d}(E)$. It is clear that $\rho_{\pi d}(E) = 0$ at $E = E_\perp$, showing that pi bands are pure p orbital in composition at $E_\perp$. In contrast to this result, at $E = E_t$, $\rho_{\pi d}(E) = \rho_\pi(E_t)$ (the total DOS at E_t), showing that the states are pure d orbital in composition at the edge of the conduction bands.

For the p-orbital partial DOS, $\rho_{\pi p_x}(E)$ and $\rho_{\pi p_y}(E)$, we have

$$-\frac{1}{\pi} \Im\left\{G(E; \vec{R}_m, \vec{R}_m)_{2,2} + G(E; \vec{R}_m, \vec{R}_m)_{3,3}\right\} = \delta(E - E_t) + \frac{|E - E_t|}{2(pd\pi)^2} \rho_\pi(\varepsilon(E)). \tag{C.56}$$

The first term of (C.56) is the partial DOS for the pure non-bonding p bands. The second term is the sum of the two p-orbital partial DOS functions for the π and π^* bands. Symmetry indicates that the two partial DOS functions are equal,

$\rho_{\pi p_x} = \rho_{\pi p_y}$, thus

$$\rho_{\pi p_x}(E) = \frac{1}{2}\delta(E - E_t) + \frac{1}{4}\frac{|E - E_t|}{(pd\pi)^2}\rho_\pi(\varepsilon(E)) = \rho_{\pi p_y}(E). \qquad (C.57)$$

As may be seen from (C.57), these partial DOS functions vanish at $E = E_t$ because the states are pure d orbital at that energy. For $E = E_\perp$, $\rho_{\pi p_x}(E) = \frac{1}{2}\rho_\pi(E_\perp)$. The partial DOS functions are obtained in Chapter 8 directly from the wavefunction amplitudes.

(e) Covalency ratio

For the π and π^* bands we define

$$\rho_{\pi p}(E) = \frac{1}{2}\frac{|E - E_t|}{(pd\pi)^2}\rho_\pi(\varepsilon(E)), \qquad (C.58)$$

the sum of the p-orbital partial densities excluding the non-bonding delta function. Figure C.3 shows $\rho_{\pi d}(E)$ and $\rho_{\pi p}(E)$ as functions of E. The parameters used for the figure are appropriate for $SrTiO_3$ or $BaTiO_3$. The total area under the two curves for each function is 1. For $\rho_{\pi d}(E)$ the conduction-band area is 0.82 and the valence-band area is 0.18. For $\rho_{\pi p}(E)$ the valence-band area is 0.82 and the conduction-band area is 0.18.

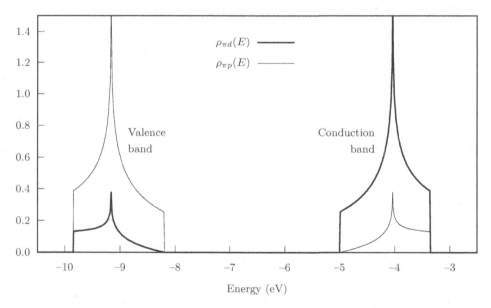

Figure C.3. Partial density of states. Parameters used are $E_\perp = -8.2\,\text{eV}$, $E_t = -5.0\,\text{eV}$, and $(pd\pi) = 1.0\,\text{eV}$.

The d-orbital mixture into the π valence band is

$$R(E) \equiv \frac{\rho_{\pi\,d}(E)}{[\rho_{\pi\,d}(E) + \rho_{\pi\,p}(E)]} = \frac{(E - E_\perp)}{2(E - E_m)} \qquad \text{for} \quad E_2 \leq E \leq E_\perp \qquad \text{(C.59)}$$

where $E_2 = E_m - \sqrt{(E_g/2)^2 + 8(pd\pi)^2}$ is the bottom of the valence band. Equation (C.59) is a measure of the covalent mixing of the d orbitals into the valence band as a function of E. If we make the substitution, $E' = E - E_m$ in (C.59) the result is

$$R(E') = \frac{(E' + E_g/2)}{2E'} \qquad \text{for} \quad E_2 - E_m \leq E \leq -E_g/2. \qquad \text{(C.60)}$$

This result was obtained in Chapter 4 by other means and $\langle R \rangle$, the value of $R(E)$ averaged over the valence band is shown in Fig. 4.5 as a function of E_g and of $(pd\pi)$. Equation (C.60) shows that for a given E', the d-orbital mixing into the valence band is dependent only on the energy gap, E_g, however, the total mixing for the entire band depends on both E_g and $(pd\pi)$ because the band width depends on both of these parameters. Symmetry demands that the probability of the p orbitals in the π^* conduction band is the same as the probability of the d orbitals in the valence band. That this requirement is met is evident in Fig. C.3.

Appendix D

Surface and bulk Madelung potentials for the ABO_3 structure

The electrostatic potentials (Madelung potentials) due to the ionic charges of the ABO_3 perovskite structure are given in this appendix for the infinite and semi-infinite lattices. The potential for an electron at the point $\vec{r}$ due to the charged ions is

$$V_M(\vec{r}) = \sum_{\vec{R}_B} \frac{e^2 q_B}{|\vec{r} - \vec{R}_B|} + \sum_{\vec{R}_A} \frac{e^2 q_A}{|\vec{r} - \vec{R}_A|} - \sum_{\vec{R}_O} \frac{e^2 q_O}{|\vec{r} - \vec{R}_O|}, \qquad (D.1)$$

where $\vec{R}_B$, $\vec{R}_A$ and $\vec{R}_O$ are the vector positions of the B, A, and O ions respectively. For the infinite lattice the sum extends over all of the lattice sites, while for the semi-infinite lattice the sum extends over the half-space bounded by a type I or type II (001) surface. In (D.1), q_B, q_A, and q_O are the magnitudes of the charges on the B, A, and O ions, respectively.

A bit of finesse is required to carry out these sums and various methods are discussed in the scientific literature [1–4]. The potentials at the sites are summarized in Table D.1 for a perovskite such as SrTiO$_3$ with $q_B = 2q_A = 2q_O$.

In Column 1, the notation "Site(z)" indicates the type of site and z, the distance below the surface for the type I and type II (001) surfaces. The convention is that a positive potential is repulsive to an electron and a negative potential is attractive. The reduction in the B-ion potential at the type I surface means that the surface site potential is less repulsive.

The potentials in Table D.1 are in units of $e^2/2a$ and the charge is $e = 4.802 \times 10^{-10}$ esu. To convert these potentials to electronvolts multiply the entry by $(14.3942/2a)$, where $2a$ is the lattice constant in angstroms. For SrTiO$_3$, for example, $2a = 3.92$ in angstroms and the conversion factor is 3.6720. Thus, the Madelung potential at the Ti site for a type I (001) surface is: $11.7045 \times 3.6720 = 42.9789$ eV. For the infinite crystal the value is $12.3775 \times 3.6720 = 45.4502$ eV. The reduction of the repulsive potential at the surface is 2.4713 eV. These values are for the full

Table D.1. *Bulk and surface electrostatic potentials.*

Ion site(z)	Infinite lattice potential	Surface potential	Change from bulk value
Type I (001)			
$B(0)$	12.3775	11.7045	-0.6730
$A(a)$	5.3872	5.4159	0.0287
$O(0)$	-6.4559	-6.4590	-0.0031
$O(a)$	-6.4559	-6.4845	-0.0029
Type II (001)			
$B(a)$	12.3775	12.4092	0.0317
$A(0)$	5.3872	4.9783	-0.4089
$O(0)$	-6.4559	-5.5127	-0.9432
$O(a)$	-6.4559	-6.4590	0.0031

ionic charges $q_{Sr} = 2$, $q_{Ti} = 4$, and $q_O = 2$. The charge may be adjusted for covalent effects, however charge neutrality must be maintained. This means that

$$q_A + q_B - 3q_O = 0. \tag{D.2}$$

For bulk SrTiO₃ it is found that due to covalent mixing the effective charge is approximately 85% of the ionic charge so that $q_A = 1.7$ for the Sr ions, $q_B = 3.4$ for the Ti ions, and $q_O = 1.7$ for the oxygen ions. With this covalency reduction, the change in the Madelung potential on the type I (001) surface is 2.1003 eV (less repulsive).

Using Table D.1, the surface unit-cell site potentials can be calculated for the type I and type II (001) surfaces. Additional tables and results may be found in [1–4].

References

[1] E. A. Kraut, T. Wolfram, and W. Hall, *Phys. Rev. B* **6**, 1499 (1972).
[2] T. Wolfram, E. A. Kraut, and F. J. Morin, *Phys. Rev. B* **7**, 1677 (1973).
[3] F. Hund, *Z. Physik* **94**, 11 (1935).
[4] Y. Sakamoto and U. Takahasi, *J. Chem. Phys.* **30**, 337 (1952).

Index

Printed in the United States
By Bookmasters